Electro-Optics Handbook

ELECTRO-OPTICS SERIES

Series Editor: Dr. Herbert Elion
Chief Executive Officer
International Communications and Energy, Inc.
Natick, Massachusetts

Volume 1 Molecular Electro-Optics: Part 1—Theory and Methods; Part 2—Applications to Biopolymers, *edited by Chester T. O'Konski*

Volume 2 Fiber Optics in Communications Systems, *by Glenn R. Elion and Herbert A. Elion*

Volume 3 Electro-Optics Handbook, *by Glenn R. Elion and Herbert A. Elion*

Other volumes in preparation

Electro-Optics Handbook

GLENN R. ELION
President
International Communications & Energy, Inc.
Natick, Massachusetts

HERBERT A. ELION
Chief Executive Officer
International Communications & Energy, Inc.
Natick, Massachusetts

MARCEL DEKKER, INC. New York and Basel

Library of Congress Cataloging in Publication Data

Elion, Glenn R   [Date]
Electro-optics handbook.

(Electro-optics series ; v. 3)
Includes index.
1. Electrooptics--Handbooks, manuals, etc.
I. Elion, Herbert A., joint author. II. Title.
III. Series.
TA1750.E46   621.36   79-20585
ISBN 0-8247-6879-5

MARCEL DEKKER, INC.
270 Madison Avenue, New York, New York 10016

Current printing (last digit):
10 9 8 7 6 5 4 3 2 1

PRINTED IN THE UNITED STATES OF AMERICA

PREFACE

This first edition contains fundamental material underlying all of electro-optic engineering. Where appropriate, a small overlap was permitted for material closely related to the specialized branches of engineering that are still substantially evolving.

This handbook covers the entire practice of electro-optic engineering; the authors have presented the data that have been useful to them in their work as practicing electro-optic engineers. Where no credits are shown, the material was prepared by either Glenn R. Elion or Herbert A. Elion.

The book had its beginnings more than 30 years ago, when one of the authors began to make notes for his own "pocket book" on electro-optics. Then, three years ago, at the publisher's request, the authors began the work of selecting from this continually growing mass of material pertaining to electro-optical engineering, and of condensing, digesting, and arranging it in a usable form for modern publication.

Much attention has been paid to the abstraction of data from recent literature, and numerous references to other data sources are given. In this respect, the present work will be found to differ from other product-oriented pocket books of components manufacturers: it is a handbook prepared as a service to the entire engineering profession. Heretofore, the basic data required for evaluation, selection, and proper application of electro-optic tests have been scattered in a multitude of technical papers, reports, and publications; others have been published by components manufacturers and specifically directed towards utilization of their products. This handbook is the first large-scale international effort to integrate this vitally needed information in a single fundamental handbook for use in industry, military practice, engineering education, and technical training.

Glenn R. Elion
Herbert A. Elion

CONTENTS

LIST OF FIGURES

LIST OF TABLES

Electro-Optics Handbook

Chapter 1

FUNDAMENTAL DEFINITIONS

The following definitions are for some of the more basic and recurrent terms used in electro-optics. For some words used in the industry, there are two different meanings for identical terms. In these cases, the more prominent usage of the word has been defined. All equations presented in this handbook use the standard accepted international symbols. To prevent any confusion of terms and symbols, Chapter 2 presents a listing of symbols that will be followed throughout.

Abaxial: Optical term indicating points or areas off the central axis

Abbe constant: The ratio of refractivity to dispersion in optical materials

Aberration: Any type of error resulting in an imperfect image

Abridged spectrophotometer: An instrument to measure or analyze light at various predetermined wavelengths

Absentee layer: A layer used to separate stacks or optical cavities without influencing the optical design properties and without changing the normal incidence reflectance

Absolute spectral sensitivity $S(\lambda)$: The quotient of the output to input radiation at a given wavelength in the narrow range of two specified wavelengths; the ratio of the generated photocurrent to the radiant power striking a given device

Absorptance: The ratio of the radiant flux absorbed to the radiant flux incident upon a given body

Absorption: $I = I_0 e^{-Kx}$ (Lambert's law), where I_0 is the original intensity, I is the intensity after passing through a material with thickness x and absorption coefficient $K = 4\pi K' n/\lambda$, where K' is the index of absorption, n is the refractive index, and λ is the wavelength of the incident radiation

Absorption spectrum: The dark line or band spectrum obtained from light passing through a gas; the broad dark-region spectrum with no resolvable lines obtained from light passing through a liquid or solid

Absorptivity: The fraction of radiant energy falling on a given body which is absorbed or transformed into heat; or the ratio of the radiation absorbed to that absorbed under identical conditions by a blackbody

Achromatic: Colorfree or gray; a term used for optical lenses indicating partial or complete chromatic aberration correction

Actinity: The ratio of the sensitivity s(Z) to a given radiation to the sensitivity s(N) to a reference radiation N, as related to a device with a known spectral sensitivity distribution

Actinoelectric effect: The change in resistance of a material with exposure to light in the presence of an applied electric current

Acutance: An objective measure of image sharpness given as the degree of steepness of the angle of the edge gradient between density boundaries in an image

Albedo: The ratio of reflected to incident radiation from an unpolished surface of a given body

Allochromy: Any type of fluorescence where the emerging radiation has a different wavelength from the incident radiation

Alloy-junction photocell: A type of photodiode where an alloy junction is made by mixing an indium disk with a thin wafer of n-type germanium

Analog communications: A basic form of telecommunications where the transmitted data are in a continuous form as opposed to digital communications which use discontinuous formats

Anastigmat: Where the astigmatic difference is zero for one or more abaxial areas in a compound optical lens

Angular aperture: The largest angular sweep of a beam of radiation that an objective can accept or transmit

Angle of half-sensitivity (α): The angle at which a detector can be rotated in both directions off the optical axis where the device output falls to half its maximum value

Angle shift: The shifting to shorter wavelengths in light transmitted through dielectric multilayer thin films as the layers are tilted from the normal incident radiation direction

Anisotropic: Any material which has different electrical or optical properties along different axes

Antireflection coating: A thin film of material applied to an optical or semiconductor surface to reduce external or internal reflections, usually applied at a thickness equal to one-fourth of the peak wavelength or more

Aperture: The maximum diameter of an entering radiation beam that can completely transmit radiation through the optical system

Aplanat: A lens which has been corrected for spherical aberration

Apochromat: The highest degree of color correction; or where three colors have been brought into common focus in microscope objectives

Apostilib: A luminance unit equal to $1/\pi$ cd/m^2

Astigmatism: An error in spherical lenses where the image of a point becomes two focal lines at right angles to each other separated by a measurable distance along the axis

Attenuation: The optical power loss per unit length along a waveguide which is the sum of the absorption and light scattering

Automatic gain control (AGC): A type of self-controlling device that maintains a constant peak transmission amplitude within specified limits in response to variations in the system attenuation, temperature, or other changes

Average noise figure: The ratio of the total output noise power to that part attributed to thermal noise in the input

Back focal length: The distance from the center of a lens to its principal focus on the side of the lens opposite the imaged object

Backscattering coefficient (β): $\beta = 4\pi\Phi r^2 W_r/W_i$, where Φr is the reflected power per unit solid angle in the source direction, W_i is the power per unit area in the incident radiation, and W_r is the power per unit area at a distance r

Band-elimination filter: A type of filter that has a single attenuation band, transmitting wavelengths on both sides of the band

Band-gap energy: The energy difference between the conduction band and the valence band of a given elemental material system

Bandpass filter: A filtering device that blocks all wavelengths except for a narrow region (band)

Band spectrum: A type of radiation spectrum produced by discharges in polyatomic gases consisting of a series of lines placed very near to each other giving the appearance of bands

Bandwidth: The complete range of frequencies over which a particular information system can function, basically determined by the maximum decodable frequency

Bandwidth %: The full width at half-maximum (FWHM) given as a percent of the center wavelength in reference to the transmittance curve of a given bandpass filter

Barrier-layer cell: A type of photovoltaic cell where radiation striking the contact layer of copper and cuprous oxide generates a potential

Barrier voltage: The voltage needed to cause electrical conduction in a pn-junction diode

Beam candlepower: The candlepower rating of a given radiation source which when located at the same distance as a given beam would produce an equivalent illumination as the beam

Beam spreader: An electrical or optical device that increases the angular divergence of an incident beam of radiation

Beam width: The angular width of a light beam; the least beam width is λ/A, where λ is the incident wavelength and A is the aperture

Bend loss: A type of increased attenuation caused by permitting high-order modes to radiate from the side of an optical waveguide when it is curved around a small radius of curvature

Blackbody: A body that absorbs all of the energy of all wavelengths of the incident radiation; a body with an emissivity of unity

Bolometer: A type of radiation detector that converts radiation into heat which causes a thermal variance in the detector material which is measured to indicate the flux of incident radiation

Bragg's law: Sin $\theta = n\lambda/2d$ where θ is the angle formed by the incident and emerging x-ray beams, d is the distance between layers of atoms, λ is the x-ray wavelength, and n is a whole number

Breakdown voltage (V_{br}): The reverse voltage at which an incremental increase in the voltage causes a sharp rise in the reverse current of a given device

Brewster's law: The tangent of the polarizing angle θ for a given material is equal to the index of refraction: r - tan θ, where θ is the angle of incidence at which the reflected polarized ray is at right angles to the reflected ray

Brightness: The radiant flux emitted per unit emissive area as projected on a plane normal to the line of sight given in units of lamberts; lambert = 1 lumen/cm^2 of projected surface area

Candela (cd): The luminous intensity perpendicular to a surface area of 1/600,000 square meter of a blackbody at the temperature of solidification of platinum (2042 K) under a pressure of 101,325 N/m^2

Cando-luminescence: A type of light phenomenon where white light is produced without going to very high source element temperatures

Carrier-current communication: The use of high-frequency alternating current (ac) on telecommunications and power lines for control and communication functions

Carrier lifetime: The time needed for excess carriers doped into a semiconductor device to recombine with opposing carriers

Carrier mobility: In semiconductor materials, the average drift velocity of the carriers per unit electric field

Cathode luminous sensitivity: The ratio of the measured photocathode current to the incident luminous flux

Cathodo-luminescence: A type of phosphor luminescence caused by electron bombardment

Cauchy's dispersion formula: $n = A + B/\lambda_2 + C/\lambda_4 + \cdots$, where n is the refractive index of the material at the incident wavelength λ and A, B, and C are empirically determined constants

Cavity: An elementary interference filter made with two reflecting stacks and a solid transparent spacer layer between them

Cavity filter: A type of filtering device in transmission lines that induces attenuation of undesired signals

Channel capacity: The total number of individual channels in a given communications system or the maximum information transmission rate through a given channel at a specified error rate

Channel electron multiplier: A photoelectric detector where the applied voltage causes photoelectric emission from the part of the device exposed to the incident radiation and accelerates the emission to the anode, creating secondary emission

Chelate laser: A laser containing a rare earth chelate material contained within a transparent solid material with high quantum efficiency, usually exhibiting narrow emission lines

Child's law: The current in a thermionic diode varies proportionally with the 3/2 power of the anode voltage and inversely with the distance squared between electrodes

Christiansen-effect filter: An optical filter containing fine, powdered crystalline substances; depending upon the incorporated material, the absorption spectrum of the filter can be predetermined

Chromatic aberration: The optical focusing error caused by the difference in the index of refraction of different wavelengths of the incident light

Circle of confusion: Disks of light on a given image formed by a lens or lens system from point sources of light on the subject

Cladding: Material usually of a low refractive index that surrounds the central core of an optical fiber to minimize surface scattering losses

Coefficient of reflection: The square root of the ratio of the reflected to incident power on a given surface area

Coherent radiation: Light propagation where the phase between any two points in the field is exactly the same or else maintains a constant difference throughout the duration of the light pulse

Coherent reference: A stable reference frequency used to compare other signals and frequencies to establish the relative coherency of a communications or electronic system

Collector dark current (I_{CO}): The current passing through a device sensitive to radiation in the absence of any illumination

Colorimetry: The measurement of radiation quantities given in terms of their effectiveness in producing the sensation of color

Color saturation: The degree of absence of white light in a color, given as a percentage

Color temperature: The temperature of a blackbody having the same visible color as that of a given nonblackbody, given in units of K

Coma: A type of lens error or aberration which causes circular patches of light at the edges of the given lens field

Complementary colors: A pair of colors that when added together in a given proportion produce white light

Continuous spectrum: A type of radiation spectrum produced by incandescent bodies consisting of wavelength regions where energy is present to some degree at every wavelength

Conversion efficiency: The ratio of the maximum available radiant or luminous flux output to the total input power of a photon-emitting device

Core: The center of an optical fiber bound by the cladding which conducts the transmitted light in the high-refractive-index region of the fiber

Coupling aperture: An opening or aperture in the wall of various waveguides which can transfer energy from or into some external circuit or waveguide

Coupling capacitance (C_k): The total capacitance between the detector and the emitter of an opto-isolator

Critical angle (θ_c): The smallest angle of incidence at which total reflection will occur at the boundary between two media of different indexes of refraction

Cross coupling: Various types of undesirable coupling between optical or electrical devices or different communications channels

Crush strength: The physical limit of an optical fiber or cable to withstand an applied force perpendicular to the axis of the fibers

Current transfer ratio (CTR): The quotient between the output and input current in a given optoelectronic coupler or photoelectric coupler device

Cutoff: The wavelength at which the transmittance falls to 5% of the peak value; usually applies to bandpass filters

Cutoff frequency (f_g): The frequency at which the incident light beam generates a photovoltage or photocurrent 0.707 times the value of incident radiation with f = 1 kHz

Czochralski technique: A process of growing large single crystals from the molten material state, usually used for Ge and Si

Dark current (I_D): The current passing through a photovoltaic device in the absence of any illumination

dBm: The abbreviation for decibels above or below 1 mW of power

dc transfer ratio: The quotient of the direct current (dc) output and input current of an opto-coupler

Decibel (dB): The ratio of the output to input power given by dB = 10 log (P_0/P_i), which indicates the level of transmission power gains or losses

Delay time (t_d): The time interval between the points at which the leading edges of the input and output pulses have reached 10% of their maximum amplitude

Demodulation bandwidth (B): The frequency interval in which the demodulated output of a photodetector device or system is not more than 3 dB below the output in the region of flat response (midband output)

Demultiplexer: A type of device or circuit which can separate two or more signals transmitted on a single channel

Dendritic growth: A process technique of growing long semiconductor material crystals in uniform optically flat ribbons

Depletion layer: A semiconductor material barrier layer where the mobile-carrier charge density is insufficient to neutralize the donor and acceptor charge density

Depletion region: The intrinsic region of a semiconductor junction where the carriers are depleted

Derating: The reduction in the rating or power output of a device or system to extend the lifetime or to permit operation at higher operating temperatures

Dichroic polarizer: A type of polarizer that transmits light that is nearly completely plane polarized with an intensity that is half that of the incident beam

Differential forward resistance (r_f): The resistance measured for small ac currents or voltages at a given point in the forward direction of the voltage–current plot

Diffracting grating: A solid that has been ruled with a large number of narrowly spaced equidistant lines; $\lambda = (s/n)(\sin i + \sin d)$, where s is the distance between rulings, i is the angle of incidence, d is the angle of diffraction, and n is the order of the spectrum

Diffraction: The phenomena produced by radiation passing objects which are comparable in size to the incident radiation wavelengths; diffraction efficiency is the ratio of incident radiant flux to the emerging diffracted flux

Diode capacitance (C_D): Total capacitance measured between diode terminals caused by the junction, parasitic, and case capacitances

Diopter: A measure of the refractive power of optical lenses or prisms

Discrimination ratio: The ratio of the passband to stopband widths of an optical filter

Dispersion: The difference between the index of refraction of a given material for any two given wavelengths

Dispersive power (W): $W = (n_2 - n_1)/(n - 1)$, where n_2 and n_1 are the indexes of refraction for wavelengths λ_2 and λ_1 and n is the mean index

Display primaries: Red, green, and blue receiver primaries which when mixed in the proper ratios can produce other colors

Doping: The addition of precisely controlled amounts of one or more n- or p-type elements (Al, As, Ga, In, Sb) to a semiconductor material to control various electrical or optical properties

Doppler effect: The altering of the apparent frequency by the motion of the source relative to the detection device; $f_o = f_s[(c + v_r)/(c - v_r)]^{1/2}$

where f_o is the observed frequency, f_s is the source frequency, c is the speed of light, and v_r is the velocity of the source relative to the observer

Drift mobility: The average drift velocity of semiconductor carriers per unit of electric field

Duplex: A method of communications where each terminal end can simultaneously transmit and receive

Dynamic sensitivity: The ratio of a device's photocurrent variations to a given small variance in the incident radiant or luminous power

Edge filter: An interference filter where the transmittance changes abruptly with a small change in wavelength

Electroluminescence: A type of phosphor luminescence caused by the presence of an electrical field

Electron optics: The control of electron speed and direction using various electric and magnetic fields similar to the control of light via geometrical optical lenses

Emissive power: Emissivity is measured by the energy radiated from a unit area of surface per unit time for a unit difference of temperature between the surface and surroundings, given in units of ergs per second per square centimeter

Epitaxial growth: A type of process for semiconductor devices where thin layers of material are grown in a controlled orientation

Excess noise: The current or $^1/f$ noise resulting from the passage of current through a semiconductor device

Extrinsic detector: A type of semiconductor photodetector whose optical and/or electrical response properties can be altered by the addition of various elemental impurities

Facsimile: The reproduction of an image or picture by scanning the image and then converting it into various types of electrical or optical signals

Fall time (t_f): The time required for the photocurrent to fall from 90% to 10% of the maximum value of a given device or system

Fiber optics: The transmission of light through transparent fibers usually by multiple reflections

Fiberscope: A fiber optic bundle with an objective lens at one end and an eyepiece at the other end to permit the observation of normally inaccessible objects, particularly useful in medical applications

Field curvature: A type of lens error or aberration which causes the plane of focus to be distorted (curved)

Field lens: A radiometer lens having a focal length such as to image the objective lens onto a detector to establish uniform illumination on the sensitive area

Filter discrimination: The difference between the minimum and maximum insertion loss at any given wavelength in the attenuation band of an optical fiber

Flicker photometer: A device to measure the intensity of a light source by comparing the given illumination to a known standard light source by viewing both alternately in rapid succession

Flip-flop: A device or circuit that has two stable operating states which can be switched by the incoming transmission signal to the desired operating condition

Fluorescence: The emission of radiation as the consequence of absorption of radiation from another source; usually of a longer wavelength than that of the absorbed (incident) radiation

Flux: The number of photons or the amount of radiant energy passing through a given surface element per unit time, given in watts or lumens

Flux density: The flux per unit area orthogonal to the given flux direction or flow path

f Number: The ratio of the focal length to the diameter of an optical lens

Focal length (f): The distance from the optical center of a lens to the principal focus

Footcandle (fc): The luminance at a surface 1 ft from a uniform radiation source with a luminous intensity of 1 candlepower

Footlambert: A unit of photometric brightness equal to $^1/\pi$ cd/ft^2, where a footcandle is equal to 1 lm/ft^2

Frequency-division multiplexing (fdm): In communications systems, a method of transmitting two or more signals on a single channel by using different frequencies for each signal

Frequency response characteristic: The relationship between the modulation frequency of the incident radiation and the radiant dynamic sensitivity

Frequency selectivity: The extent to which a circuit or device can differentiate between the center or signal frequency and other frequencies

Fresnel number: The ratio of the square of the radius of a lens aperture to the products of the focal length and wavelength, giving an indication of the lens diffraction effects

Gain bandwidth product (G_B): The product of the voltage dependent photocurrent gain M times the frequency at which the measurement is made when the device is biased for maximum attainable gain

Graded index: A type of optical fiber where the refractive index of the core decreases radially outward toward the cladding achieving high bandwidth capacities and coupling efficiencies

Graybody: A type of radiant body whose spectral emissivity is constant at all wavelengths and in a constant ratio to a totally radiant body at the same temperature

Guard ring: A small metal ring put on a terminal of a semiconductor device to uniformly distribute the surface charge

Half-duplex (hdx): A communications system which can receive or transmit in one direction at a time, but not simultaneously between both end terminals

Hefner lamp: A standard radiation source that emits a luminous intensity of 0.9 candlepower

Hexadecimal display: Any solid-state device which can display the numbers 0 to 9 and the alphabetical characters A to F

High-pass filter: An optical filter with a single transmission band that starts at the cutoff frequency and extends to infinite frequency

Hot cathode: A thermionic cathode that generates electrons by thermionic emission

Hue: The dominant optical wavelength which determines a color other than gray

Hybrid communications network: A system network with the ability to operate with analog and digital signals or other types of dissimilar modes

Illuminance (E_V): The ratio of the luminous radiant flux incident on an element of the surface of a given body to the area of that element

Illumination sensitivity: The ratio of the output current to the incident illumination in a photosensitive device held at constant electrode potentials

Incandescence: The state when any given body is at such an elevated temperature that it emits light, usually in the visible light region

Induction furnace: A type of oven used in the preparation of optical materials that is heated by electromagnetic induction

Information feedback system: A type of system where an echo is used to check the transmission accuracy in the telecommunications line

Injection laser: A forward-biased semiconductor diode that emits radiation during the recombination process, usually modulated by altering the input current

Insertion loss: The ratio of the power delivered to a particular component or device to the power measured directly after the component, device, or coupling, usually given in decibels for optical and electrical transmissions

Integrating photometer: A type of light-measuring device that gives the average emitted candlepower from a given source from all directions or from all angles in a single plane

Intrinsic brightness: The ratio of the luminous intensity to the projected unit area from a given direction

Intrinsic concentration: The number of minority carriers in a semiconductor material or device above the average equilibrium number

Intrinsic detector: A type of photoconductive material that generates a potential without the addition of dopants, such as Si and CdS

Ionoluminescence: A type of phosphor luminescence caused by bombardment of positive ions

Irradiance (E_e): The ratio of the radiant flux incident on a given surface element to the area of the irradiated element, given in units of watts per square meter (W/m^2)

Isolation voltage (V_{is}): The maximum allowable operating voltage between the input and output of an opto-isolator

Junction capacitance (C_j): The total pn-junction diode capacitance which increases with decreasing applied reverse voltage

Junction filtering: The combined use of high- and low-pass filters to split two or more frequency bands for transmission over electrical or optical communications lines

Kerr effect: The phenomena where certain transparent materials can become double refracting when placed in an electric field perpendicular to the incident beam of radiation

Kirchoff's law: The ratio of the emissive power to the absorptivity of any body is the same for all bodies and is equal to the emissive power of a blackbody at the same absolute temperature

Lambert: A unit of luminance equal to $1/\pi$ cd/cm or 1 L/cm^2

Lambertian emitter: An optical light source where the transmitted radiation is distributed uniformly in all directions

Lambert's law: The illumination on a surface element from a point source is inversely proportional to the square of the distance to the source for radiation perpendicular to the source; for an off-axis element the illumination is proportional to the cosine of the source to surface off-axis angle

Lateral chromatic aberration: The optical error affecting abaxial image sharpness caused by magnification variations of different colors

Leakage current (I_R): The current which passes through a semiconductor junction when reverse bias is applied

Leaky waveguide: An optical waveguide containing a narrow longitudinal slot intended to allow a predetermined continuous amount of energy to escape from the waveguide

Lens speed: The ratio of the lens focal length to the diameter, giving an indication of the amount light the lens passes

Lens transmission: The ratio of intensities of a collimated beam of light before entering and after passing through an optical lens

Light dispersion: The process in which white light is separated into its various component wavelengths by traveling at different angles from the central incident beam

Light source efficiency: The ratio of the total luminous flux to the total power consumed by a given light source, given in lumens per watt

Limit of resolution: The minimum angle of resolution of an optical lens is given by $\theta = 1.22\lambda/\alpha$, where α is the aperture diameter and λ is the wavelength of the incident light

Linear detector: Any electrical or optical detection device that gives a response or signal output directly proportional to the input

Line spectrum: A type of radiation spectrum produced by discharges in monatomic gases consisting of lines of narrow wavelength intervals or single lines separated by regions or intervals of little or no energy

Longitudinal chromatic aberration: A type of image error caused by the focusing of different colors at different distances from the center of the lens

Lossev effect: The radiation emitted in a semiconductor material when the charge carriers recombine

Lumen (lm): The luminous flux emitted within a unit solid angle (1 sr) by a point source of radiation having a uniform intensity of 1 cd

Luminance: The luminous flux given from a radiation source given in units of lumen per steradian per square meter, candela per square foot, meter-lambert, or footlambert

Luminescent spectrum: A type of radiation spectrum produced by fluorescent or phosphorescent emissions consisting of broad continuous bands over limited regions of the spectrum

Luminosity factor (K): The ratio of the luminous flux to the radiant flux measured at a given wavelength, given in units of lumens per watt

Luminous intensity (I_V): The ratio of the luminous flux from a given source within an element of solid angle to the element of solid angle, given in units of candela

Luminous sensitivity: The ratio of the device photocurrent to the incident luminous flux in A/lm or the illuminance in A/lx

Lux (lx): Illuminance made by a luminous flux of 1 lm uniformly distributed over 1 m

Magnifying power: The ratio of the angle from the principal axis made by the image of an object seen through an optical device to the angle of the object to the principal axis when directly observed

Majority carrier: The carrier type occurring in the largest number in semiconductor materials; in p-type semiconductors, it is holes; and in n types, it is electrons

Mandrel test: A common procedure to determine the relative flexibility or the minimum bend radius of a fiber or cable; results usually given in centimeters radius

Matched filter: An optical filter which can separate a desired transmission band from undesired noise or signals

Matching factor (m): The quotient of the minimum to maximum radiant flux measured on devices which can be considered to be an array

Mean spherical intensity: The average measured intensity of a radiation source integral to all spherical directions

Minimum deviation: The smallest change in the direction of light passing through a prism occurs when the angle of incidence is the same as the angle of emergence; $n(\lambda) = \sin (1/2)(\theta_p + \theta_D)/\sin (1/2)\theta_p$, where n is the prism refractive index at wavelength λ, θ_D is the minimum deviation angle, and θ_p is the angle of the prism

Minority carrier: The carrier type occurring in the smallest number in semiconductor materials; in p-type semiconductors, it is electrons, and in n types, it is holes

Mired: An abbreviation for microreciprocal degree, which is a color temperature measurement scale; the mired value of a given light source is given by the ratio of 10^6 to the source color temperature in Kelvin

Mismatch loss: The ratio of the optimal power that can be absorbed to the actual measured absorbed power from a given light source or transmission line

Mobility (u): The degree of ease with which carriers can travel through a semiconductor material in the presence of an applied potential

Modal dispersion: The component of pulse spreading caused by differential optical path lengths in a multimode optical waveguide

Mode filter: A type of waveguide filter that can separate signals of different transmission modes at the same frequency

Modulation capability: The ability of a device or circuit to attain the highest degree of modulation without the introduction of undesirable distortion or noise

Monochromatic sensitivity: The ability of an optical device to respond to light of a single wavelength or color

Mosaic detector: A type of detector consisting of a number of individual elements arranged in a two- or three-dimensional pattern

Multiplexer: A type of device or circuit which can simultaneously combine and transmit two or more signals on a single channel

Neutral density filter: A gray filter which decreases the light intensity passing through the filter without altering the relative spectral distribution

Nodal points: Two points on the axis of a lens or lens system such that a ray entering the lens in the direction of one point leaves as if from the other end parallel to the original direction

Noise equivalent bandwidth (NEB): The equivalent bandwidth of a sharp-cutoff noise spectrum having the same noise power and maximum value as the measured output noise of a given device, circuit, or system, given in units of hertz

Noise equivalent power (NEP): The ratio of the broadband output noise voltage or current to the responsivity, given in units of watts

Numerical aperture: The measure of the degree of light acceptance of a fiber defined by $NA = (N_{core}^2 - N_{clad}^2)^{1/2}$, where N is the refractive index

Off-state collector current [$I_C(off)$]: The measured output current when the input current is zero for an opto-coupler

On-state collector current [I_C(on)]: The measured output current when the input current level is above the threshold for an opto-coupler

Optical cement: A permanent transparent adhesive, usually epoxy or methacrylate, capable of handling extreme temperatures

Optical communications: Communications that use radiation beams (visible), ultraviolet (uv) and infrared (ir), transmitted either through the atmosphere or through transparent fibers or cables

Opto-coupler: Any device designed to be used to transform electrical signals using optical energy to attain a coupling with electrical isolation between the output and input; usually made with a GaAs diode and a silicon phototransistor

Optoelectronic transistor: A type of transistor that has a transparent base with a photoelectric collector and an electroluminescent emitter

Packing fraction: The ratio of the active fiber core area to the total cross-sectional area of a fiber bundle

Parallax: A type of optical error or visual illusion that causes an object to appear to be physically displaced when viewed from nonorthogonal angles

Passband: The range of wavelengths in a bandpass filter between the cut-on and cutoff

Passband ripple: The difference between the minimum and maximum loss points in a filter within a given bandwidth

Period: A given sequence of thin-film layers on optical lenses which is repeated to build up a stack reflector

Phosphorescence: The emission of light from a material after the electrical stimulus has been removed

Photocathode luminous sensitivity: The ratio of the photoelectric emission current to the incident luminous flux, which is usually a collimated beam orthogonal to the photocathode

Photocurrent: The difference between the light and dark currents, $I_L - I_D$, in a given photodetector

Photocurrent gain (M): The quotient of the photocurrent at a certain reverse voltage to the photocurrent at a 10-V bias

Photodielectric effect: The change in the dielectric constant of phosphors under illumination

Photoelastic: Types of transparent isotropic dielectrics that change one or more optical properties under an applied stress

Photoelectric material: A type of substance that can emit electrons when illuminated in a vacuum

Photoelectric sensitivity: The rate at which electrons are emitted from a material element per unit of radiant flux at a specified wavelength

Photoluminescence: A type of phosphor luminescence caused by visible or uv radiation

Photovoltaic effect: The generation of an electric field in a substance in the presence of a radiant source

Polarized light: A type of light propagation that exhibits different energy properties in different directions orthogonal to the line of travel

Preform: A rod of optical material, usually various types of carefully doped glasses, which is subsequently drawn into optical fibers

Primary colors: A basic set of colors from which all other colors can be made by mixing the primaries in various proportions

Priming illumination: A small beam of light incident to a photoelectric device to make it more sensitive to the radiation being measured

Pulse duty factor: The average pulse duration times the pulse-repetition rate

Pulse spreading: The increase in pulse width within a given length of fiber due to the effects of modal and material dispersion

Quantum efficiency (QE): The ratio of the number of emitted photoelectrons to the number of incident photons at a given wavelength, given by $QE = hcs(\lambda)/e\lambda$ where $s(\lambda)$ is the spectral sensitivity in amperes per watt at wavelength λ

Quantum noise: The random signal variations caused by changes in the average quantum radiation incident upon a detector

Quarter-wave stack: A type of multilayer dielectric film where each layer thickness is 1/4 of the given wavelength; successive film layers usually alternate between high- and low-index-of-refraction materials

Quenching: Any process in phosphor luminescence which reduces the radiant emission

Radiance (L_e): The ratio of the radiant flux at an element of a given body or device and the product of the solid angle of the cone and the area of the orthogonal projection of the element, given in units of watts per steradian per square meter

Radiant exitance (M_e): The ratio of the actual measured radiant power leaving an element of the surface of a given body by the area of the element, given in units of watts per square meter

Radiant sensitivity: The ratio of the device photocurrent to the incident radiant power in amperes per watt or the incident irradiance in amperes per watt per square meter

Radio luminescence: A type of phosphor luminescence caused by high-energy radiation such as χ, β, and γ rays

Rayleigh scattering: A scattering of the incident radiation through a fiber inversely proportional to the fourth power of the wavelength, caused by various heterogeneities in the fiber

Recombination radiation: The radiation emitted in a semiconductor material when the holes in the valence band recombine with the electrons in the conduction band

Refractive index: The ratio of the velocity of light in a vacuum to its velocity in the given material

Rejection ratio: The quotient of the power transmitted within the passband to the power outside the passband

Relative luminosity: The ratio of the actual measured luminosity to the maximum luminosity at a specified wavelength

Relative spectral sensitivity: The ratio of the radiant sensitivity at a given wavelength to the radiant sensitivity at a reference wavelength usually taken as the wavelength of maximum response

Resolving power: The degree to which an optical lens or lens combination can separate two points close together or the distance two points can be separated before the lens can distinguish two separate images

Rise time (t_r): The time required for the photocurrent to rise from 10 to 90% of the maximum value of a given device or system

Roentgeno luminescence: A type of phosphor luminescence caused by x-ray excitation

Saturation current: The output current at which little or no change occurs by an increase in the operating voltage under constant irradiance, or in the irradiance under constant operating conditions

Saturation voltage: The minimum operating voltage which causes little or no change in the photocurrent when the voltage is increased under conditions of constant radiation.

Signal-to-noise ratio (SNR): The ratio of the power in the transmitted signal to the undesirable noise present in the absence of any signal

Single mode fiber: An optical fiber of core diameter d whose value V is <2.4, where $V = \pi d(NA)/\lambda$, where λ is the wavelength of the transmitted light

Slope: The rate at which the transmittance changes in the transition region, defined as slope = $[\lambda(80) - \lambda(5)]/\lambda(50)$, where $\lambda(x)$ is the wavelength where the transmission is x% of peak transmission

Smoothing filter: A ripple filter used to remove ac components or random fluctuations of a given transmission signal

Snell's law: The ratio of the sine of the angle of incidence to the sine of the angle of refraction is equal to the index of refraction

Spectral bandwidth ($\Delta\lambda$): The range of wavelengths where the spectral concentration of a given photometric or radiometric quantity is greater than half of its maximum value

Spectral flatness: The variation of the transmittance over a given range of wavelengths

Spike filter: A special type of interference filter that has a full width at half-maximum bandwidth of less than 1% of the given center wavelength

Stack: A type of multilayer reflecting film used in the construction of an interference filter cavity and other devices

Standard noise temperature: The reference temperature for noise measurements specified at 290° K

Stefan-Boltzmann law: The total emitted radiant energy from a surface element of a blackbody is proportional to the fourth power of its absolute temperature, with the proportionality constant equal to 5.672×10^{-5} ergs/cm^2/T^4/sec

Step-index fiber: An optical fiber whose core has a uniform refractive index up to the inner edge of the cladding

Stopband: That part of the wavelength spectrum that is subjected to a specified signal attenuation by a filter or some other interference device

Surface reflection: The fraction of incident light reflected (r) from the surface of a transparent refractive material given by $r = [(n - 1)/(n + 1)]^2$, where n is the refractive index

Temperature coefficient (TK): The quotient of the relative change of a given electrical quantity and the incremental change in temperature which caused it

Temperature derating: The lowering of the current or voltage applied to a device or circuit which is subject to elevated temperature operation

Thermal noise: The resistance noise caused by the random thermal motion of charged particles

Total emissivity: The ratio of the radiation emitted from the surface element of a given body to that omitted by a blackbody under identical conditions

Transmittance: The ratio of the radiant power transmitted through a given material to the incident radiant power

Visibility: The ratio of the luminous flux to the total radiant energy given in units of lumens per watt

Waveguide critical dimension: The width of a waveguide cross which determines the critical wavelength for propagation along the waveguide

Wedge filter: A type of optical filter where the density increases from one end to the other or with increasing angles about a circular disk

YAG laser: The solid-state laser of small surface area and volume used in fiber optics, composed of yttrium aluminum garnet

Zero-order transmission grating: A type of optical grating that transmits long wavelengths and diffracts shorter wavelengths in various predetermined directions

Chapter 2

SYMBOLS, DESIGNATIONS, AND CONVERSIONS

This chapter contains the internationally accepted symbols for semiconductor and electro-optics devices, symbols for various terms and units of measure, and a conversion factor table. The derived units of measure for most quantities are based on the international symbols for the various units of measurement as listed.

Measurement	Unit	Symbol
Electric current	Ampere	A
Length	Meter	m
Luminous intensity	Candela	cd
Mass	Kilogram	kg
Temperature	Kelvin	K
Time	Second	s

Table 1 Physical Constants

Quantity	Symbol	Value
Atomic mass unit	amu	$1.6605655 \times 10^{-27}$ kg 9.3148125×10^{8} eV
Bohr magneton		$9.2740966 \times 10^{-24}$ JT^{-1} 5.7883811×10^{-5} eVT^{-1} 1.3996108×10^{10} HzT^{-1} 4.6685981×10^{-1} $cm^{-1}T^{-1}$ 6.7173329×10^{-1} KT^{-1}
Bohr radius	α_0	$5.2917706 \times 10^{-11}$ m
Boltzmann constant	K	$1.3806624 \times 10^{-23}$ JK^{-1}
Electron Compton wavelength	λ_{ce}	$2.4263089 \times 10^{-12}$

Table 1-Continued

Electron magnetic moment	μ_e	$9.2848516 \times 10^{-24}$ JT^{-1}
Electron radius (classical)	r_e	$2.8179380 \times 10^{-15}$ m
Electron rest mass	M_e	$9.1095344 \times 10^{-31}$ kg 5.1100411×10^{5} eV
Electron volt	eV	$1.6021917 \times 10^{-19}$ J 2.4179659×10 Hz 8.0654653×10 m^{-1} 1.1604855×10 K
Elementary charge	e	$1.6021892 \times 10^{-19}$ C
Faraday constant	F	9.6484562×10^{4} C $mole^{-1}$
Gas constant	R_o	8.3143435×10^{3} Jk $mole^{-1}$-K^{-1} 8.2056235×10^{-2} m -atm^3K $mole^{-1}$ K^{-1}
Gravitational acceleration	g_e	9.80621 ms^{-2}
Gravitational constant	G	6.672041×10^{-11} Nm^2 kg^{-2}
Magnetic flux quantum	Φ_o	$2.0678506 \times 10^{-15}$ Wb
Neutron Compton wavelength	λ_{cn}	$1.3196217 \times 10^{-15}$ m
Neutron rest mass	M_n	$1.6749543 \times 10^{-27}$ kg 9.3955275×10^{8} eV
Nuclear magneton	μ_n	$5.0509515 \times 10^{-27}$ JT^{-1} 3.1525262×10^{-8} eVT^{-1} 7.6227004×10^{6} HzT^{-1} 2.5426591×10^{-2} $m^{-1}T^{-1}$ 3.6584616×10^{-4} KT^{-1}
Permeability of vacuum	μ_o	$1.25663706 \times 10^{-6}$ Hm^{-1}
Permittivity of vacuum	ϵ_o	$8.8541878 \times 10^{-12}$ Fm^{-1}
Planck constant	h	$6.6261763 \times 10^{-34}$ JHz^{-1}
Proton Compton wavelength	λ_{cp}	$1.3214409 \times 10^{-15}$ m
Proton magnetic moment	μ_p	$1.4106203 \times 10^{-26}$ JT^{-1}

Table 1-Continued

Proton rest mass	M_p	$1.6726485 \times 10^{-27}$ kg 9.3825925×10 eV
Radiation constant (first)	C_1	$3.7418322 \times 10^{-16}$ Wm^2
Radiation constant (second)	C_2	1.4387864 mK
Rydberg constant	R_∞	1.0973732×10^7 m^{-1} $2.1799142 \times 10^{-18}$ J 1.3605826×10^1 eV 3.2898423×10^{15} Hz 1.5789367×10^5 K
Speed of light in vacuum	c	2.99792458×10^8 ms^{-1}
Stefan-Boltzmann constant	σ	$5.67037271 \times 10^{-8}$ Wm^{-2} K^{-4}
Wavelength of krypton-86, (orange-red line)	λ(Kr86)	6057.80211 Å

Table 2 Prefix Names for Powers of 10

Prefix	Symbol	Power of 10
Peta	P	+18
Exa	E	+15
Tera	T	+12
Giga	G	+9
Mega	M	+6
Kilo	k	+3
Hecto	h	+2
Deca	da	+1
Deci	d	-1
Centi	c	-2
Milli	m	-3
Micro	μ	-6
Nano	n	-9
Pico	p	-12
Femto	f	-15
Atto	a	-18

Table 3 Semiconductor Device Type Designation Code

The type designation code for semiconductor devices consists of two letters followed by a serial number signifying material, function, and serial number. The first letter gives information about the material used for the active part of the device.

A Materials with a band gap of 0.6 to 1.0 eV, such as germanium
B Materials with a band gap of 1.0 to 1.3 eV, such as silicon
C Materials with a band gap greater than 1.3 eV, such as GaAs
D Materials with a band gap less than 0.6 eV, such as InSb
R Compound materials as employed in photoconductive cells, such as CdS

The second letter indicates the circuit function or main application and device construction.

A Diode: detection, mixer, switching
B Variable capacitance diode
C Transistor for audio frequency (a.f.) applications
D Power transistor for a.f. applications
E Tunnel diode
F Transistor for high frequency (h.f.) applications
G Oscillator diode or multiple of dissimilar devices, miscellaneous
H Magnetic sensitive diode, field probe
K Hall effect device in an open magnetic circuit
L Power transistor for h.f. applications
M Hall effect device in a closed magnetic circuit
N Photocoupler
P Radiation sensitive diode
Q Radiation generating diode
R Low-power thyristor
S Low-power switching transistor
T Power thyristor
U Power switching transistor
X Multiplier diode, e.g., step recovery, varactor
Y Diode: booster, efficiency, rectifying
Z Diode: transient suppressor, voltage reference or regulator

The serial number consists of three figures ranging from 100 to 999 for semiconductor devices primarily for use in domestic equipment; and one letter and two figures ranging from 10 to 99 for devices primarily used in professional equipment.

A version letter can be used to indicate a deviation of a single electrical or mechanical characteristic. The letter never has a fixed meaning, except for the letter R indicating reversed voltage, i.e., collector to case.

Table 4 Symbols for Semiconductor Devices

Term	Symbol[a]
Capacitance	C, c
Current	I, i
Power	P, p
Resistance	R, r
Time	T, t
Voltage	V, v

The following subscripts can be used with the above quantities to describe numerous quantities in semiconductor device.

Anode terminal	A, a
Average value	AV, AVG, av, avg
Base terminal	B, b
Breakdown	BR, br
Collector terminal	C, c
Delay; drain terminal; diode	D, d
Emitter terminal	E, e
Forward	F, f
Gate	G, g
Junction	J, j
Cathode terminal	K, k
Peak value	M, m
Reverse	R, r (first subscript)
Repetitive	R, r (second subscript)
Root-mean-square value	RMS, rms
Source terminal	S, s (first subscript)
Nonrepetitive	S, s (second subscript)
Specified circuit	X, x
Actual working voltage	Z, z

[a]Lower case letters are usually used to signify instantaneous values.

Table 5 Symbols for Quantities and Terms

Quantity	Symbol
Absorption coefficient	a
Absorption factor	α
Acceptable quality level	AQL
Actinity of radiation	α
Admittance	γ
Angle of half-intensity	α
Anode terminal	A
Attenuation coefficient	α
Avalance photodiode	APD
Capacitance	C
Charge density	p
Circular wavenumber	k
Color temperature	T_c
Conductance	G
Conductivity	γ
Coupling coefficient	k
Current transfer ratio	CTR, K
Cutoff frequency	f_g
Demodulation bandwidth	B
Dielectric polarization	P
Electric current density	j
Electric dipole moment	P
Electric displacement	D
Electric field strength	E
Electric flux	ψ
Electric force	E
Electric potential	V
Electric susceptibility	χ_e
Electromagnetic moment	m
Extinction coefficient	μ
Frequency	ν
Gain bandwidth product	$G\beta$
Illuminance	E_v
Illumination	E_v
Impedance	Z
Irradiance	E_e
Light-emitting diode	LED
Loss angle	δ
Luminance	L_v
Luminous emittance	M_v
Luminous flux	Φ_v
Luminous intensity	I_v
Magnetic field strength	H
Magnetic flux	Φ

Table 5–Continued

Magnetic flux density	B
Magnetic induction	B
Magnetic polarization	J
Magnetic susceptibility	χ_m
Magnetic vector potential	A
Magnetization	M
Matching factor	m
Modulus of elasticity	E
Mutual inductance	M
Noise-equivalent bandwidth	Δf
Noise-equivalent power	NEP
Normal stress	σ
Permeability	μ
Permeability of vacuum	μ_0
Permittivity	ϵ
Permittivity of vacuum	ϵ_0
Phase coefficient	β
Phase number	m
Polarizability	α
Propagation coefficient	γ
Radiance	L_e
Radiant emittance	M_e
Radiant energy	Q_e
Radiant flux	Φ_e
Radiant intensity	I_e
Radiant power	Φ_e
Radiant sensitive area	A
Reactance	X
Reflection factor	ρ
Refractive index	n
Relative permeability	μ_r
Relative permittivity	ϵ_r
Relaxation time	τ
Resistance	R
Resistivity	ρ
Self-inductance	L
Sensitivity	s
Spectral bandwidth	$\Delta\lambda$
Speed of light	C
Surface charge density	σ
Transmission factor	τ
Wavelength	λ
Wavenumber	σ
Work function	Φ

Table 6 Greek Alphabet

Greek letter Small	Capital	English name	Common symbol usage
α	A	Alpha	Absorption factor, actinity of radiation, Angles, area, attenuation factor, polarizability
β	B	Beta	Angles, coefficients, phase constant
γ	Γ	Gamma	Angles, electrical conductivity, propagation coefficient
δ	Δ	Delta	Angles, density, determinant, increments, permittivity
ϵ	E	Epsilon	Dielectric constant, electric intensity, Napierian logs, permittivity
ζ	Z	Zeta	Coefficients, spacial coordinates
η	H	Eta	Charge density, efficiency, intrinsic impedance, spacial coordinates
θ	Θ	Theta	Angles, phase displacements, time constants
ι	I	Iota	Nominal quantity, unit vector
κ	K	Kappa	Coupling coefficient, susceptibility
λ	Λ	Lambda	Attenuation constant, permeance, wavelength
μ	M	Mu	Amplification factor, extinction coefficient
ν	N	Nu	Frequency, reluctivity
ξ	Ξ	Xi	Spatial coordinates
o	O	Omicron	Reference subscripts
π	Π	Pi	Circumference to diameter ratio
ρ	P	Rho	Density, reflection factor, resistivity
σ	Σ	Sigma	Charge density, conductivity, summation
τ	T	Tau	Relaxation time, volume intensity, transmission factor
υ	Υ	Upsilon	Equation symbol

Table 6–Continued

ϕ	Φ	Phi	Angles, luminous flux, magnetic flux, work function
χ	X	Chi	Angles, electric susceptibility
ψ	Ψ	Psi	Angles, electric flux, phase difference
ω	Ω	Omega	Angular velocity, resistance, spacial coordinates

Table 7 Common Basic Units of Measure

Measurement	Basic unit	Symbol
Absorbed dose	Gray	Gy
Conductance	Siemens	S
Electric capacitance	Farad	F
Electric charge	Coulomb	C
Electric current	Ampere	A
Electric potential	Volt	V
Electric resistance	Ohm	Ω
Energy	Joule	J
Force	Newton	N
Frequency	Hertz	Hz
Illumination	Lux	lx
Inductance	Henry	H
Luminous flux	Lumen	lm
Magnetic flux	Weber	Wb
Magnetic flux density	Tesla	T
Power	Watt	W
Pressure	Pascal	Pa
Radioactivity	Becquerel	Bq

Table 8 Common Derived Units of Measure

Quantity	Derived unit	Unit symbol
Acceleration	Meter per second squared	m/s^2
Angular acceleration	Radian per second squared	rad/s^2
Angular velocity	Radian per second	rad/s
Current density	Ampere per square meter	A/m^2
Electric charge density	Coulomb per cubic meter	C/m^3
Electric field strength	Volt per meter	V/m
Electric flux density	Coulomb per square meter	C/m^2
Energy density	Joule per cubic meter	J/m^3
Irradiance	Watt per square meter	W/m^2
Luminance	Candela per square meter	cd/m^2
Magnetic field strength	Ampere per meter	A/m
Permeability	Henry per meter	H/m
Permittivity	Farad per meter	F/m
Radiance	Watt per square meter-steradian	W/m^2-sr
Radiant intensity	Watt per steradian	W/sr
Thermal conductivity	Watt per meter-Kelvin	W/m-k
Velocity	Meter per second	m/s
Wavenumber	Unity per meter	l/m

Table 9 Conversion Factors

To convert from	To	Multiply by
Abamperes	Amperes	1.0000×10
	E.S. cgs units	2.9979×10^{10}
	Faradays per second	1.03610×10^{-4}
	Statamperes	2.9979×10^{10}
Abcoulombs	Ampere-hours	2.7778×10^{-3}
	Coulombs	1.0000×10
	Electron changes	6.2420×10^{19}
	E.S. cgs units	2.9979×10^{10}
	Faradays	1.0361×10^{-4}
	Statcoulombs	2.9979×10^{10}
Abfarads	E.S. cgs units	8.9876×10^{20}
	Farads	1.0000×10^{9}
	Statfarads	8.9876×10^{20}
Abhenries	E.S. cgs units	1.1126×10^{-21}
	Henries	1.0000×10^{-9}

Table 9-Continued

Abmhos	E.S. cgs units	8.9876×10^{10}
	Mhos	1.0000×10^{9}
	Statmhos	8.9876×10^{20}
Abohms	Ohms	1.0000×10^{-9}
	Statohms	1.1126×10^{-21}
Abvolts	Volts	1.0000×10^{-8}
Abvolts per centimeter	E.S. cgs units	3.3356×10^{-11}
Amperes	Abamperes	1.0000×10^{-1}
	Faradays per second	1.0361×10^{-5}
	Statamperes	2.9979×10^{9}
Ampere-turns	E.M. cgs units	1.2566
	E.S. cgs units	3.7673×10^{10}
	Gilberts	1.2566
Ångstroms	Centimeters	1.0000×10^{-8}
	Inches	3.9370×10^{-9}
	Millimicrons	1.0000×10^{-1}
Apostilbs	Candela per square foot	2.957×10^{-2}
	Footlamberts	9.290×10^{-2}
	Lamberts	1.000×10^{-4}
	Nits	3.183×10^{-1}
	Stilbs	3.183×10^{-5}
Atmospheres	Centimeter of Hg (0° C)	7.6000×10
	Centemeters of H_2O (4° C)	1.0333×10^{3}
	Dynes per square centimeter	1.4696×10^{6}
	Pounds per square inch	1.4696×10
	Torrs	7.6000×10^{2}
Btu	Calories (g)	2.5200×10^{2}
	Ergs	1.0544×10^{10}
	Foot pounds	7.7765×10^{2}
	Gram-cm	1.0751×10^{7}
	Horsepower-hours	3.9275×10^{-4}
	Joules	1.0544×10^{3}
	Kilowatt hours	2.9288×10^{-4}
	Liter-atm	1.0405×10
Btu per second	Horsepower	1.4139
	Watts	1.0544×10^{3}
Cable lengths	Fathoms	1.2000×10^{2}
	Feet	7.2000×10^{2}
	Meters	2.1946×10^{2}

Table 9–Continued

Calories (g)	Btu	3.9683×10^{-3}
	Cubic feet-atm	1.4582×10^{-3}
	Ergs	4.1840×10^{7}
	Foot pounds	3.0860
	Kilogram-meters	4.2665×10^{-1}
	Watt-hours	1.1622×10^{-3}
Candles per square foot	Apostlibs	3.3816×10^{1}
	Footlamberts	3.1416
	Lamberts	3.3816×10^{-3}
	Nits	1.0764×10
	Stilbs	1.0764×10^{-3}
Circumferences	Degrees	3.600×10^{2}
	Minutes	2.1600×10^{4}
	Quadrant	4.0000
	Radians	6.2832
	Seconds	1.2960×10^{6}
Coulombs	Abcoulombs	1.0000×10^{-1}
	Ampere-hours	2.7777×10^{-4}
	Electron charges	6.2420×10^{18}
	E.S. cgs units	2.9979×10^{9}
	Faradays	1.0361×10^{-5}
	Statcoulombs	2.9979×10^{9}
Degrees	Circumferences	2.7778×10^{-3}
	Minutes	6.0000×10
	Quadrants	1.1111×10^{-2}
	Radians	1.7453×10^{-2}
	Seconds	3.6000×10^{3}
Dynes	Grams	1.0197×10^{-3}
	Newtons	1.0000×10^{-5}
Dynes per square centimeter	Atmospheres	9.8692×10^{-7}
	Centimeters of Hg (0° C)	7.5006×10^{-5}
	Centimeters of H_2O (0° C)	1.0197×10^{-3}
	Pascals	1.0000×10^{-1}
Dyne-centimeter	Ergs	1.0000
	Foot pounds	7.3756×10^{-8}
	Newton-meters	1.0000×10^{-7}
Electron volts	Ergs	1.6022×10^{-12}
	Grams	1.7825×10^{-33}
	Joules	1.6022×10^{-19}

Table 9-Continued

Electron charges	Abcoulombs	1.6021×10^{-20}
	Coulombs	1.6021×10^{-19}
	Statcoulombs	4.8030×10^{-10}
Ergs	Btu	9.4845×10^{-11}
	Calories (g)	2.3901×10^{-8}
	Cubic centimeter-atm	9.8692×10^{-7}
	Electron volts	6.2415×10^{11}
	Kilogram-meters	1.0197×10^{-8}
	Liter-atm	9.8692×10^{-10}
Farads	Abfarads	1.0000×10^{-9}
	E.S. cgs units	8.9876×10^{11}
	Statfarads	8.9876×10^{11}
Fathoms	Centimeters	1.8288×10^{2}
	Inches	7.2000×10
	Miles (naut.)	9.8747×10^{-4}
	Miles (stat.)	1.1364×10^{-3}
Footcandles	Lumens per square foot	1.0000
	Lumens per square meter	1.0764×10
	Lux	1.0764×10
	Phots	1.0764×10^{-3}
Footlamberts	Apostlibs	1.0764×10
	Candela per square centimeter	3.4263×10^{-4}
	Candela per square foot	3.1831×10^{-1}
	Lamberts	1.0764×10^{-3}
	Lumens per square foot	1.0000
	Nits	3.4263
	Stilbs	3.4263×10^{-4}
Gauss	E.S. cgs units	3.3356×10^{-11}
	Maxwells per square centimeter	1.0000
	Lines per square inch	6.4516
Gilberts	Amperes-turns	7.9577×10^{-1}
	E.S. cgs units	2.9979×10^{10}
Gilberts per centimeter	Ampere-turns per inch	2.0213
	Oersteds	1.0000
Grams	Carats	5.0000
	Dynes	9.8067×10^{2}
	Grains	1.5432×10
	Ounces	3.5274×10^{-2}
	Pounds	2.2046×10^{-3}

Table 9-Continued

Henries	Abhenries	1.0000×10^{9}
	E.S. cgs units	1.1126×10^{-12}
	Stathenries	1.1126×10^{-12}
Henries/meter	E.M. cgs units	7.9577×10^{5}
	E.S. cgs units	8.8542×10^{-10}
	Gauss per oersted	7.9577×10^{5}
Horsepower	Btu per hour	2.4572×10^{3}
	Calories (g) per second	1.7830×10^{2}
	Ergs per second	7.4600×10^{9}
	Joules per second	7.4600×10^{2}
	Kilowatts	7.4600×10^{-1}
Joules	Btu	9.4845×10^{-4}
	Calories (g)	2.3901×10^{-1}
	Ergs	1.0000×10^{7}
	Foot pounds	7.3756×10^{-1}
	Gram-cm	1.0197×10^{4}
	Horsepower-hours	3.7251×10^{-7}
	Liter-atm	9.8689×10^{-3}
	Watt-second	1.0000×10^{4}
Lamberts	Apostlibs	1.0000×10^{4}
	Candela per square foot	2.9572×10^{2}
	Footlamberts	9.2903×10^{2}
	Nits	3.1831×10^{3}
	Stilbs	3.1831×10^{-1}
Lumens	Candlepower	7.9577×10^{-2}
	Watts	1.4706×10^{-3}
Lumens per square meter	Footcandles	9.2903×10^{-2}
	Lumens per square foot	9.2903×10^{-2}
	Phots	1.0000×10^{-4}
Lux	Footcandles	9.2903×10^{-2}
	Lumens per square meter	1.0000
	Phots	1.0000×10^{-4}
Maxwells	E.S. cgs units	3.3356×10^{-11}
	Gauss-square centimeter	1.0000
	Volt-seconds	1.0000×10^{-8}
	Webers	1.0000×10^{8}
Meters	Angstroms	1.0000×10^{10}
	Fathoms	5.4681×10^{-1}
	Feet	3.2808

Table 9–Continued

Meters	Inches	3.9370 X 10
	Miles (naut.)	5.3961 X 10^{-4}
	Miles (stat.)	6.2137 X 10^{-4}
Mhos	Abmhos	1.0000 X 10^{-9}
	E.S. cgs units	8.9876 X 10^{11}
	Statmhos	8.9876 X 10^{11}
Millimeters of Hg (0° C)	Atmospheres	1.3158 X 10^{-3}
	Dynes per square centimeter	1.3332 X 10^{3}
	Pounds per square inch	1.9337 X 10^{-2}
	Torrs	1.0000
Minutes	Circumferences	4.6296 X 10^{-5}
	Degrees	1.6667 X 10^{-2}
	Quadrants	1.8519 X 10^{-4}
	Radians	2.9089 X 10^{-4}
	Seconds	6.0000 X 10
Newtons	Dynes	1.0000 X 10^{5}
	Pounds	2.2481 X 10^{-1}
Newton-meters	Gram-centimeter	1.0197 X 10^{4}
	Pound-feet	7.3756 X 10^{-1}
Oersteds	Ampere-turns per meter	7.9577 X 10
	E.S. cgs units	2.9979 X 10^{10}
	Gilberts per centimeter	1.0000
Ohms	Abohms	1.0000 X 10^{9}
	Statohms	1.1126 X 10^{-12}
Phots	Footcandles	9.2902 X 10^{-1}
	Lux	1.0000 X 10^{4}
Quadrant	Circumferences	2.5000 X 10^{-1}
	Degrees	9.0000 X 10
	Minutes	5.4000 X 10^{3}
	Radians	1.5708
	Seconds	3.2400 X 10^{5}
Radians	Circumferences	1.5915 X 10^{-1}
	Degrees	5.7296 X 10
	Minutes	3.4377 X 10^{3}
	Quadrants	6.3662 X 10^{-1}
	Revolutions	1.5915 X 10^{-1}
	Seconds	2.0626 X 10^{5}

Table 9-Continued

Revolutions	Degrees	3.6000×10^{2}
	Radians	6.2832
Seconds	Circumferences	7.7161×10^{-7}
	Degrees	2.7778×10^{-4}
	Minutes	1.6667×10^{-2}
	Quadrants	3.0864×10^{-6}
	Radians	4.8481×10^{-6}
Statamperes	Abamperes	3.3356×10^{-11}
	E.M. cgs units	3.3356×10^{-11}
Statcoulombs	Ampere-hours	9.2657×10^{-14}
	Coulombs	3.3356×10^{-10}
	Electron charges	2.0821×10^{9}
	E.M. cgs units	3.3356×10^{-11}
Statfarads	E.M. egs units	1.1126×10^{-21}
Stathenries	Abhenries	8.9876×10^{20}
Statohms	Abohms	8.9876×10^{20}
Statvolts	Abvolts	2.9979×10^{10}
	Volts	2.9979×10^{2}
Steradians	Hemispheres	1.5915×10^{-1}
	Solid angles	7.9577×10^{-2}
	Spheres	7.9577×10^{-2}
	Square degrees	3.2828×10^{3}
Stilbs	Apostlibs	3.1416×10^{4}
	Candela per square foot	9.2902×10^{2}
	Footlamberts	2.9192×10^{3}
	Lamberts	3.1416
	Nits	1.0000×10^{4}
Volts	Abvolts	1.0000×10^{8}
	Statvolts	3.3356×10^{-3}
Watts	Btu per hour	3.4144
	Calories (g) per hour	8.6042×10^{2}
	Ergs per second	1.0000×10^{7}
	Horsepower	1.3596×10^{-3}
	Joules per second	1.0000
	Liter-atm per hour	3.5528×10
Watt-hours	Foot pounds	2.6552×10^{3}
	Horsepower hours	1.3410×10^{-3}

Table 9-Continued

Watt-hours	Joules	3.6000×10^3
	Kilogram meters	3.6710×10^2
Webers	E.M. cgs units	1.0000×10^8
	Volt-seconds	1.0000
Webers per square centimeter	Gauss	1.0000×10^8
	Lines per square inch	6.4516×10^8

Chapter 3

PHOTOMETRIC AND RADIOMETRIC QUANTITIES, STANDARDS, AND UNITS

Photometry

Photometry relates specifically to light in the wavelength region visible to the human eye, or between the 3800- and 7600-Å bands, and involves the measuring of those effects of radiant energy upon the retina of the eye that require interpretation by the observer, these measurements being made in terms of brightness and color. The standardized response of the human eye to light in this region is given in Table 10 and shown in Fig. 1. The response peaks at about 5550 Å in the green visible zone, and falls rapidly on either side to the red (infrared) or blue (ultraviolet) regions.

The basic terms used to describe and define light energy in photometry are as follows:

Illuminance: The ratio of the total luminous flux at a given distance from a light source to the receiving surface area, given in units of lumens per square centimeter (= $1m/cm^2$ = phot = ph) or lumens per square foot (= $1m/ft^2$ = footcandle)

Luminance: The ratio of the luminous intensity to the receiving surface area, given in units of lumens per square foot-steradian (= $1m/ft^2$-sr)

Luminous exitance: The ratio of the luminous flux emitted from a given light source to the surface area of the source, given in the same units as illuminance

Luminous flux: The total luminous power emitted from a light source, given in units of lumens (= 1m)

Luminous intensity: The ratio of the luminous flux emitted by a light source to the solid angle of emittance of the source, given in units of lumens per steradian (= 1m/sr = candela = cd)

Table 10 Spectral Response of Standard Human Eye

Wavelength (nm)	Normalized luminous efficiency	Luminous efficiency (lm/W)
380	0.0000	0.05
390	0.0001	0.13
400	0.0004	0.27
410	0.0012	0.82
420	0.0040	2.73
430	0.0116	7.91
440	0.0230	15.7
450	0.0380	25.9
460	0.0600	40.9
470	0.0910	62.1
480	0.1390	94.8
490	0.2080	142
500	0.3230	220
510	0.5030	343
520	0.7100	484
530	0.8620	588
540	0.9540	650
550	0.9950	679
555	1.0000	685
560	0.9950	679
570	0.9520	649
580	0.8700	593
590	0.7570	516
600	0.6310	430
610	0.5030	343
620	0.3810	260
630	0.2650	181
640	0.1750	119
650	0.1070	73.0
660	0.0610	41.4
670	0.0320	21.8
680	0.0170	11.6
690	0.0082	5.59
700	0.0041	2.78
710	0.0021	1.43
720	0.0010	0.716
730	0.0005	0.355
740	0.0003	0.170
750	0.0001	0.820
760	0.0001	0.041

Source: CIE Standard Tables.

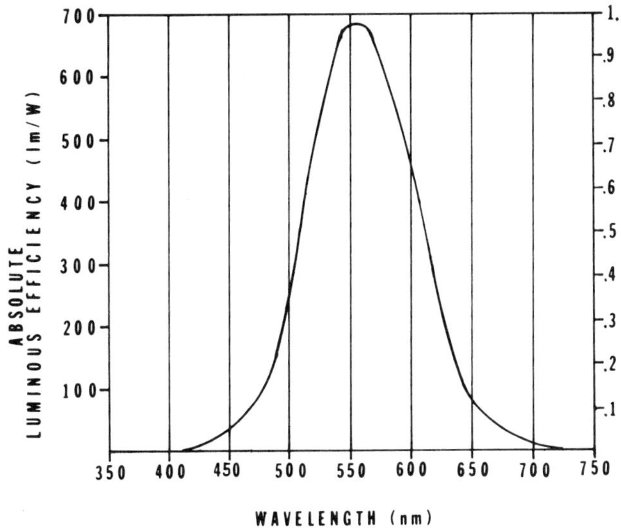

Fig. 1 Spectral response of standard human eye.

Each of these photometric quantities can be derived from the other given the details of the source and the space geometry from the source to the measuring device or impinging surface area. In all of the units for photometry and radiometry, it is implied that the source of radiant energy is defined with regard to size, shape, intensity, and distance to the given surface element or device. When considering spherical sources of light, it can become effectively a point source when the distance to the object of concern is at least 20 times the radius of the confining sphere of the source. For a simple planar source, the effective minimum point source distance is 10 times the factor $(\text{length}^2 + \text{width}^2)^{1/2}$, or its equivalent. For a line or tube source, the distance is at least 10 times the factor $(\text{length}^2 + \text{diameter}^2)^{1/2}$, or its equivalent. These are approximations used for the purpose of visualizing spatial relationships. For accurate work, the source to object distances must be measured very accurately for any type of radiation source. For source-object distances less than the above given minimum path lengths, care must be taken to define the radiation and object surface directions, as well as spatial distances.

Table 11 lists some of the commonly used luminosity color temperatures.

Table 12 lists the various photometric quantities with the appropriate units and symbols. Note that when there is any doubt as to whether a given quantity is photometric or radiometric, a subscript is used to differentiate between the two. For photometric terms the subscript is v, and for radiometric terms the subscript is e.

Table 11 Luminosity Standards

Standard source	Color temperature (K)
Hefner	1880
Harcourt pentane	1920
Standard British candle	1930
Melting point of platinum	2042
Incandescent carbon (4.0 W/candela)	2080
Incandescent tungsten (1.25 W/candela)	2400
Acetylene flame	2415

Table 12 Photometric Quantities

Quantity	Symbol	Equation definition[a, b, c]	Common units	Symbol
Absorptance	α, α_V	$\alpha = \Phi a/\Phi i$	Numerical	
Reflectance	ρ, ρ_V	$\rho = \Phi r/\Phi i$	Numerical	
Transmittance	τ, τ_V	$\tau = \Phi t/\Phi i$	Numerical	
Luminous energy	Q, Q_V	$Qv = \int_{380}^{760} K(\lambda)Q_e\lambda\, d\lambda$	Lumen-second	lm s
Luminous density	w, w_V	$w = dQ/dV$	Lumen-second per cubic meter	lm s/m^3
Luminous flux	Φ, Φ_V	$\Phi = dQ/dt$	Lumen	lm
Luminous exitance	M, M_V	$M = d\Phi/dw$	Lumen per square meter	lm/m^2
Illuminance	E, E_V	$E = d\Phi/dA$	Lux	lx
Luminous intensity	I, I_V	$I = d\Phi/dw$	Candela	cd
Luminance	L, L_V	$L = dl/(dA \cos \theta)$	Nit	nt
Luminous efficacy	K	$K = \Phi v/\Phi e$	Lumen per watt	lm/W
Luminous efficiency	V	$V = K/K_{max}$	Numerical	

[a] $\alpha + \rho + \tau = 1.0$.
[b] Subscripts: a = absorbed, i = incident, r = reflected, t = transmitted, max = maximum function value.
[c] Symbols: w = solid angle through which flux from a point source is radiated; θ = angle between line of sight and normal to emitting surface being measured.

Radiometry

The essential difference between radiometry and photometry is that in radiometry, measurements are not restricted to the visible region of the spectrum and commonly extend into the infrared and ultraviolet areas. Another difference is that radiometric quantities are based on watts and photometric quantities on lumens. Radiometry involves the measurement of electromagnetic radiation, which has numerous measurable entities such as frequency, intensity, modulation, coherence, and direction. The radiation source also has numerous measurable quantities, such as size, shape, temperature, material composition, and radiant output. Table 10 gives the conversion factor of lumens per watt for the visible (photometric) region. Outside of this region, no matter how high the radiant output, the photometric value in lumens is zero, since it is beyond the range of the standardized human eye response.

The basic terms used to describe and define light energy in radiometry are as follows:

Irradiance: The ratio of the total radiant flux density at a given distance from the light source to the projected surface area, given in units of watts per square centimeter (= W/cm^2)

Radiance: The ratio of the radiant intensity of the light source to the surface area of the source, given in units of watts per square centimeter-steradian (= W/cm^2-sr)

Radiant exitance: The ratio of the total radiant flux to the surface area of the source, given in units of watts per centimeter

Radiant flux: The total radiant power emitted from a light source, given in units of watts

Radiant intensity: The ratio of the radiant flux of the light source to the solid angle through which it is spread, given in units of watts per steradian (= W/sr)

Table 13 lists the various radiometric quantities with the appropriate units and symbols. Note that in Table 11, the radiance equation definition is given as $dI(dA \cos \theta)$. This can also be expressed as $d^2\Phi/dW(dA \cos \theta)$ for both the radiance and the luminance equations.

When a specific wavelength is measured or referred to in radiometry, it is often convenient to express quantities in their spectroradiometric terms. Table 14 lists the important quantities used to define spectral radiant quantities and lists the appropriate units and symbols. Note that the measurements and quantities are very similar to those in radiometry except for the use and reference to a specific unit frequency or wavelength.

Table 13 Radiometric Quantities

Quantity	Symbol	Equation definition	Common units	Symbol
Absorptance	α, α_e	$\alpha = \Phi_a/\Phi_i$	Numerical	
Reflectance	ρ, ρ_e	$\rho = \Phi_r/\Phi_i$	Numerical	
Transmittance	τ, τ_v	$\tau = \Phi_t/\Phi_i$	Numerical	
Radiant energy	Q, Q_e		Joule	J
Radiant density	W, W_e	W = dQ/dV	Joule per cubic meter	J/m^3
Radiant flux	Φ, Φ_e	$\Phi = dQ/dt$	Watt	W
Radiant exitance	M, M_e	$M = d\Phi/dA$	Watt per square meter	W/m^2
Irradiance	E, E_e	$E = d\Phi/dA$	Watt per square meter	W/m^2
Radiant intensity	I, I_e	$I = d\Phi/dw$	Watt per steradian	W/sr
Radiance	L, L_e	$L = dI/(dA \cos \theta)$	Watt per steradian-square meter	W/sr m^2
Emissivity[a]	ϵ	$\epsilon = M/M_{blackbody}$	Numerical	

[a]M = radiant exitance of measured object; $M_{blackbody}$ = radiant exitance of a blackbody at the same temperature as the measured object.
[b]Subscripts: a = absorbed, i = incident, r = reflected, t = transmitted, max = maximum function value.
[c]Symbols: W = solid angle through which flux from a point source is radiated; θ = angle between line of sight and normal to emitting surface being measured.

Photometric and Radiometric Instrumentation

To determine which type of instrumentation should be used for various photometric and radiometric measurements, many factors must be considered. The basic consideration is the wavelength (frequency) of the radiation in question. Figure 2 shows the wavelength and frequency spectrum from x-ray to sonic waves given in angstrom units for wavelength and cycles per second (hertz) for frequency. This simple nomograph shows the relationships between wavelength and photon energy, given in electron

Table 14 Spectroradiometric Quantities

Quantity	Symbol[a]	Equation definition	Common units	Symbol
Spectral radiant flux	Φ_λ	$\Phi_\lambda = d\Phi/d\lambda$	Watt per micrometer	$W/\mu m$
Spectral radiant flux	Φ_ν	$\Phi_\nu = d\Phi/d\nu$	Watt per hertz	W/Hz
Spectral radiant exitance	M_λ	$M_\lambda = dM/d\lambda$	Watt per square meter-micrometer	$W/m^2\,\mu m$
Spectral irradiance	E_λ	$E_\lambda = dE/d\lambda$	Watt per square meter-micrometer	$W/m^2\,\mu m$
Spectral radiant intensity	I_λ	$I_\lambda = dI/d\lambda$	Watt per steradian-micrometer	$W/sr\,\mu m$
Spectral radiance	L_λ	$L_\lambda = dL/d\lambda$	Watt per square meter-steradian-micrometer	$W/m^2\,sr\,\mu m$

[a]Symbols: λ = wavelength or unit wavelength; ν = frequency or unit frequency.

volts (eV). Note that the spread in wavelengths, frequencies, and photon energies for the ultraviolet-to-infrared region covers a span of four orders of magnitude, which accentuates the need for the proper choice of instrumentation depending upon where in the electromagnetic spectrum the required radiation measurement lies. The two basic considerations for light-measuring equipment may be summarized as spectral range (wavelength region) and the radiant intensity of the radiation.

The other considerations mostly concern the light source. These include the distance to the source and the size of the source, both of which determine the type of optics needed to couple the radiant output into the measuring device. Also included here is the type of light source and whether its emission is pulsed or continuous, which determines the tone span of the measurement and the required frequency range needed to examine and attain accurate absolute, relative, and repeatable measurements.

The measurement of the total flux (Φ_S) of a given light source is often made using an integrating sphere, which is a large (in comparison to the source) spherical shell with an inner surface area S_S. The inside shell is uniformly covered, usually with a white paint having a high diffuse reflectance ρ_S,

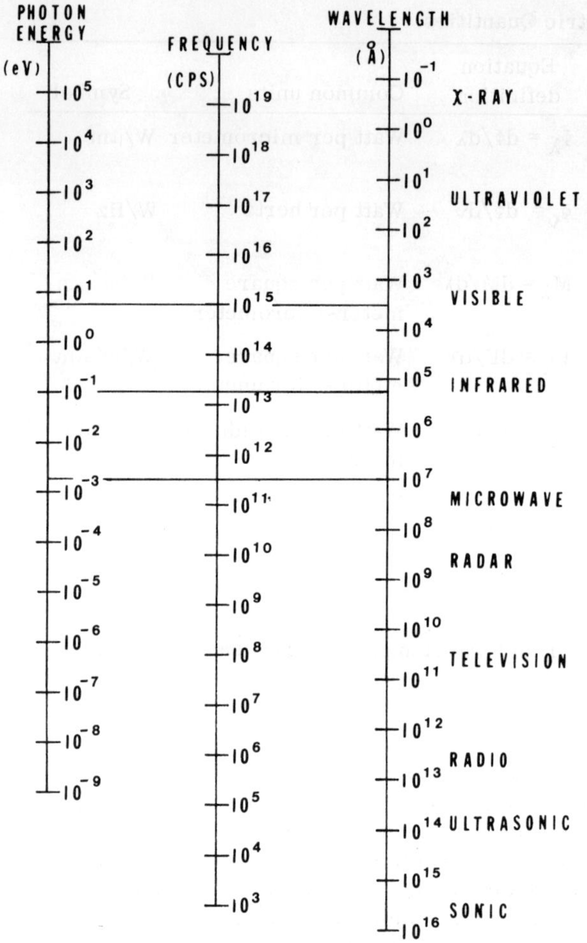

Fig. 2 Wavelength, frequency, energy conversion chart.

which generates a uniform illumination within the sphere. A small hole with an area of S_h is cut into the sphere. Once the flux (Φ_h) through the hole or the luminance of the inner surface (L) is measured, the total source flux can be estimated by

$$\Phi_s = \frac{1 - \rho_s}{\rho_s} \frac{S_s}{S_h} \Phi_h \quad \text{or} \quad \Phi_s = (1 - \rho_s)\frac{LS_s}{\rho_s}$$

The basic types of instrumentation used to measure radiation are photometers, radiometers, and spectroradiometers. Radiometers use wavelength-calibrated photodetectors, such as photoconductive cells, photodiodes, phototubes, or various types of photomultipliers, and attain the necessary wavelength region using narrow bandpass filters. Spectroradiometers use some sort of optical device to attain single frequencies (monochromators) for measurements and use the same types of photodetectors as radiometers. Photometers employ a similar detection scheme with special optical filters to currect the device's response to the standard luminosity curve to give a direct reading of photometric units, such as footlamberts or a similar unit.

Another group of instruments used for sources with relatively high intensity consists of bolometers, pyrometers, and thermopiles. These devices generate voltage (potentials) or currents from changes in the temperature of a given device element caused by the absorption of radiant (heat) energy from the source. The amplified signal voltages can be directly related to the incident radiant power for a defined optical system.

Standards and Calibrations

The calibration of most instruments used for light measurements requires the use of some sort of standard which can be used to compare experimentally determined values to attain some absolute measure of various quantities. Many devices are calibrated by using standard tungsten lamps such as the FEL tungsten-halogen lamp standard of spectral irradiance or by using calibrated thermopiles. A common standard light source is the spectral distribution of an incandescent tungsten-filament lamp operated at a color temperature of 2856 K. Other standards are often derived from tungsten lamps by using various filters with accurately determined spectral attenuations.

The accuracy of any given measurement is only as good as the accuracy and stability of the lamp standard, which leads to final absolute accuracies of about ±10% in the ultraviolet region and ±8% in the visible and infrared regions. The relative accuracy and reproducibility of measurements in many devices are very high (±1%). Thus, the major source of error lies in the definition and reproducibility of the standards.

The calibration of photometric devices which dive direct readings of photometric values is complicated by the fact that their spectral sensitivity $[K_{d\lambda}(\lambda)]$ does not usually match the spectral luminous efficiency $[K_\lambda(\lambda)]$ of the standardized human eye. The calibration of the detector thus changes with the spectral distribution $[e_\lambda(\lambda)]$ of the illumination impinging on the device. Illumination-dependent recalibration can sometimes be eliminated by computing a calibration factor (C), given by

$$C = \frac{\int K_\lambda e_\lambda \, d\lambda \quad \int K_{d\lambda} e_{o\lambda} \, d\lambda}{\int K_\lambda e_{o\lambda} \, d\lambda \quad \int K_{d\lambda} e_\lambda \, d\lambda}$$

where $e_{o\lambda}$ is the original illumination source spectral distribution.

BIBLIOGRAPHY

Bernstein, Neil S., Light Measurement Terminology and Techniques, Light Instrumentation Application Note L2001A-3, EG&G, Electro-Optics Division, 1976.

Cornsweet, Tom N., Visual Perception, Academic Press, New York, 1970.

Smith, Warren J., Modern Optical Engineering, McGraw-Hill, New York, 1966.

Chapter 4

SOURCES OF RADIATION

There are numerous types of radiation sources available, including extraterrestrial sources such as the sun, moon, stars, planets, and sky; man-made lamps, flashtubes, and discharge sources; and blackbody, nonblackbody, and graybody radiators. Table 15 lists a variety of common light sources in a relative luminance order, starting with the sun and ending with a moonless sky, and differing by more than 12 orders of magnitude of the given candela per square centimeter units of luminance. These average figures give an indication of the vast number and types of radiation sources available for use and study in electro-optics.

Solar Radiation

The sun approximates a blackbody radiator with a color temperature of about 4870 K for noon sunlight and 6600 K for average daylight. The *radiant exitance* of the sun is defined by the solar constant, the radiant power density received by an imaginary spherical surface at a radius equal to the mean earth-to-sun distance and concentric with the sun's center. The average value of the solar constant is about 0.138 W/cm^2 or 8.1×10^7 erg/min cm^2, which corresponds to a solar emittance of about 6340 W/cm^2. The solar illumination averages about 9.1 lm/cm^2 on clear days, corresponding to a solar luminance of about 1350 cd/mm^2.

Table 16 shows the solar spectral radiance and irradiance at sea level with the sun at the zenith. Although the sun emits radiation along the entire wavelength spectrum, the spectrum observed on the earth's surface has numerous gaps caused by various atmospheric absorptions. The infrared regions of absorption centered about 1.3, 1.9, 2.7, and 4.2 μm are caused by the presence of water vapor and carbon dioxide in the atmosphere. The sun's spectrum contains a number of absorption lines

Table 15 Luminance of Common Light Sources

Source description	Average luminance (cd/cm^2)
Sun (seen without earth's atmosphere)	225,000
Sun (at meridian seen from earth)	150,000
Carbon arc lamp (rotating)	80,000
Xenon arc lamp (150 W)	30,000
Mercury arc lamp (200 W)	26,000
Photoflash bulbs	25,000
Blackbody radiator (4000 K)	25,000
Carbon arc lamp (nonrotating)	17,500
Zirconium arc lamp (100 W)	9,600
Tungsten filament (750 W projection lamp)	7,500
Metal halide (deluxe-white, 1500 W)	1,600
Tungsten filament (gas-filled lamp)	1,200
Mercury discharge lamp (clear, 400 W)	970
Metal halide (color-improved, 1000 W)	930
Metal halide (clear, 400 W)	810
Sodium high pressure lamp (400 W)	780
Sun (at horizon, seen from earth)	600
Tungsten (3310 K color temperature)	210
Tungsten filament (vacuum lamp)	200
Blackbody (2040 K)	60
Mercury discharge (deluxe-white, 1000 W)	17
Mercury discharge (color-improved, 1000 W)	15
Acetylene flame	10
Fluorescent lamp (96T12, 800 mA)	1.1
Candle flame	0.90
Kerosene lamp	0.85
Average clear sky	0.75
Moon (seen from earth, clear sky)	0.30
Snow (average sunlight)	0.17
Moonlight (clear sky)	1.7×10^{-7}
Moonless sky (clear night)	4.5×10^{-8}

known as Fraunhofer lines, which are due to various solar elements and to water vapor and oxygen in the earth's atmosphere. Table 17 lists the Fraunhofer lines with the source or origin of the given line. These lines along with flame lines of various elements are commonly used to calibrate spectrometers. Table 18 lists the major lines used for this purpose.

Figure 3 shows the solar spectral irradiation at sea level on an average clear day with the sun at the zenith.

Table 16 Sea Level Solar Spectral Radiance and Irradiance

Wavelength (nm)	Solar radiance (W/sr cm^2 nm)	Solar irradiance (W/cm^2 nm) X 10^{-4}
380	0.736	0.500
400	1.103	0.750
420	1.562	1.062
440	1.773	1.205
460	2.030	1.380
480	2.140	1.458
500	2.160	1.465
520	2.086	1.418
540	2.046	1.391
560	2.043	1.389
580	2.039	1.386
600	2.033	1.382
620	2.001	1.360
640	1.961	1.333
660	1.934	1.315
680	1.896	1.289
700	1.839	1.250
720	1.745	1.186
740	1.650	1.122
760	1.565	1.064
780	1.488	1.012
800	1.412	0.960
820	1.344	0.914
840	1.276	0.868
860	1.216	0.826
880	1.161	0.789
900	1.106	0.752
920	1.066	0.724
940	1.025	0.696
960	0.991	0.673
980	0.962	0.654
1000	0.934	0.635
1100	0.791	0.538
1200	0.663	0.451
1300	0.559	0.380
1400	0.463	0.315
1500	0.394	0.268
1600	0.347	0.236
1700	0.309	0.210
1800	0.266	0.181
1900	0.228	0.155
2000	0.182	0.124

Source: P. Moon, Scientific Basis for Illuminating Engineering, McGraw-Hill, New York, 1936.

Table 17 Wavelengths of Fraunhofer Lines

Line	Origin	Wavelength (Å)	Line	Origin	Wavelength (Å)
Y		8990	G	H	4340
Z		8220	G^1	Fe	4326
A	O	7621	G	Ca + Fe	4308
A	O	7596	g	Ca	4227
a	H_2O	7185	h	H	4102
B	O	6870	H	Ca	3968
C	H	6563	K	Ca	3943
D_1	Na	5895	L	Fe	3820
D_2	Na	5890	M	Fe	3728
D_3	H	5876	N	Fe	3581
E	Ca	5270	O	Fe	3441
b_1	Mg	5184	P	Ti	3361
b_2	Mg	5172	Q	Fe	3287
b_3	Fe	5169	R	Ca	3180
b_4	Fe + Mg	5167	S_{1+2}	Fe	3100
C	Fe	4958	s	Fe	3048
F	H	4861	T	Fe	3021
d	Fe	4668	t	Fe	2994
e	Fe	4384	U	Fe	2948

Stellar, Sky, and Planetary Radiation

Stars on a clear night with no moon contribute about 30% of the light in the sky. The rest is essentially generated by emissions from molecules in the upper atmosphere. Sources of light outside the earth's atmosphere are often characterized by their illuminance relative to a zero-magnitude star with an illuminance of 2.65×10^{-6} lm/m^2. The term magnitude is an arbitrary scale defined by the ratio of two illuminances differing by one magnitude and equal to $(100)^{1/5} \simeq 2.512$. Thus, given two extraterrestrial light sources of magnitudes m_1 and m_2, the ratio of their brightness or illuminances is given by

$$\frac{E_{m_1}}{E_{m_2}} = (2.512)^{m_2 - m_1}$$

The log-log curves of the spectral irradiance for several stars and planets are given in Fig. 4. The spectral irradiance of all the planets depends to

Table 18 Typical Spectroscope Calibration Wavelengths

Source	Wavelength (Å)	Source	Wavelength (Å)
K flame (red)	7699	b_2 solar	5173
K flame (red)	7665	Cd (green)	5086
He (red)	7282	He (green)	5048
He (red)	7065	He (green)	5016
Hg arc	6907	Hg arc	4960
B solar	6869	He (green)	4922
Li flame (red)	6708	Hg arc	4916
He (red)	6678	H (blue-green)	4861
H (red)	6563	Cd (blue)	4800
Cd (red)	6438	He (blue)	4713
Hg arc	6234	St (blue)	4607
Na (yellow)	5896	He (blue)	4472
D_2 solar	5890	He (blue)	4388
He (yellow)	5876	Hg arc (blue)	4358
Hg arc	5791	H (blue)	4341
Hg arc	5769	Hg arc	4079
Hg arc (green)	5461	Hg arc (violet)	4047
Tl flame	5351	He (violet)	4026
E solar	5270	H_1 solar	3969
Mg flame (green)	5184	K solar	3934

some extent on their positions relative to earth at any given time as well as on the clarity of the sky and the presence (or concentration) of dust, water vapor, and other scattering and absorbing materials. The approximate value of the visual magnitude m_V is determined by the intersection of the spectral irradiance curves at the peak visual response wavelength of 555 nm. Under the best possible conditions the unaided eye can perceive a star or celestial object with a magnitude of 8.5. Table 19 lists some visual magnitudes of selected stars along with their spectral classification.

The sky on a clear day contributes about one-fifth of the total illuminance of the earth's surface by scattered sunlight and has a color temperature of about 22,000 K. The brightness of the clear sky is about 0.75 candela/ cm^2.

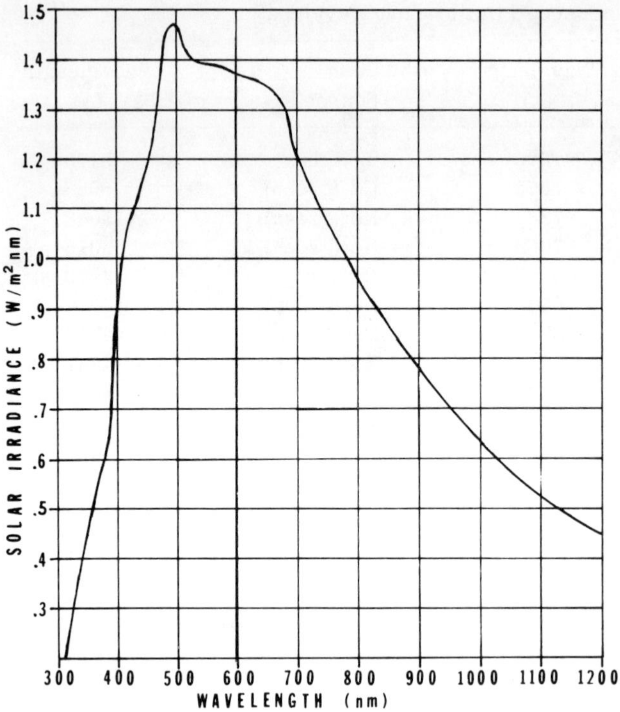

Fig. 3 Solar spectral irradiation at sea level with sun at the zenith.

Incandescent Lamps

One of the most common incoherent light sources found in electro-optics is the incandescent filament lamp, usually made with a tungsten-wire filament in a hermetically sealed glass or quartz bulb. The power ratings for individual filament lamps range from less than 1 W to 10^4 W. The heated filament is usually coiled to increase the overall luminance and to increase the emissivity. The actual shape and dimensional configurations vary with the particular application. The filament ratio of length to cross-sectional area is determined by the power and voltage requirements. The filament surface area is determined by the radiant power and temperature. The life expectancy of incandescent lamps is mostly a function of the evaporation rate of the metal from the filament. The reduction in lamp efficiency due to the coating of metal on the inside of the bulb can range from 15 to 50% before bulb failure. The addition of halogens, particularly iodine, into the bulb space extends the filament lifetime and causes the tungsten deposits on the bulb wall to return to

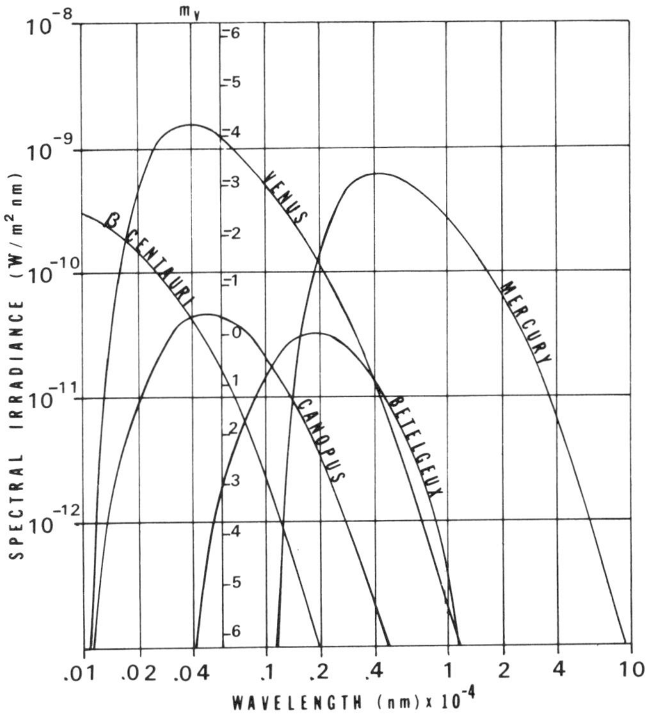

Fig. 4 Spectral irradiance and visual magnitudes of stars and planets. Source: R. C. Ramsey, Spectral irradiance from stars and planets, above the atmosphere from 0.1 to 100.0 microns, Appl. Opt., vol. 1, no. 4, July, 1962.

the hot filament where the tungsten-halogen compound dissociates and tungsten metal is redeposited on the filament. This effectively increases either the bulb's lifetime or its illuminance, the latter, as the bulb can be operated at a higher temperature but with a lifetime equivalent to a non-halogen-filled bulb.

The spectral distribution of incandescent sources is predominantly in the visible and ultraviolet regions and depends on whether glass or quartz is used for the bulb, since glass absorbs some of the ultraviolet emission. The filament radiant power (in watts) is approximately proportional to the filament surface area and to the fourth power of the filament temperature. The luminous output (in lumens) mostly depends on the filament temperature (in kelvin) and power (in watts). The lumens per watt mostly depends on the filament temperature.

Table 19 Selected Stellar Visual Magnitudes

Star	Spectral type	Visual magnitude
Sirius	A0	-1.58
Canopus	F0	-0.86
α Centauri	G0	0.06
Vega	A0	0.14
Capella	G0	0.21
Arcturus	K0	0.24
Procyon	F5	0.48
Achemar	B5	0.60
β Centauri	B1	0.86
α Crucis	B1	1.05
Spica	B2	1.21
Fomalhaut	A3	1.29
Regulus	B8	1.34
β Crucis	B1	1.50
Castor	A0	1.58
λ Crucis	M3	1.61
λ Orionis	B2	1.70
β Tauri	B8	1.78
β Carniae	A0	1.80
α Persei	F5	1.90
λ Geminorium	A0	1.93
ϵ Sagitarii	A0	1.95
β Canis Majoris	B1	1.99

Source: American Institute of Physics Handbook, McGraw-Hill, New York, 1957, pp. 6-82.

The radiation from a tungsten incandescent filament in the visible light region is very similar to a graybody with an emissivity of about 45%. The luminance, spectral emissivity, color emissivity, total emissivity, bright temperature, color temperature, and radiant emittance of tungsten are given in Table 20. Table 21 compares tungsten with some other metals, listing the total normal emissivity at six different temperatures for several polished metals. The spectral characteristics of incandescent bulbs are in part a function of the spacial configuration of the filament. Compact coiled-coil filaments are closer to blackbody than to graybody radiators. The spectral radiant emittance of tungsten is shown in Fig. 5 for several temperatures. Figure 6 shows the spectral luminous emittance of tungsten for the same temperatures as in Fig. 5.

Table 20 Radiant Properties of Tungsten

Temperature (K)	L (cd/cm^2)	Spectral emissivity 467 nm	Spectral emissivity 650 nm	Color emissivity	Total emissivity	Brightness T (K; 650 nm)	Color T (K)	M_e W/cm
1100	0.001	484	456	0.392	0.118	1055	1108	1.06
1200	0.006	482	454	0.390	0.138	1151	1210	1.68
1300	0.029	480	452	0.387	0.158	1242	1312	2.56
1400	0.11	478	450	0.385	0.175	1332	1414	3.82
1500	0.33	476	448	0.382	0.192	1422	1516	5.51
1600	0.92	475	446	0.380	0.207	1511	1619	7.71
1700	2.3	473	444	0.377	0.222	1599	1722	10.51
1800	5.1	472	442	0.374	0.236	1687	1825	14.07
1900	10.4	470	440	0.371	0.249	1774	1928	18.45
2000	20.0	469	438	0.368	0.259	1861	2032	23.6
2100	36	467	436	0.365	0.270	1946	2136	29.8
2200	61	466	434	0.362	0.278	2031	2241	37.2
2300	107	464	432	0.359	0.288	2115	2345	46.0
2400	157	463	430	0.356	0.296	2198	2451	55.8
2500	240	462	428	0.353	0.301	2280	2556	67.1
2600	347	460	426	0.349	0.309	2362	2662	80.6
2700	500	459	424	0.346	0.315	2443	2769	96.0
2800	694	458	422	0.343	0.321	2523	2876	112.9
2900	950	456	420	0.340	0.329	2602	2984	132.1
3000	1257	455	418	0.336	0.334	2681	3092	153.8
3100	1650	454	416	0.333	0.337	2760	3200	176.6
3200	2110	452	414	0.330	0.341	2837	3310	203
3300	2700	451	412	0.326	0.344	2913	3420	232
3400	3370	450	410	0.323	0.348	2989	3530	264
3500	4200	449	408	0.320	0.351	3063	3642	299
3600	5200	447	406	0.317	0.354	3137	3754	338

Source: W. F. Roeser and H. T. Wensel, Temperature, Vol. 1, Reinhold, New York, 1941.

Table 21 Total Normal Emissivity of Polished Metal at Various Temperatures

Material	300 K	500 K	800 K	1400 K	3000 K	5000 K
	Total normal emissivity					
Chromium	0.07	0.15	0.23	0.36	0.45	0.49
Copper	0.04	0.05	0.17	0.18	0.26	0.50
Gold	0.02	0.02	0.05	0.28	0.51	0.62
Iron	0.06	0.08	0.15	0.29	0.35	0.45
Steel	0.07	0.10	0.14	0.23	0.37	0.45
Nickel	0.04	0.06	0.11	0.18	0.28	0.40
Platinum	0.04	0.06	0.10	0.19	0.30	0.45
Silver	0.01	0.02	0.03	0.04	0.14	0.11
Tin	0.08	0.11	0.13	0.19	0.30	0.39
Tungsten	0.02	0.04	0.10	0.16	0.35	0.45
Zinc	0.02	0.04	0.05	0.06	0.48	0.46

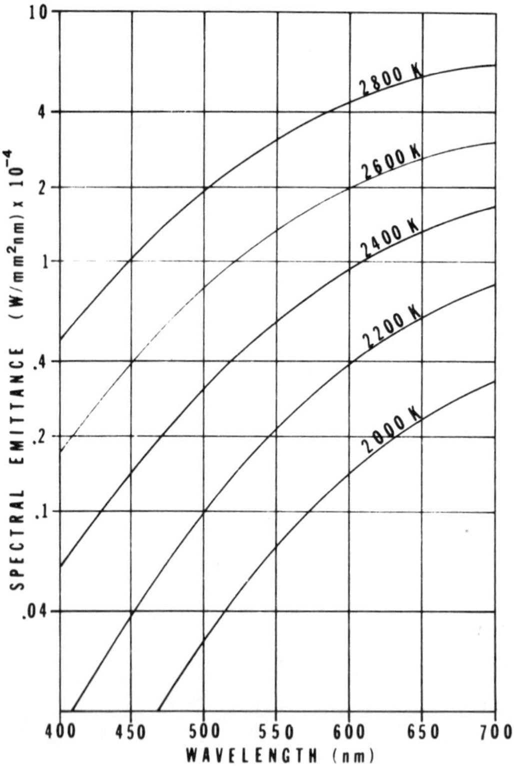

Fig. 5 Spectral radiant emittance of tungsten.

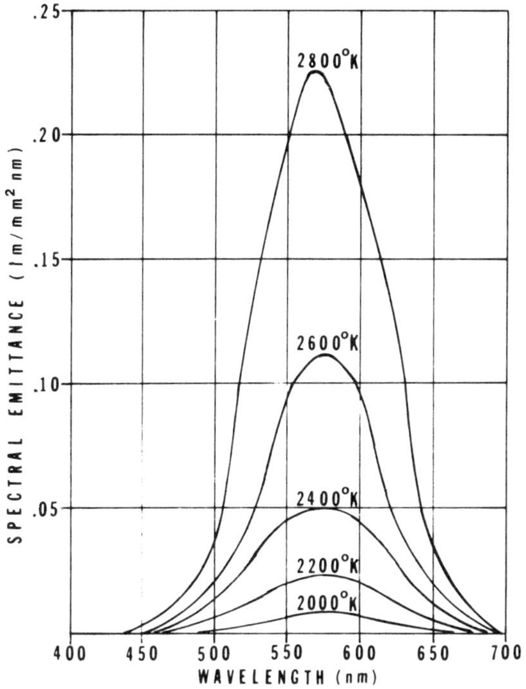

Fig. 6 Spectral luminous emittance of tungsten.

One disadvantage of tungsten incandescent lamps is that a temperature beyond 3655 K cannot be exceeded for very long before the tungsten evaporates or melts. This makes it difficult to exceed color temperatures of 3760 K or luminances of 52 cd/mm^2.

Arc Sources

Compact electrical discharge source arcs commonly found in electro-optics are grouped into five types depending upon the compound enclosed in the quartz bulb: mercury (Hg), xenon (Xe), mercury-xenon (Hg-Xe), sodium, and metal halide. The other types are less common and include those containing gaseous mixtures consisting of argon (Ar) or krypton (Kr). Arc sources (gaseous discharges) can be used to generate essentially monochromatic radiation. Sodium arc lamps operated at low current density and low temperature and pressure are good monochromatic sources. The luminance of the sodium arc lamp is about 1 cd/mm^2. Another means of

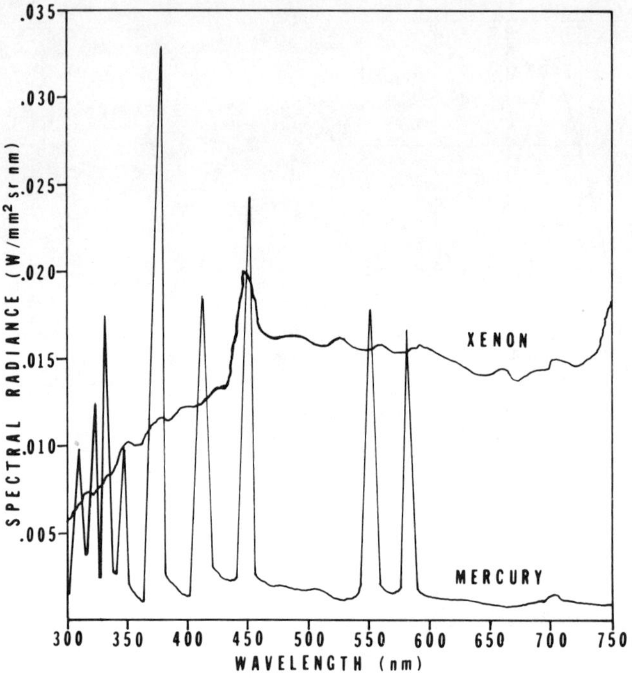

Fig. 7 Spectral radiance of xenon and mercury arc lamps.

generating monochromatic light is by using low-pressure mercury discharge lamps in combination with various optical filters to select specific lines or narrow spectral regions.

Carbon arc sources are usually two carbon rods with a given air gap between them, differentiated by the rotating and nonrotating types. When higher luminances are desired, the rotating type is used. These arcs approximate graybody radiators at a color temperature of about 6000 K, well beyond the range of incandescent lamps. By the use of air jets to constrict the high-temperature arc, the luminance may be increased over stationary arcs which are dominated by air-plasma convection. The color of the light emitted from carbon arcs can be altered by using various material dopants in the electrodes.

Mercury lamps have a spectral distribution that varies with the bulb temperature and the electrical potential at the arc gap. When a mercury lamp is at low pressure (about 0.01 nm Hg), about 62% of the emitted radiation is at 253.7 nm at a current of about 0.25 A. At a current of about 1 A, about 40% of the emitted radiation is given off at 253.7 nm. The visible

Table 22 Spectral Luminance Distributions of Ar, Kr, and Xe Gas Arcs

Spectral range (nm)	Percent luminance Ar	Kr	Xe
400-420	0.042	0.020	0.021
420-440	0.42	0.30	0.21
440-460	0.98	0.74	0.70
460-510	11.6	9.5	10.9
510-560	40.7	38.2	38.7
560-610	32.0	37.5	36.0
610-660	12.7	12.6	12.4
660-720	1.60	1.09	1.05

Source: J. N. Aldington, The gas arc, Trans. Ill. Eng. Soc. (London), vol. 14, 1949.

radiation of the mercury lamp increases when higher pressures and current densities are used. The lamp color is blue-green with an approximate color temperature range from 7000 to 10,000 K.

The radiant intensity of high-pressure mercury arc sources varies as the 1.5 power of the current and is approximately proportional to the mercury vapor density. Depending upon the desired radiant output and the ability to rapidly restart a hot lamp, the lamp may be convection, forced air, or water cooled. Various dopants added to the mercury vapor can be used to alter the light output efficiency, such as by the use of thallium iodide. The spectral radiance of a 200-W high-pressure mercury arc lamp is shown in Fig. 7. High-pressure mercury lamps are a combination of a continuum and a spectral line spectrum source as can be seen in Fig. 7. For applications where a smoother continuum is desired, xenon can be used. The spectral luminosity of the gas arc lamps Ar, Kr, and Xe are all very similar as shown in Table 22. The line portion of the gas arc spectrum is approximately proportional to the current, and the continuous portion varies approximately with the 1.7 power of the current.

The addition of xenon to the mercury lamp adds a few lines to the near-infrared region, but the spectral distribution is essentially the same as the mercury lamp. Xenon lamps have a color temperature of about 6000 K. Unlike the mercury lamp, the spectral distribution is relatively stable with variations in the applied operating voltage. The xenon lamp color is similar to sunlight, and its spectrum is almost continuous throughout the infrared, visible, and ultraviolet regions, with several strong lines in the near infrared. Sodium lamps have a spectral distribution highly dependent upon the operating pressure. Low-pressure lamps have essentially two principal lines at 598.0 and 589.6 nm. Increasing the operating pressure broadens the spectral output. Metal halide lamps are commonly sodium halides in mercury or xenon operated at high pressures and input voltages. The addition of the metal halide adds to the total spectral output of the mercury or xenon alone.

Zirconium arc lamps are sometimes used to simulate a point source of light, using an argon atmosphere in the bulb. The emission spectra of these lamps are almost free of lines in the visible portion of the spectrum and are dominated by lines in the 720 to 900 nm region.

Flash Tubes

In most flash tubes an electric spark generated by capacitor energy storage techniques is discharged through a gaseous medium as a high-intensity arc. Some types have the ability to achieve luminances up to 4×10^5 cd/mm^2, compared to 1.5×10^3 cd/mm for the sun. The discharge occurs over a short period of time, and is governed by a time constant that is approximately proportional to the tube gas vapor pressure and arc length and inversely proportional to the applied voltage. Gaseous discharge sources can be modulated using various types of current modulation. Arc discharge lamps can be modulated up to 15 kHz and glow discharges up to 1 MHz.

The efficiency of xenon flash tubes varies approximately proportionally to the arc length and exponentionally with the flash duration time. Figure 8 shows the typical pulse duration responses of a xenon flash tube at 2.5 atm as a function of arc length and voltage. The repetition rate of flash tubes is limited mostly by the thermal energy dissipation characteristics of the envelope and electrodes. The emission spectrum of most flash tubes is significantly affected by current density. Usually as the current decreases, the shorter wavelengths decrease more rapidly than the longer wavelengths.

Fluorescent Lamps

Most fluorescent lamps are low-pressure (1.5×10^{-5} atm at 18° C) mercury lamps which operate at low currents (0.15 to 1.5 A) with the dominant emission at 253.7 nm. By coating the inside of the lamp with the various mixtures of phosphors, the ultraviolet radiation can be converted into various visible colors. The coatings in common white lamps are made from a mixture of zinc beryllium silicate and magnesium tungstate. The start-up problem of fluorescent lamps is usually solved by using an inert gas mixture with mercury to initiate the discharge and by using an electrical ballast to control the current and provide the initial discharge voltage pulse.

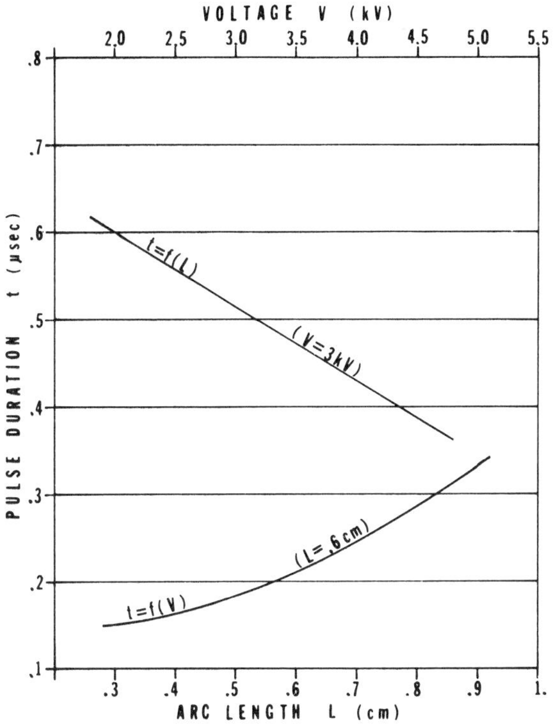

Fig. 8 Typical pulse duration response of xenon flashtube at 2.5 atm. Source: I. S. Marshak and L. I. Shchoukin, Physical and technical parameters of flashtubes, J. Soc. Motion Picture Television Engrs., vol. 70, 1961.

Table 23 lists some of the more common fluorescent lamps and some of their electrical and optical properties.

Blackbody Radiation

The spectral emittance of a blackbody in a vacuum as a function of wavelength is given by Planck's law of radiation:

$$M_{e\lambda} = \frac{C_1}{\lambda^5\left(e^{C_2/\lambda T} - 1\right)} \tag{1}$$

Table 23 Common Fluorescent Lamp Characteristics

	8 W	15 W	20 W	30 W	40 W	90 W
Nominal length (inches)	12	18	24	36	48	60
Average voltage (V)	56	56	56	98	102	63
Average current (A)	0.17	0.30	0.38	0.36	0.43	1.5
Lumens at 100 h						
Standard cool white	370	730	1000	1860	2650	5200
Deluxe cool or warm white	245	540	740	1380	1890	3850
Daylight	325	665	910	1700	2320	4750
White or standard warm white	375	760	1040	1930	2660	5400
Lamberts at 100 h						
Standard cool white	3.38	2.66	1.70	3.08	2.17	2.50
Deluxe cool or warm white	2.30	1.97	1.26	2.28	2.46	1.83
Daylight	3.06	2.40	1.55	2.81	1.91	2.26
White or standard warm white	3.52	2.77	1.77	3.20	2.18	2.57

Source: Illuminating Engineering Society, IES Lighting Handbook, 3rd ed., New York, 1959.

where $C_1 = 2\pi c^2 h = 3.7405 \times 10^{-16}\ \mathrm{Wm}^2 = 3.7405 \times 10^8\ (\mathrm{W/m}^2)\mu^4$ and $C_2 = ch/k = 1.43879 \times 10^4 \mu\ \mathrm{K}$ where k is Boltzmann's constant; $k = 1.38054 \times 10^{-23}$ J/K. The spectral radiance $L_{e\lambda}$ is equal to $M_{e\lambda}$ divided by π. The spectral radiance expressed per unit frequency is given by

$$L_{e\nu} = \frac{2h}{c^2}\ \frac{\nu^3}{e^{h\nu/kt} - 1}$$

where ν is the frequency of radiation. $L_{e\nu}$ is usually expressed in watts per square meter-steradians-hertz.

A blackbody is assumed to be in thermal equilibrium as a perfect energy absorber. A blackbody can be simulated in practice by having a small hole or entrance leading into a large cavity. When a blackbody is heated to 900 K its apparent color is red, at about 3000 K its color is white, and at 9000 K turns to light blue.

When Eq. (1) is integrated over all wavelengths, the total emittance of a blackbody is given by the Stefan–Boltzmann law:

$$M_e = C_3 T^4$$

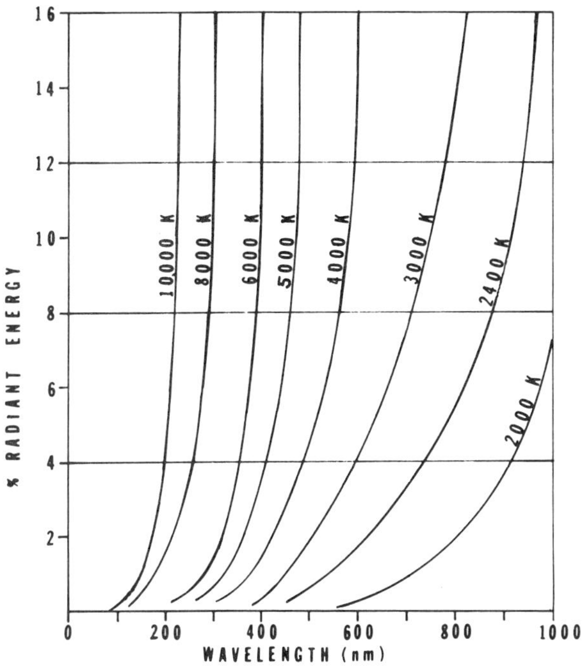

Fig. 9 Percent of blackbody radiant energy emitted between wavelengths $0 \rightarrow \lambda$ at various temperatures.

where

$$C_3 = \frac{\pi^4}{15} \frac{C_1}{C_2^4} = \frac{2\pi^5}{15} \frac{k^4}{c^2 h^3} = 5.6697 \times 10^{-8} \text{ W/m}^2 \text{ K}^4$$

C_3 is the Stefan-Boltzmann constant. Figure 9 shows the percent of blackbody radiant energy emitted between wavelengths 0 and λ for various temperatures. The luminous efficiency of a blackbody as a function of temperature is shown in Fig. 10 and is seen to reach a maximum of about 6600 K, with a luminous efficiency of about 0.14.

To find where the maximum spectral emittance occurs, Eq. (1) is differentiated with respect to λ, and the derivation is set equal to 0. The maximum spectral emittance λ_{max} given by Wien's displacement law is

$$\lambda_{max} = \frac{C_2}{C_x T} = \frac{2897.8\mu}{T}$$

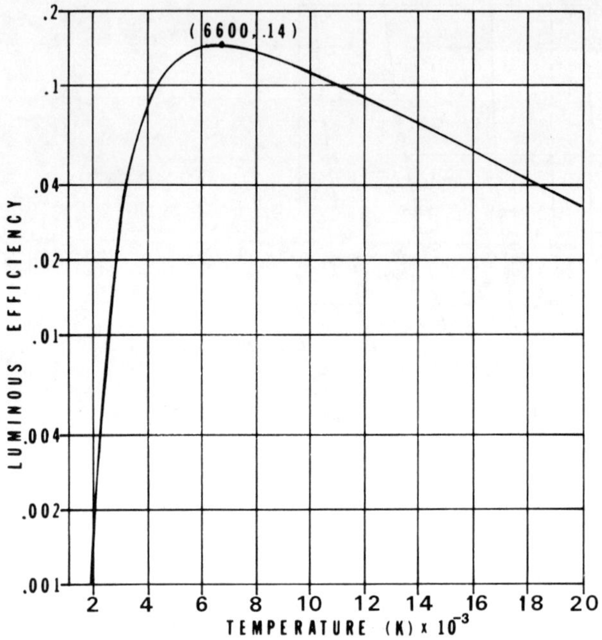

Fig. 10 Blackbody luminous efficiency as a function of temperature.

where C_X = 4.96511 [the real root of the equation $5 - x = 5e^{-x}$] and T is the temperature in kelvin. Table 24 lists several of the peak wavelengths using this relationship.

By substituting Wien's displacement laws into Eq. (1), the peak spectral emittance is given by $M_{e\lambda_{max}} = C_4 T^5$, where

$$C_4 = \frac{C_1 \left(C_X / C_2\right)^5}{e^{cx} - 1} = 1.2862 \times 10^{-4}\ \mathrm{W/m^2\,\mu\,K^5}$$

As the temperature increases, the spectral emittance increases and the peak shifts toward the shorter wavelengths. Figure 11 shows several spectral emittance curves at various temperatures. Figure 12 shows the maximum spectral emittance, the total emittance, and the peak wavelengths of a blackbody as a function of the thermal equilibrium temperature. Figure 13 is a nomograph used to calculate the relationships between total blackbody radiation, temperature, wavelength, and energy distribution.

Table 24 Peak Blackbody Radiant Emittance Wavelengths as a Function of Temperature

Peak wavelength (nm)	Temperature (K)
2897.8	1,000
1448.9	2,000
965.9	3,000
724.4	4,000
597.6	5,000
483.0	6,000
414.0	7,000
362.2	8,000
322.0	9,000
289.8	10,000
193.2	15,000
145.0	20,000
115.9	25,000
96.6	30,000
82.8	35,000
72.4	40,000
64.4	45,000
58.0	50,000
52.7	55,000
46.6	60,000
44.6	65,000

Nonblackbody Radiators

For any given absolute temperature the spectral radiance of a given nonblackbody radiator emitting only thermal radiation is always less than that of the spectral radiance of a total blackbody radiator. To relate the given nonblackbody spectral radiance L_λ' to the blackbody spectral radiance L_λ an absorption factor $a_\lambda = f(\lambda, T)$ is used, given by $L_\lambda' - a_\lambda L_\lambda$, where $a < 1$, and depends on the given radiant direction if the given body does not obey Lambert's law. This absorption factor is used to define the diffuse reflector factor ρ_λ by the equation $\rho_\lambda - (-a_\lambda)$, where $\rho_\lambda = 0$ for a full radiator and $\rho_\lambda = 1$ for a perfect reflector. If a_λ is independent of wavelength, the given body is defined as a graybody; in other words, a graybody has constant emissivity for all wavelengths. The general physical laws of graybody radiation are the same as those for a blackbody radiator except that all energies are reduced by the a_λ factor.

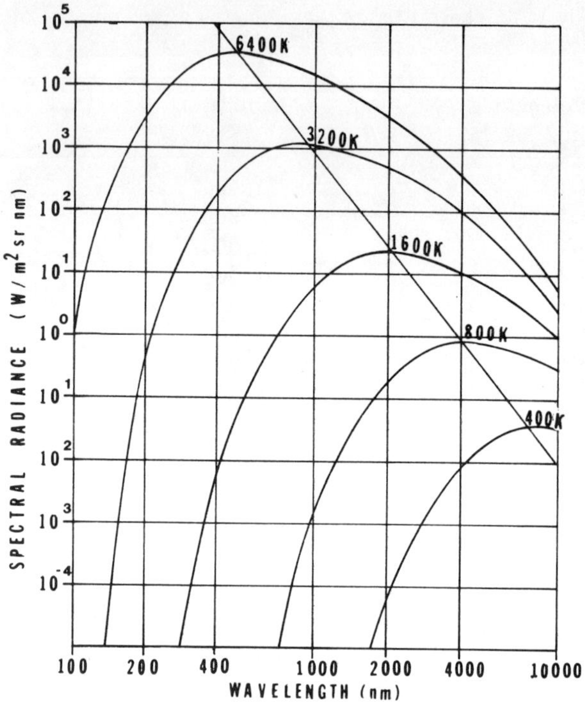

Fig. 11 Blackbody spectral radiance as a function of temperature.

Phosphors

The types of phosphors are divided into two basic groups, ionic and covalent, depending upon the chemical bonding of the components. Most common phosphors are made by combining a host crystal and various chemical activators or dopants by heating the mixture into solution or by physically grinding the materials together. Often, the individual components of the phosphor are nonluminescent. Extraneous materials (usually metals) in the final phosphor combination often act to reduce or quench the total potential luminescence, making material purity and preparation techniques important factors in achieving superior and efficient phosphors.

Phosphor screens can be formed by material settling in a gravitational field or by evaporation. The settling method requires the use of a binder which can keep the screen intact under various types of thermal stresses and physical shocks. Evaporation methods can produce relatively transparent screens with high contrast and resolving power, but with lower

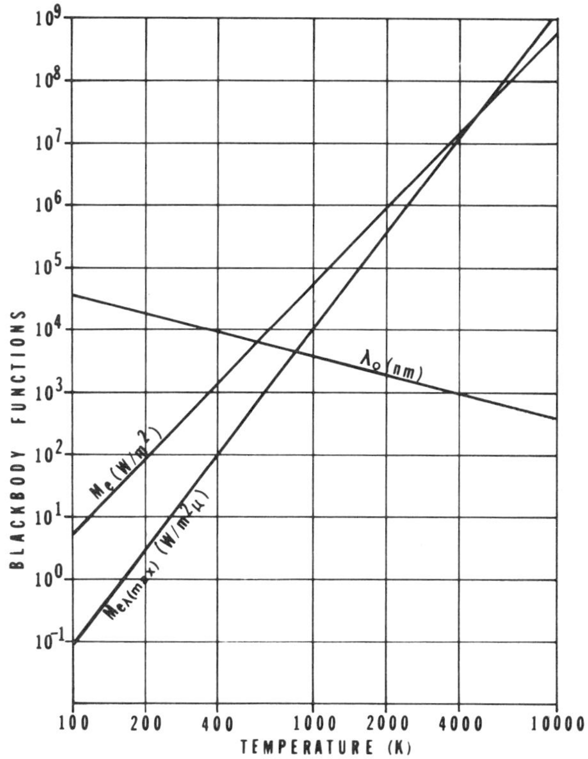

Fig. 12 Blackbody peak wavelengths and total and maximum spectral emittance.

luminance levels than thicker granular-type phosphor screens. In phosphor screens the emitted radiation has a longer wavelength than the impinging radiation, resulting in a relatively small amount of absorption by the phosphor of the emitted radiation. Table 25 lists some phosphor compositions and designations as given by the Joint Electron Devices Engineering Council (JEDEC), as well as the average decay times at the 10% point, the peak wavelengths, and the upper and lower 35% point wavelengths.

The common types of luminescence encountered in electro-optics are <u>cathodoluminescence</u> which is due to electron bombardment, <u>electroluminescence</u> which is due to an electric field, <u>ionoluminescence</u> which is due to positive ion bombardment, <u>photoluminescence</u> which is due to visible or ultraviolet radiation, and <u>roentgenoluminescence</u> caused by x-rays. Zinc sulfides are the most common phosphors used in electroluminescence.

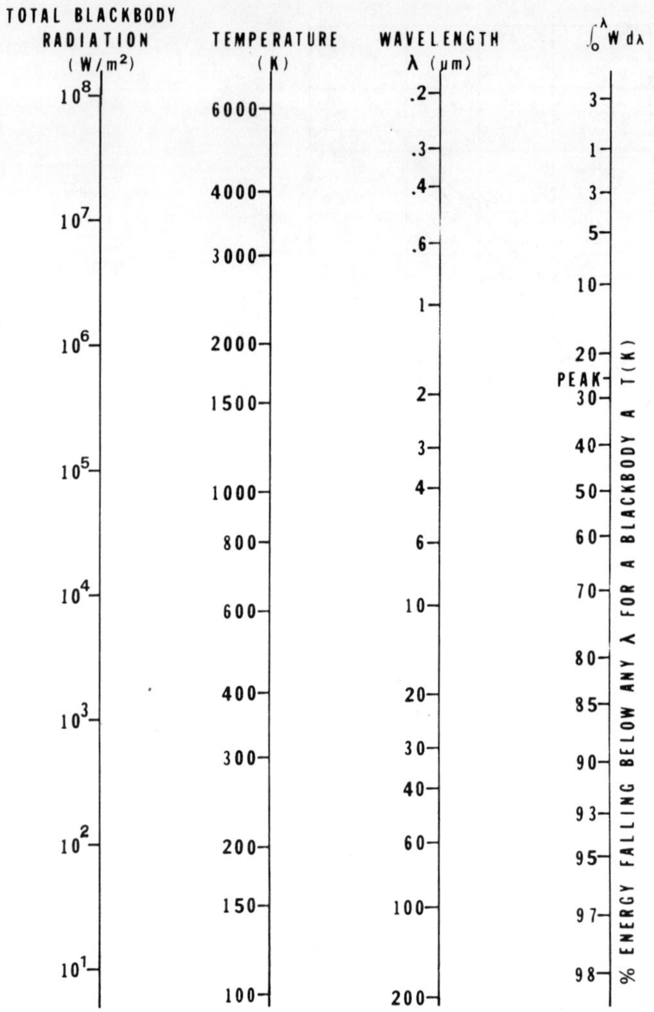

Fig. 13 Nomograph to calculate intensity and spectral distribution of radiation.

The efficiency of a phosphor system is usually described in terms of the quantum, energetic, luminous, or luminescence efficiency. The quantum efficiency is defined as the ratio of the number of emitted quanta to the absorbed quanta. In photoluminescence the quantum efficiency can approach unity. In roentgenoluminescence due to secondary electron emissions the quantum efficiency can range from 10^{-1} to 10^{3} depending on the

Table 25 Phosphor Compositions, Wavelength Ranges, and Decay Times

JEDEC code no.	Material composition	Lower 35% point (nm)	Peak wavelength (nm)	Upper 35% point (nm)	Average decay time, 10% point
P1	Zn_2SiO_4:Mn	500	525	555	24 ms
P2	ZnS:Cu, Ag	490	535	585	50 μs
P3	$(Zn, Be)_2SiO_4$:Mn	530	600	665	19 ms
P4	ZnS:Ag	425	457	485	22 μs
	(Zn, Cd)S:Ag	505	565	615	60 μs
P5	$CaWO_4$	365	415	500	25 μs
P7	ZnS:Ag	400	435	480	50 μs
	(Zn, Cd)S:Cu	490	555	610	390 ms
P11	ZnS:Ag, Ni	420	460	520	35 μs
P12	$(Zn, Mg)F_2$:Mn	555	590	645	210 ms
P14	ZnS:Ag	405	440	490	28 μs
	(Zn, Cd)S:Cu	540	600	675	5 ms
P15	ZnO	450	505	575	2.8 μs
P16	$(Ca, Mg)SiO_3$:Ce	355	303	425	0.12 μs
P17	ZnO	400	450	480	5.2 μs
	(Zn, Cd)S:Cu	490	550	610	420 ms
P18	$(Ca, Mg)SiO_3$:Ti	350	410	490	53 μs
	$(Zn, Be)SiO_4$:Mn	450	530	670	13 ms
P19	$KMgF_3$ Mn	565	590	640	220 ms
P20	(Zn, Cd)S:Ag	505	560	630	60 μs
P21	MgF_2:Mn	580	605	630	170 ms
P22	ZnS:Ag	420	450	490	22 μs
	$ZnSiO_4$:Mn	500	515	555	60 μs
	$Zn\ (PO_4)_2$:Mn	595	650	700	700 μs
P24	ZnO	450	510	585	1.5 μs
P25	$CaSiO_3$:Pb, Mn	550	610	675	45 ms
P26	ZnF_2:Mn	560	595	635	17 s
P27	$Zn_3(PO_4)_2$:Mn	600	635	685	27 ms
P28	(Zn, Cd)S:Cu, Ag	500	550	605	510 ms

Table 25-Continued

P29	ZnS:Cu, Ag	490	535	585	50 μs
	$CaSiO_3$:Pb, Mn	550	610	675	45 ms
P31	ZnS:Cu	430	522	570	38 μs
P32	(Ca, Mg)SiO_3:Ti	415	555	620	700 ms
	(Zn, Cd)S:Cu	540	600	675	420 ms
P33	MgF_2:Mn	555	585	635	8 s
P34	ZnS:Pb, Cu	510	535	600	140 s
P35	ZnS, ZnSe:Ag	430	485	610	600 μs

x-ray intensity and phosphor system. For phosphors that follow Stoke's rule, the energetic efficiency (η_e) is lower than the quantum efficiency and is given by

$$\eta_e - \eta_q \frac{\lambda_a}{\lambda_e}$$

where

η_q = quantum efficiency
λ_a = wavelength of the absorbed radiation
λ_e = wavelength of the emitted radiation

The luminous efficiency is in essence a measure of the phosphor efficiency using the human eye as the photodetector and is given as the ratio of the lumens of light emitted by the phosphor to the absorbed or incident power of light impinging on the phosphor surface. The luminescence efficiency (η_m) is given by

$$\eta_m = P_r/P_r + P_o$$

where P_r is the probability of a radiating transition relative to P_o, the sum of probabilities for all transitions.

The reflectance p_e of a phosphor system is the fraction of the incident radiation that is reflected by the phosphor surface and subsurface layers and materials. It varies over an order of magnitude (0.04 to 0.40) depending upon the materials used and the preparation method. Table 26 lists several phosphors with the reflectance and luminescence efficiencies.

Table 26 Phosphor Luminescence Efficiencies and Reflectance

Phosphor	Reflectance (p_e)	Luminescence photo Ne	Nq	Efficiency cathodo Ne
$ZnSiO_4$:Mn	0.36	0.35	0.70	0.08
$BaSi_2O_5$:Pb	0.35	0.55	0.75	0.04
$CaSiO_3$:Pb, Mn	0.17	0.29	0.68	0.07
$Ca_3(PO_4)_2$:Ti	0.15	0.49	0.56	0.002
$MgWO_4$	0.07	0.44	0.84	0.025
$CaWO_4$:Pb	0.05	0.42	0.75	0.03

Source: H. P. Kallman and G. M. Sprach, Luminescence of Organic and Inorganic Materials, Wiley, New York, 1962.

BIBLIOGRAPHY

Cortorillo, S. F., H. Kee, H. S. Strauss, and W. E. Thouret, High-brightness xenon lamps with liquid-cooled electrodes, Illum. Eng., vol. 60, 1965.

Cotton, H., Principles of Illumination, Wiley, New York, 1960.

Friedman, Herbert, The sun and the upper atmosphere, U.S. Naval Research Laboratory, Washington, D.C., Astronautics, July, 1955.

Gaydon, A. G., The Spectroscopy of Flames, Wiley, New York, 1957.

Gibbs, J. E., and G. W. Gordon-Smith, The color temperature of an incandescent helical filament, Brit. J. Appl. Phys., vol. 12, 1961.

Goncz, J. H., and P. B. Newell, Spectra of pulsed and continuous xenon discharges, J. Opt. Soc. Amer., vol. 56, 1966.

Harvey, E. N., A History of Luminescence, American Philosophical Society, Philadelphia, 1975.

Henderson, S. T., and A. M. Marsden, Lamps and lighting, Crane, Russak, New York, 1972.

Ivey, H. F., Color and efficiency of luminescent light sources, J. Opt. Soc. Amer., vol. 53, 1963.

Kingslake, R. (Ed.), Applied Optics and Optical Engineering, Vol. 1, Academic Press, New York, 1975.

Kuiper, G. P. (Ed.), The Solar System, University of Chicago Press, Chicago, 1954.

Larabee, S. D., Spectral emissivity of tungsten, J. Opt. Soc. Amer., vol. 49, 1959.

Levi, Leo, Applied Optics, Vol. 1, Wiley, New York, 1968.

Leverenz, H. W., An Introduction to Luminescence of Solids, Wiley, New York, 1950.

Marshak, I.S., and L. J. Shchoukin, Physical and technical parameters of flashtubes, J. Soc. Motion Picture Television Engrs., vol. 70, 1961.

Mauro, J. A. (Ed.), Optical Engineering Handbook, General Electric Co., Syracuse, N.Y., 1966.

Middleton, W. E. K., Vision Through The Atmosphere, University of Toronto Press, Toronto, Canada, 1958.

Optical Characteristics Of Cathode Ray Tube Screens, JEDEC Electron Tube Council, JEOEC Publication No. 16, Newark, N.J., 1960.

Page, L., Optics And Spectroscopy, D. Van Nostrand, New York, 1959.

Ramsey, R. C., Spectral irradiance from stars and planets, Appl. Opt., Vol. 1, No. 4., July, 1962.

RCA Electro-Optics Handbook, RCA Corporation, Technical Series EOH-11, Harrison, N.J., 1974.

Segal, S. M., A study of high-intensity light sources, Illum. Engr., Vol. 50, 1955.

Thomas, Woodlief (Ed.), SPSE Handbook of Photographic Science And Engineering, Wiley, New York, 1973.

Chapter 5

LASERS

The four basic elements essential to laser systems are a lasing material, an excitation mechanism, a feedback mechanism, and an output coupler. The lasing material or active lasing medium is usually a solid (crystalline) or a gas enclosed in a transparent container. The lasing material will emit and amplify light when properly excited, usually over a narrow spectral region. The excitation mechanism for gas lasers is essentially the electrodes connected inside the transparent container (glass tube). The excitation mechanism for solid lasers is optical pumping, light being absorbed by the crystalline material from an external source, usually a gas-filled arc lamp for continuous operation (cw) or gas-filled flash lamps for pulsed operation (P). The feedback mechanism for most lasers is achieved by using mirrors of various configurations to make the lasing radiation oscillate in the material for a period long enough to achieve the desired output power level. The output coupling for most lasers is made by having one of the mirrors partially reflective so that the radiation can escape, while having the other mirror close to completely reflective.

The most common laser parameters used to characterize a given laser system are the peak output power, efficiency, spectral range, beam divergence, and pulse repetition rate (for pulsed modes). Table 27 lists some common laser characteristics for gas, solid-state, and semiconductor lasers.

Some of the areas where lasers have been used with distinct advantages are range finding, welding, plasma diagnostics, Raman spectroscopy, high-speed photography, Schlieren photography, chemical emission spectroscopy, retina welding, cancer research, optical weather radar systems, narrow beam illumination, photographic holography, atmospheric transmissions, optical instrumentation, optical communications, mechanical measurements, standards of length, seismographs, tracking systems, microscopic spectroscopy, flow-rate measurements, rotation-rate sensing, power transmission, display devices, holography, optical memory systems, light sources, and computers.

Table 27 Comparison of Laser Characteristics

Laser characteristic	Gas laser CW	Gas laser Pulsed	Solid state CW	Solid state Pulsed	Semiconductor CW	Semiconductor Pulsed
Peak pulsed power		>200 kW (CO_2)		6 X 10^9 W		100 W
Peak cw power	1000 W (CO_2)		250 W		10 W	
Output efficiency	15% (CO_2)			5% (glass)	25% (GaAsP)	
Spectral range (um)	0.25-774	0.27-10.6	0.49-2.7	0.265-2.7	0.25-22	0.27-2
Spectral width (full power)	50 MHz (CO_2)	50 MH_2 (CO_2)		1 Å (ruby)	18 Å (GaAsAl)	
Beam divergence (radians)	2 X 10^{-4}	8 X 10^{-4}	10^{-3}	10^{-3}	0.2	0.2
Pulse repetition rate (pps)		10^5		5 X 10^3		10^3

Ruby Lasers

One of the most commonly known lasers is constructed with Al_2O_3 doped with approximately 0.05% chromium ions (Cr^{3+}), and is known as the ruby laser. It is usually a cylindrical rod with mirrored ends perpendicular to the cylinder axis. The ruby laser uses a three-level energy system with a maximum emission power output at about 0.6943 μm. The overall efficiency of a typical ruby laser can be as high as 3% depending on numerous factors, particularly the design used to couple the external optical pump to the laser rod. The ruby laser can usually only withstand several optically pumped pulses a second and, therefore, is usually powered by flashtubes instead of continuous sources. Depending on the size of the crystalline or glass rod and the power and the power source used, the energy of a single pulse can range from milliwatts (mW) to several hundred watts. The lifetime of ruby lasers is mainly determined by the usable lifetime of the mirrored ends, assuming the rods are not cracked by overheating or excessive thermal cycles. When the mirrored surfaces become damaged, the ends can often be repolished and resurfaced (coated) to restore the rod to original quality.

Using Q-switching, ruby lasers can attain peak powers of 10^9 to 10^{10} W. The power levels attainable are high enough to damage the laser; thus, the pulse length is often kept to the order of tens of nanoseconds and the energy in the order of tens of joules. The wavelength of ruby lasers varies from 0.6943 μm at 295 K to 0.6934 μm at 77 K. Commonly used ruby lasers utilize water cooling and a Pockels cell to Q-switch the laser. Typical pulse repetition rates are 1 to 3 pulses per second (pps) with individual emission spikes of 30- to 50-nsec durations with 10- to 100-MW peak powers.

Typical optical pumping of ruby lasers consists of using a linear flashlamp (usually xenon) inside of a polished elliptical cavity with the flashlamp at one longitudinal axis and the center of the ruby cylinder at the other. Another method is to place the rod in the center of a polished cylindrical mirrored surface and use a coiled flashlamp for the external optical pumping.

In general, an optically pumped solid-state laser system should have broad adsorption bands to permit efficient coupling of the external optical pump energy into the lasing material. It is also preferable for a solid-state laser system to exhibit a narrow fluorescent emission line to attain some flexibility in controlling the pulse width. For high-frequency emissions, it is more efficient to generate the energy at a lower frequency and then frequency multiply in a nonlinear optical material.

Table 28 Lasers Used for Material Removal

Laser type	Emission λ (μm)	Peak power (W/cm^2)	Repetition rate (pps)	Typical applications
CO_2	10.6	10^5	100–1000	Metal cutting, cable insulation, stripping, plastic and rubber drilling, scribing of ceramic substrates
Nd-YAG	1.06	10^5	100–5000	Machine tool alignment scribing of silicon wafers
Nd-glass	1.06	10^6	<3	Gyroscope balancing
Ruby	0.6943	10^6	<3	Hole drilling in gemstones

Source: Laser Institute of America, Guide for Material Processing by Lasers, The Paul M. Harrod Co., Baltimore, Md., 1977, pp. 8-9.

Like other solid-state lasers, ruby types have problems of multifrequency output, multimoding, and spiking. For certain applications such as metal working and material removal, these limitations and problems are not important. Table 28 lists several laser types used for material removal, and lists several of the more important physical parameters, including the emission wavelength, peak power, and pulse repetition rate.

Neodymium-Doped Lasers

Host matrices such as glass, yttrium aluminum garnet (YAG), and calcium tungstate can be doped by neodymium to achieve a four-level energy system with a maximum output at about 1.06 μm. The neodymium-YAG (Nd-YAG) laser is similar to the neodymium glass laser. Here the laser rod is made from a single yttrium aluminum garnet crystal doped with 1 to 3% neodymium ions (Nd^{2+}). The pulsing rate of solid lasers is in part determined by the lasing material thermal conductivity and the ability of a given system to cool the laser during continuous or pulsed operation. The Nd-YAG laser has a relatively high thermal conductivity and can be operated in either a pulsed or continuous mode. When operated in the continuous (cw) mode, the optical pump is usually a xenon arc lamp. The output power in the cw mode ranges from less than a watt to several hundred watts.

The neodymium glass laser is similar to the ruby laser in terms of shape, optical pumping methods, and the range of pulse repetitions that can be used. These lasers are usually made by doping an optical glass rod with 2 to 5% neodymium ions (Nd^{2+}). They can be made larger than ruby rods and, in general, when properly made and properly driven, are more efficient as well. As the size and power of the neodymium lasers increase, the potential pulse rate decreases.

CO_2 Lasers

The CO_2 laser is usually a mixture of carbon dioxide, helium, and nitrogen. The addition of the helium and nitrogen enhances the energy transfer between the two CO_2 vibrational states and can lead to a laser efficiency of up to 20%. High-power CO_2 lasers used increased gas pressure in the laser chamber to attain more excited molecules per unit volume. This is commonly known as the transverse excited atmospheric (TEA) CO_2 laser. The efficiency of lasers is defined as the ratio of the electrical power input (to drive the lasing material) to the laser output energy. CO_2 lasers can develop continuous output powers in the range of 150 kW at 10.6- or 9.3-μm wavelengths, giving them a practical importance for various types of materials processing such as welding and cutting operations.

Table 29 lists some of the more commonly known solid laser systems with their appropriate emission wavelengths. The materials are listed by the host lasing material (the majority or bulk material) and the dopant used in the bulk. In some cases, pure crystalline materials are used without dopant ions. Table 30 lists some of the gas laser systems that have been examined along with their emission wavelengths. It should be noted that for both of these tables the wavelengths of the reported emissions have varying degrees of accuracy depending upon the measurement technique used by the individual system. Details of a given system or measurement technique can be found by checking the appropriate reference material.

Dye Lasers

External optical pumping dye lasers can be used to obtain coherent light over a large wavelength region almost continuously by proper selection of the laser dye and solvent. Continuous mode lasing action can be attained in dye lasers by pumping the dye solution from a reservoir solution and subsequently cooling the dye-solvent system. The dye solution can also be cooled by removing heat directly from the dye cuvette to attain continuous or rapid-pulse operation. The laser wavelength of a particular dye system can be tuned over a significant range by varying the solution or cavity parameter or by using an intracavity wavelength selector such as a diffraction grating.

Table 29 Solid Laser Systems

Host	Dopant	Wavelengths of emission (μm)			Refs.
Al_2O_3	Cr^{3+}	0.6929	0.6934	0.6943	37, 38
		0.6953	0.7009	0.7041	
BaF_2	Nd^{3+}	1.05	1.06		44, 45
	U^{3+}	2.20	2.556	2.60	
CaF_2	Dy^{2+}	2.36			41, 42
	Er^{3+}	1.617	2.69		44, 46
	Ho^{3+}	2.092	2.094		48, 55
	Nd^{3+}	1.046			
	Sm^{2+}	0.708			
	Tm^{2+}	1.116	1.189	2.36	
	U^{3+}	2.223	2.438	2.490	
		2.511	2.613		
$CaMoO_4$	Ho^{3+}	2.055	2.071	2.074	44, 68
	Nd^{3+}	1.061	1.067		
	Tm^{3+}	1.906	1.911		
$Ca(Nb_2O_3)$	Er^{3+}	1.61			53
	Ho^{3+}	2.047			
	Nd^{3+}	1.06			
	Pr^{3+}	1.04			
	Tm^{3+}	1.19			
$Ca(VO_4)_2$	Nd^{3+}	1.037			69
$CaWO_4$	Er^{3+}	1.612			37, 54
	Ho^{3+}	2.046	2.059		61, 69
	Nd^{3+}	0.914	0.918	1.006	
		1.057	1.058	1.063	
		1.064	1.065	1.350	
	Pr^{3+}	1.047			
	Tm^{3+}	1.916	1.918		
CeF_3	Nd^{3+}	1.064			70
Chelate	Eu^{3+}	0.613			48, 67
	Tb^{3+}	0.545			
Er_2O_3	Tm^{3+}	1.934			71
$Gd_3Ga_5O_{12}$	Nd^{3+}	1.063			
Gd_2O_3	Nd^{3+}	1.074	1.079		62

Table 30 Gas Laser Systems

Gas	Wavelengths of emission (μm)				Refs.
Ar	0.26249	0.27539	0.28842	0.29129	7, 17, 28
	0.29262	0.30026	0.30240	0.30548	84-86
	0.33614	0.33448	0.33585	0.35111	
	0.35142	0.35769	0.36379	0.37052	
	0.37953	0.38583	0.41466	0.41829	
	0.43707	0.44820	0.45450	0.45794	
	0.46096	0.46580	0.47269	0.47649	
	0.48799	0.48891	0.49651	0.49926	
	0.50172	0.51453	0.52860	0.52870	
	0.55022	0.79931	0.8780	1.0923	
	1.0935	1.3472	1.618	1.694	
	1.793	2.016	2.1339	2.142	
	2.2045	2.3139	2.3973	2.5014	
	2.5421	2.5494	2.5512	2.5364	
	2.5668	2.6843	2.7356	2.8202	
	2.8238	2.8663	2.8783	2.8843	
	2.9272	2.9788	3.0453	3.0996	
	3.1333	3.1346	4.7330	4.9160	
	4.9213	5.1216	5.1218	5.4680	
	5.4770	5.8477	5.8666	6.050	
	6.9429	6.9448	7.2150	7.8063	
	12.140	15.038	26.936	26.944	
Ar-Br	0.8446				22
$Ar-CO_2$	0.2896	0.6328			29
Ar-Hg	0.5225	1.222	1.246	1.276	6
Ar-Kr	0.3336		0.7990		
$Ar-N_2$	0.3577				36
$Ar-O_2$	0.8446				25
Br	0.47427	0.50546	0.51824	0.51850	84, 87
	0.52383	0.53320	0.53341	0.84463	
	0.84464	0.84467	0.84468	2.2845	
	2.3511	2.8375			
CH_3CN	119.0	334.4	334.8		88
CH_4	0.4942	0.5773	0.8703		1
Cl_2	0.26327	0.31914	0.33929	0.33935	85, 89
	0.35300	0.35607	0.36379	0.36455	
	0.36692	0.37205	0.37488	0.41325	

Table 30-Continued

	0.47404	0.47687	0.47813	0.48969	
	0.49047	0.49177	0.50783	0.51031	
	0.52179	0.52213	0.53922	0.60947	
	0.9451	1.3859	1.3891	1.9755	
	2.0199	2.4466			
C_2N_2	194.7013				90
CO	0.55906	0.55934	0.55960	0.55983	8, 91, 92
	0.56004	0.56025	0.56038	0.60629	
	0.60657	0.60682	0.60705	0.60725	
	0.60742	0.65955	0.65995	0.66031	
	0.66064	0.66091	0.66115	0.66135	
	5.03705	5.04750	5.05755	5.06773	
	5.07807	5.08845	5.09905	5.10410	
	5.10985	5.11418	5.12445	5.13485	
	5.14530	5.15595	5.16666	5.17220	
	5.17765	5.18250	5.18865	5.19290	
	5.20345	5.21410	5.22498	5.23600	
	5.24195	5.24710	5.25250	5.26310	
	5.27380	5.28465	5.29570	5.30695	
	5.36820	5.37950	5.4080	5.41961	
	5.42682	5.42991	5.44247		
CO_2	9.5691	9.5862	9.6063	9.6211	93
	9.6391	9.6570	9.6762	10.5135	
	10.5326	10.5518	10.5713	10.5912	
	10.6118	10.6324	10.6534	10.6748	
	10.6965	10.7194	10.7415	10.7748	
	10.7880				
Cs	3.20	7.18			27
D_2	0.4905	0.5748	0.5778	0.5807	1
	0.8760				
DCN	189.9490	190.0080	194.7027		94
	194.7644	204.3872			
Diethylamine	126.24	128.75	130.95	135.03	95
	201.19	211.14	223.25	309.94	
	311.08	336.83	372.80		
D_2O	33.896	35.090	36.319	36.524	90, 96
	37.791	40.994	56.845	71.965	
	72.429	72.747	73.337	74.545	
	76.305	84.111	84.291	107.71	
	171.60	189.95			

Table 30-Continued

F	0.27596	0.28261	0.30798	0.31216	85, 89
	0.31742	0.32027	0.40248	0.45526	
	0.45678	0.63472	0.63715	0.8446	
H_2	0.1098	0.1613	0.3240	0.3721	1, 34, 97
	0.4402	0.5388	0.83496	0.88762	98
	0.88988	0.9755	1.11621	1.1222	
	1.1680	1.3058	1.31623	1.8751	
HCl	3.808	3.810	3.840		
HCN	126.164	128.629	130.838	134.932	99-101
	309.731	310.908	311.309	336.579	
	336.70	372.547			
He	1.7880	2.0603	2.1019	2.5420	28
He-Cl_2	1.9754	2.020			23, 32
He-CO	1.0689	1.454			2, 31
HF	2.680				
He-Hg	0.5677	0.6150	0.7346	1.0583	5, 19, 20
	1.5295	1.8101			
He-Ne	0.5939	0.6046	0.6118	0.6293	2, 4, 9-14
	0.6328	0.6351	0.6401	1.0621	24
	1.0798	1.0844	1.1143	1.1177	
	1.1180	1.1390	1.1409	1.1523	
	1.1525	1.1601	1.1614	1.1617	
	1.1767	1.1770	1.1985	1.1988	
	1.2066	1.2069	1.2689	1.2887	
	1.2912	1.5231	1.5235	1.8281	
	1.8287	1.8309	1.8408	1.8595	
	1.8602	2.0351	2.1041	2.1708	
	2.3951	2.3957	3.3913	31.928	
	34.679	35.602	37.231	41.741	
	53.486	54.019	54.117	57.355	
He-NO	1.3583				22
HeSe	0.5228				
HeSF_6	1.0455	1.0628			22
He-Xe	2.026	3.508			30
Hg	0.4797	0.5678	0.7346	0.8547	20, 21, 85
	0.8628	0.8677	0.9396	1.0586	86
	1.1181	1.2545	1.2981	1.3655	

Table 30-Continued

	1.5295	1.5550	1.6918	1.6921	
	2.6942	1.7073	1.7110	1.8130	
	3.93	5.88	6.49		
H_2O	16.931	23.365	26.666	27.974	103
	28.054	28.273	28.356	32.929	
	33.033	35.000	35.841	36.619	
	37.859	38.094	39.698	40.629	
	45.523	47.251	47.469	47.693	
	48,677	53.906	55.077	57.660	
	67.177	73.402	78.455	79.106	
	89.775	115.42	118.65	120.08	
I	0.4986	0.5216	0.5407	0.5419	33, 85, 104
	0.5625	0.5678	0.5760	0.6127	
	0.6585	0.6904	0.7032	0.8250	
	0.8800	0.980	1.01	1.03	
	1.04	2.5987	2.7571	3.236	
	3.431				
Kr	0.2440	0.26494	0.26645	0.27415	17, 28, 84-86
	0.30497	0.31244	0.32394	0.33750	
	0.35074	0.35642	0.40674	0.41313	
	0.41545	0.41718	0.42265	0.43181	
	0.43870	0.44433	0.45722	0.45830	
	0.46152	0.46192	0.46339	0.46502	
	0.46804	0.46950	0.47624	0.47657	
	0.48252	0.48467	0.50164	0.50221	
	0.51260	0.52083	0.53087	0.56819	
	0.57530	0.61688	0.64710	0.65700	
	0.67646	0.68710	0.75251	0.79311	
	0.79930	0.8589	1.690	1.694	
	1.7842	1.8190	1.9211	2.1160	
	2.1891	2.5230	2.6267	2.6288	
	2.8618	2.9845	2.9878	2.0536	
	3.0672	3.1515	3.3409	3.3419	
	3.4680	3.4883	3.4898	4.3748	
	4.8773	4.8832	5.3000	5.3919	
	5.5700	5.5863	5.6306	7.0581	
KrF	0.2501				3
N_2	0.33673	0.33710	0.34010	0.34787	3, 84, 85 105, 106
	0.34830	0.39950	0.40973	0.41033	
	0.45105	0.45144	0.46303	0.56666	

Table 30-Continued

	0.56760	0.56795	0.86840	0.86911	
	0.8698	0.8704	0.8710	0.8844	
	0.8847	0.8852	0.8856	0.8862	
	0.8871	0.8879	0.8886	0.8892	
	0.8898	0.8909	1.0449	1.0461	
	1.0472	1.0480	1.0491	1.0495	
	1.0505	1.2303	1.2312	1.2319	
	1.2334	1.2347	1.3585	1.4544	
	3.28804	3.29522	3.30756	3.31240	
	3.31616	3.31892	3.45212	3.45852	
	3.45877	3.46804	3.47127	8.18384	
	8.21106				
N_2-Cs_2	11.4823	11.4893	11.5031	11.5099	107
	11.5166	11.5237	11.5307	11.5376	
	11.5446	11.5962			
Ne	0.23581	0.24350	0.26780	0.26787	15, 16, 22
	0.27775	0.28688	0.33198	0.33238	84-86, 108
	0.33244	0.33275	0.33290	0.33311	
	0.33455	0.33783	0.33929	0.33934	
	0.54010	1.0295	1.1788	1.1790	
	1.2459	1.4544	1.7162	1.8210	
	1.5576	1.9577	2.0350	2.0354	
	2.0380	2.1012	2.3260	2.3951	
	2.5400	2.5524	2.7581	2.7840	
	2.8641	2.9496	2.9670	2.9813	
	3.0268	3.0278	3.3182	3.3342	
	3.3362	3.3813	3.3849	3.3912	
	3.3922	3.4481	3.5845	3.7746	
	3.9817	5.4048	5.6620	7.3300	
	7.4221	7.4237	7.4799	7.4954	
	7.6163	7.6461	7.6510	7.6925	
	7.7015	7.7407	7.7655	7.780	
	7.8368	8.0088	8.0621	8.8337	
	8.8413	9.0896	10.0633	10.960	
	10.981	11.861	11.902	12.835	
	13.739	13.759	14.930	16.638	
	16.668	16.893	16.947	17.158	
	17.189	17,802	17.841	17.888	
	18.396	20.480	21.752	22.836	
	25.423	28.053	31.928	34.550	
	34.679	35.602	37.231	41.741	

Table 30-Continued

	50.70	52.39	53.47	53.486	
	54.019	54.117	55.68	57.355	
	68.329	72.15	84.047	85.047	
	86.90	88.46	89.93	93.02	
	106.02	124.4	126.1	132.8	
Ne-CO	1.4539				22
Ne-Hg	1.529	1.153			33
Ne-O_2	0.8446	1.153			22, 26
NH_4	21.471	22.542	22.563	22.675	109
	24.918	26.282	31.951		
NO	5.9550	5.9639	5.9750	6.0008	110
	6.0266	6.032	6.0386	6.0403	
	6.0629	6.0885	6.0934	6.1417	
	6.1545	6.1972	6.2752		
N_2O	10.7698	10.7799	10.7901	10.8005	108
	10.8107	10.8212	10.8312	10.8416	
	10.8523	10.8628	10.8736	10.8843	
	10.8951	10.9061	10.9170	10.9280	
	10.9390	10.9510	10.9612	10.9724	
	10.9838	10.9951	11.0067	11.0184	
	11.0296	11.0416			
O_2	0.29839	0.30472	0.30635	0.33813	85
	0.33855	0.37547	0.37599	0.43474	
	0.43513	0.44149	0.44170	0.46055	
	0.46491	0.55924	0.67214		
Xe	0.24772	0.26918	0.29837	0.30798	17, 18, 28
	0.32469	0.33059	0.33308	0.33500	85, 86, 89
	0.34542	0.34830	0.36455	0.36692	111, 112
	0.37457	0.37810	0.38033	0.39729	
	0.40605	0.42141	0.42403	0.42746	
	0.42859	0.43058	0.44342	0.46030	
	0.46737	0.46836	0.47231	0.47489	
	0.48620	0.48695	0.48870	0.49541	
	0.49650	0.50077	0.50449	0.51590	
	0.52389	0.52602	0.52617	0.53140	
	0.53529	0.53946	0.54009	0.54192	
	0.55244	0.56591	0.57270	0.57511	
	0.59557	0.59711	0.6094	0.62709	
	0.65287	0.7828	0.7989	0.8330	

Table 30-Continued

	0.8408	0.8443	0.8569	0.8582
	0.8714	0.9063	0.9265	0.9287
	0.9697	1.0634	1.0950	2.0261
	2.0267	2.3200	2.6276	2.6518
	2.6601	3.0078	3.1101	3.2739
	3.3676	3.4343	3.4400	3.5080
	3.6219	3.6518	3.6856	3.8690
	3.8941	3.9966	4.1600	4.6021
	5.5754	7.314	9.004	9.702
	11.29	12.266	12.972	18.506
	18.514			
Xe-F	0.351	0.353		
Xe-He	5.575	9.007		

1. Appl. Phys. Lett., vol. 3, no. 10, 1963, p. 181.
2. Nature, vol. 197, January, 1963, p. 173.
3. Nature, vol. 200, no. 4907, 1963, p. 667
4. Aviation Week and Space Technology, vol. 79, no. 8, August, 1963, p. 34.
5. Appl. Phys. Lett., vol. 4, no. 2, January, 1964, p. 34.
6. Proc. IEEE, vol. 52, no. 4, April, 1964, p. 414.
7. Appl. Phys. Lett., vol. 4, no. 7, April, 1964, p. 128
8. Phys. Lett., vol. 7, no. 3, November, 1963, p. 194.
9. Appl. Phys. Lett., vol. 4, no. 1, 1964, p. 20
10. Appl. Phys. Lett., vol. 4, no. 1, 1964, p. 18.
11. Proc. IEEE, vol. 52, no, 2, February, 1964, p. 206.
12. C.R. Acad. Sci., vol. 258, no. 5, 1964, p. 1452.
13. Appl. Phys. Lett., vol. 4, no. 7, April, 1964, p. 123.
14. Appl. Phys. Lett., vol. 2, no. 11, June, 1963, p. 211.
15. Phys. Rev. Lett., vol. 133, no. 6, 1964, p. A1476.
16. Appl. Phys. Lett., vol. 4, no. 18, 1964, p. 134.
17. Appl. Phys. Lett., vol. 4, no. 10, 1964, p. 178.
18. Appl. Opt., vol. 3, no. 6, June, 1964, p. 789.
19. J. Appl. Phys., vol. 34, no. 10, 1963, p. 3148
20. C.R. Acad. Sci., vol. 258, no. 13, 1964, p. 3259.
21. Proc. IEEE, vol. 52, no. 4, 1964, p. 414.
22. Phys. Rev., vol. 133, no. 5, 1964, p. A1244.
23. Appl. Phys. Lett., vol. 3, no. 9, 1963, p. 154.
24. Appl. Phys. Lett., vol. 2, no. 5, 1963, p. 101.
25. Phys. Rev. Lett., vol. 8, 1962, p. 470.
26. J. Appl. Phys., vol. 33, 1962, p. 3194.

Table 30-Continued

27. Appl. Opt., vol, 1, 1962, p. 513.
28. Phys. Rev. Lett., vol. 9, 1962, p. 2.
29. Electronic Design, vol. 10, 1962, p. 30.
30. Appl. Phys. Lett., vol. 2, no. 5, 1963, p. 99.
31. Nature, vol. 197, January, 1963, p. 173.
32. Appl. Phys. Lett., vol. 3, no. 9, 1963, p. 154.
33. Nature, vol. 198, no. 4882, 1963, p. 774.
34. Phys. Rev., vol. 9, 1974, p. A2635.
35. Appl. Phys. Lett., vol., 25, 1974, p. 498.
36. J.E.T.P. Lett., vol. 20, 1974, p. 53.
37. Proc. IEEE, vol. 51, 1963, p. 4.
38. Phys. Rev. Lett., vol. 6, 1961, p. 96.
39. Phys. Rev., vol. 126, 1962, p. 14.
40. Electronic Design, vol. 11, 1963, p. 31.
41. Proc. IEEE, vol. 51, no. 9, 1963, p. 1236.
42. Appl. Phys. Lett., vol. 3, no. 2, 1963, p. 23.
43. Phys. Rev. Lett., vol. 3, no. 2, July, 1963, p. 23.
44. Electronic Design, vol. 10, February, 1962, p. 91.
45. Electronic Design, vol. 10, October, 1962, p. 31.
46. J. Appl. Phys., vol. 33, 1962, p. 757.
47. Proc. IRE, vol. 50, 1962, p. 1532.
48. Scientific American, vol. 209, July, 1963, p. 43.
49. Proceedings of the Electrochemical Society Meeting, Abstract No. 108, 1962, p. 260.
50. Phys. Rev. Lett., vol. 7, 1961, p. 444.
51. Appl. Opt., vol. 1, 1962, p. 534.
52. Appl. Phys. Lett., vol. 1, 1962, p. 25.
53. J. Appl. Phys., vol. 34, no. 11, 1963, p. 3155.
54. Phys. Rev., vol. 131, no. 5, 1963, p. 2038.
55. Appl. Phys. Lett., vol. 2, no. 5, 1963, p. 101.
56. Phys. Rev. Lett., vol. 11, no. 7, 1963, p. 318.
57. Appl. Phys. Lett., vol. 3, no. 8, 1963, p. 135.
58. Appl. Phys. Lett., vol. 4, no. 1, 1964, p. 22.
59. J. Opt. Soc. Amer., vol. 53, no. 12, 1963, p. 1447.
60. Proc. IEEE, vol. 52, no. 1, 1964, p. 93.
61. Phys. Rev., vol. 131, no. 5, 1963, p. 2038.
62. Appl. Phys. Lett., vol. 4, no. 6, 1964, p. 113.
63. Appl. Phys. Lett., vol. 3, no. 8, 1963, p. 135.
64. Phys. Rev. Lett., vol. 11, no. 7, 1963, p. 318.
65. Phys. Lett., vol. 6, no. 1, 1963, p. 32.
66. Appl. Opt., vol. 3, no. 1, 1964, p. 153.
67. Nature, vol. 200, no. 4902, 1963, p. 157.
68. Phys. Rev. Lett., vol. 133, 1964, p. A494.
69. J. Electrochem. Soc., vol. 112, 1965, p. 303.

Table 30–Continued

70. Appl. Phys. Lett., vol. 4, 1964, p. 208.
71. Appl. Phys. Lett., vol. 6, 1965, p. 200.
72. Appl. Opt., vol. 2, 1963, p. 87.
73. Appl. Phys. Lett., vol. 6, 1965, p. 45.
74. J. Appl. Phys., vol. 36, 1965, p. 323.
75. Phys. Rev. Lett., vol. 11, 1963, p. 318.
76. Appl. Phys. Lett., vol. 5, 1964, p. 21.
77. Appl. Phys., vol. 34, 1963, p. 897.
78. Appl. Phys. Lett., vol. 7, 1965, p. 127.
79. Phys. Lett., vol. 21, 1966, p. 37.
80. IBM Research Report RC5165, Yorktown Heights, N.Y., 1974.
81. Appl. Phys. Lett., vol. 24, no. 8, 1974, p. 372.
82. Laser Technology and Applications, McGraw-Hill, New York, 1968.
83. Appl. Phys. Lett., vol. 6, 1975, p. 202.
84. Proc. IEEE, vol. 54, 1966, p. 543.
85. Proc. IEEE, vol. 52, 1964, p. 843.
86. Phys. Rev., vol. 133, 1963, p. A1476.
87. Appl. Phys. Lett., vol. 7, 1965, p. 263.
88. Phys. Lett., vol. 23, 1966, p. 443.
89. Appl. Phys. Lett., vol. 7, 1965, p. 202.
90. Phys. Lett., vol. 24A, 1967, p. 690.
91. Phys. Lett., vol. 7, 1963, p. 194.
92. Appl. Phys. Lett., vol. 5, 1964, p. 81.
93. Phys. Rev., vol. 136, 1964, p. A1187.
94. Appl. Phys. Lett., vol. 12, 1968, p. 124.
95. Electronics Lett., vol. 1, 1965, p. 45.
96. Phys. Lett., vol. 13, 1964, p. 35.
97. J. Appl. Phys., vol. 34, 1963, p. 897.
98. J. Opt. Soc. Amer., vol. 56, 1966, p. 1260.
99. Appl. Phys. Lett., vol. 8, 1966, p. 217.
100. Appl. Phys. Lett., vol. 12, 1968, p. 122.
101. Appl. Phys. Lett., vol. 11, 1967, p. 62.
102. J. Opt. Soc. Amer., vol. 55, 1965, p. 1051.
103. Nature, vol. 201, 1964, p. 250.
104. Proc. IEEE, vol. 52, 1964, p. 1350.
105. Appl. Phys. Lett., vol. 3, 1963, p. 16.
106. IEEE J. Quantum Electronics, vol. QE-2, 1966, p. 229.
107. Appl. Phys. Lett., vol. 8, 1966, p. 217.
108. Appl. Phys. Lett., vol. 6, 1965, p. 12.
109. Phys. Lett., vol. 14, 1965, p. 33.
110. Appl. Phys. Lett., vol. 9, 1966, p. 94.
111. Appl. Phys. Lett., vol. 1, 1962, p. 85.
112. IEEE J. Quantum Electronics, vol. QE-2, 1966, p. 70.

Table 31 lists some of the typical dye laser parameters including wavelength, tuning, spectral width, beam divergence, optical efficiency, output energy, power, pulse repetition rate, and pulse duration. Table 32 lists some of the dyes used along with a typical solvent. A typical or average emission wavelength for each dye system is given which can depend, to some extent, on the external pumping frequency and method used.

One of the major problems with some dye laser systems is the instability of the dye, either in the presence of the solvent over an extended period of time, by thermal degradation, or by photochemical reactions during use. Dye lasers can be used to generate a wide variety of coherent frequencies for special display systems, underwater illumination, and optical frequency amplification.

Dye lasers can be used to attain frequencies in the visible and ultraviolet wavelength regions. To attain a single-frequency tunable ultraviolet source, two tunable dye lasers can be used as shown in Fig. 14, where laser action is produced in Cs, Rb, or Sr vapors. Here, the λ_0 is given by $\lambda_1 \lambda_2/2\lambda_2 + \lambda_1)$, where λ_1 and λ_2 are the respective emission wavelengths of the dye lasers as shown.

Q-Switches and Modulators

To obtain sharp, intense laser pulses with large peak powers, a Q-switch is used, which is essentially an optical shutter to control lasing feedback and oscillations. The Q-switch shutter for some applications must open in a few nanoseconds. Mechanical Q-switching can be accomplished by using a rotating mirror or roof prism at one end of the laser aligned perpendicular to the centerline of the resonator. The speed of rotation of the reflective optics varies with the application and depends upon the laser type and desired output power. Typical rotational speeds are in the area of 400 to 500 revolutions per second. Since in solid-state lasers there is a time delay before maximum inversion is achieved after the flashlamp is triggered, the tripping timing must be carefully controlled using various optical or magnetic pick-up triggers to phase-balance lamp flashing and Q-switch rotational speed. Mechanical Q-switches are naturally limited by the rotational speed of the reflective optics.

Table 33 lists some commercially available laser Q-switches and a few of their more important parameters including the useful spectral range and the minimum required rise time.

Table 31 Dye Laser Properties

Property	Typical values	Condition	Comments
Wavelength	3400-11,750 Å[a]	Flashlamp and/or laser pumped	Dye-solvent combination cover the wavelength span nearly continuously.
Tuning	80-400 Å	Prism, filter, grating, Q-switch	Length, concentration, and temperature of active medium can also tune.
Spectral width	0.01-15 Å	Grating and etalon-mirrors	Dependent on method used.
Beam divergence	0.5-5 mrad	Etalon in cavity-flashlamp	Dependent on pumping uniformity.
Efficiency	0.4-25%	Flashlamp-laser pumped	Dependent on optical pumping.
Output energy	0.05-2.5 J	Flashlamp pumped	Varies with pumping intensity
Output power	0.5-2.5 MW	Flashlamp pumped	Depends on dye-solvent system
Pulse repitition rate	1-2 pps 10-60 pps 50-200 pps	Annular flashlamp Linear flashlamp Laser pumped	Dependent on laser cooling, dye cooling, and failure of optical components
Pulse duration	1-150 μsec 10-20 nsec	Flashlamp pumped Laser pumped	(Shorter than pump duration) (Follow pump pulse)
Mode locked	Pulses $<10^{-9}$ sec Pulses $<10^{-11}$ sec	Flashlamp pumped with intracavity absorber Mode-locked pump	Pump cavity length integral multiple of dye cavity observed by two-photon fluorescence

[a]Lower wavelengths possible by dual-dye laser techniques.
Source: A. K. Levine and A. J. DeMaria (Eds.), Lasers, Vol. 3, Marcel Dekker, New York, 1971, p. 271.

Table 32 Dye Laser Compounds and Emission Wavelengths

Organic dye compound	Typical solvent	Average center wavelength (nm)
p-Terphenyl	Ayclohexane	341
2, 5-Diphynyl-1, 3, 4-oxadiazole	Dioxane	348
Isopropyl-2-phenyl-5(4-biphenylyl)-1, 3, 4-oxdaiazole	Ethanol	370
2, 5-Diphenylfuran	Dioxane	371
p-Quater phenyl	DMF	374
2, 5-Diphenyloxazole	Dioxane	381
Diphenylbutadiene	Toluene	383
2-Biphenyl-5-styryl-1, 3, 4-oxadiazole	Toluene	390
2, 5-Di (naphthyl)-1, 3, 4-oxadiazole	Toluene	391
2-Phenyl-5-naphthyl-1, 4-oxazole	Toluene	400
1, 2-Di-4-biphenyl-ethylene	Toluene	408
2, 5-Dibiphenylyloxazole	Benzene	409
p, p^1 -Diphenylstilbene	Benzene	409
1, 4-Distyrylbenzene	Toluene	415
1, 4-Di[2-(5-phenyloxazolyl)]-benzene	Toluene	417
p-Bis (o-methylstyryl)-benzene	Ethanol	419
4, 4^1 -Dichloro-1, 4-distyryl-benzene	Toluene	420
Aquafluor	Ethanol	420
p-Bis[2-(5-phenyloxazolyl)]benzene	Ethanol	421
Liquiflor	Ethanol	422
1, 4-Bis-2-(4-methyl-5-phenyloxazolyl) benzene	Cyclohexane	423
p-Bis-(o-methylstyryl)-benzene	Toluene	424
2^1 -Methoxy-1, 4-distyryl-benzene	Toluene	425
1, 2-Di-(naphthyl)-ethylene	Toluene	426
2, 2^1 -Dimethoxy-1, 4-distyryl benzene	Toluene	430
1-Styryl-4[w-vinyl-(n-biphenylyl)]-benzene	Toluene	432
9, 10-Diphenylanthracene	Cyclohexane	433
Acridone	Ethanol	437
2, 5-Bis-[5-tert-butylbenzoxazolyl(2)] thiphene	Benzene	437
3-Ethylaminopyrene-5, 8, 10-trisulfuric acid	Water	441
4-Methyl-7-hydroxycoumarin	Aqueous alkaline	450
9-Aminoacridine hydrochloride	Ethanol	459
7-Diethylamino-4-methyl coumarin	Ethanol	459
Esculin	Aqueous alkaline	460
7-Hydroxycoumarin	Aqueous alkaline	460
Benzyl B-methyl umbelliferane	Ethanol	460

Table 32-Continued

Calcein blue	Ethanol	470
2, 4, 6-Triphenyl-pyrilium fluoroborate	Methanol	485
3-Aminophthalimide	Isoamyl alcohol	500
Trypaflavin	Ethanol	505
Acriflavin hydrochloride	Ethanol	510
Fluorescein	Aqueous alkaline	518
Na-Fluorescein	Water	527
Eosin	Ethanol	540
Lachs	Glycerol	540
2, 7-Dichloro-fluorescein	Ethanol	545
Rhodamine 6G	Ethanol	555
Monobromofluorescein	Glycerin	560
Uranine	Ethanol	560
Pina Corthol	Ethanol	565
Dibromofluorescein	Glycerin	568
Rhodamine B	Ethanol	577
Rhodamine C	Glycerin	580
Pyronin B	Acetone	580
Acridine red	Ethanol	580
Rhodamine G	Ethanol	585
Rhodamine S	Ethanol	586
Pyronin G	Isoamyl alcohol	590
Lucegenin	Aqueous acid	600
Eosin	Ethanol	600
Sophranine-T	Ethanol	610
Isoquinoline red	Water	620
Rhodamine 3B	Isoamyl alcohol	620
3, 3^1 -Diethyloxadicarbocyanine iodide	Methanol	658
Cresyl violet	Ethanol	660
3, 3^1 -Diethyl-2, 2^1 -thiadicerbocyanine iodide	Acetone	711
3, 3^1 -Diethyl-10-chloro-2, 2^1 -(5, 6, 5^1, 6^1 -dibenzo)thiadicarbocyanine iodide	Acetone	714
3, 3^1 -Diethyl-2, 2^1 -(5, 5^1 -dimethyl) thiazolinotricarbocyanine iodide	Glycerin	717
3, 3^1 -Diethyloxytricarbocyanine iodide	Ethanol	728
3, 3^1 -Diethyl-5, 5^1 -dimethoxy-6, 6^1 -bis (methylmercapto)-10-, methyl thiadicarbocyanine bromide	Ethanol	733
3, 3 -Diethyl-2, 2^1 -oxatricarbocyanine iodide	Acetone	744
1, 1^1 -Dimethyl-11-bromo-2, 2^1 -quinodicarbocyanine iodide	Glycerine	745

Table 32-Continued

Cryptocyanine	Glycerin	745
1, 1^1 -Dimethyl-4, 4^1 -quinocarbocyanine iodide	Glycerin	749
1, 1^1 -Diethyl-2, 2^1 -dicarbocyanine iodide	Glycerin	750
Echtblau B	Glycerin	753
1, 1^1 -Diethyl-4, 4^1 -quinocarbocyanine bromide	Glycerin	754
Dicyanine	Glycerin	756
Naphthalene green	Glycerin	756
Rhodulinc blue 6G	Glycerin	758
Magnesium phthalocyanine	Quinoline	759
Malachite green G	Isoamyl alcohol	760
1, 1^1 -Diethyl- -cyano-2, 2-dicarbocyanine-tetrafluoroborate	Pyridine	760
Chloroaluminum phthalocyanine	Dimethyl sulfoxide	762
3-Ethyl-3^1-methylthiathiazolinotricarbocyanine iodide	Ethanol	770
3, 3^1 -Diethyl-10-chloro-2, 2^1 -(4, 5, 4^1, 5^1-dibenzo)thiadicarbocyanine iodide	Acetone	774
3, 3 -Diethyl-11-methoxythiatricarbocyanine iodide	Ethanol	786
1, 3, 3, 1^1 , 3^1 , 3^1 -Hexamethylinodotricarbocyanine iodide	Ethanol	794
1, 1^1 -Diethyl-λ-acetoxy-2, 2^1 -dicarbocyanine tetrafluoroborate	Methanol	797
1, 1^1 -Diethyl-λ-nitro-4, 4^1 -dicarbocyanine tetrafluoroborate	Pyridine	800
3-3^1 -Diethylthiatricarbocyanine iodide	Ethanol	807
Victoria blue	Glycerin	809
Victoria blue R	Glycerin	814
1, 1^1 -Diethyl-11-bromo-2, 2^1 -quinodicarbocyanine iodide	Glycerin	815
1, 3, 3, 1^1 3^1 , 3^1 -Hexamethyl-2, 2 -indotricarbocyanine iodide	Acetone	819
3, 3^1 -Diethylthiatricarbocyanine iodide	Dimethyl sulfoxide	822
Methylene green	Sulfuric acid	823
1, 3, 3, 1^1 , 3^1, 3^1-Hexamethyl-4, 5, 4^1, 5^1-dibenzoindoctricarbocyanine perchlorate	Ethanol	824
3, 3^1-Diethyl-2, 2^1-selenatricarbocyanine iodide	Acetone	826
Methylene blue	Sulfuric acid	824
1, 1^1-Diethyl-11-bromo-4, 4^1-quinodicarbocyanine iodide	Methanol	830
3, 3^1-diethyl-6, 7, 6^1, 7^1, -dibenzothiatricarbocyanine	Ethanol	839

Table 32-Continued

Toluidine blue	Sulfuric acid	848
Thionin	Sulfuric acid	850
3, 3^1-Diethyl-2, 2^1-(5, 6, 5^1, 6^1-tetramethoxy) thiatricarbocyanine	Acetone	853
3, 3^1-Diethyl-6, 7, 6^1, 7^{1}dibenzo-11-methyl-thiatricarbocyanine iodide	Ethanol	856
3, 3^1-Diethyl-2, 2^1-(e, r, 4 , 5 -dibenzo) thiatricarbocyanine	Acetone	860
Phthalocyanine	Sulfuric acid	863
1, 1^1-Diethyl-2, 2^1-quinotricarbocyanine iodide	Acetone	898
1, 1^1-Diethyl-4, 4^1-quinotricarbocyanine iodide	Acetone	999

Source: Anthony J. DeMaria and Albert K. Levine, Eds., Lasers, Vol. 3, Marcel Dekker, New York, 1971.

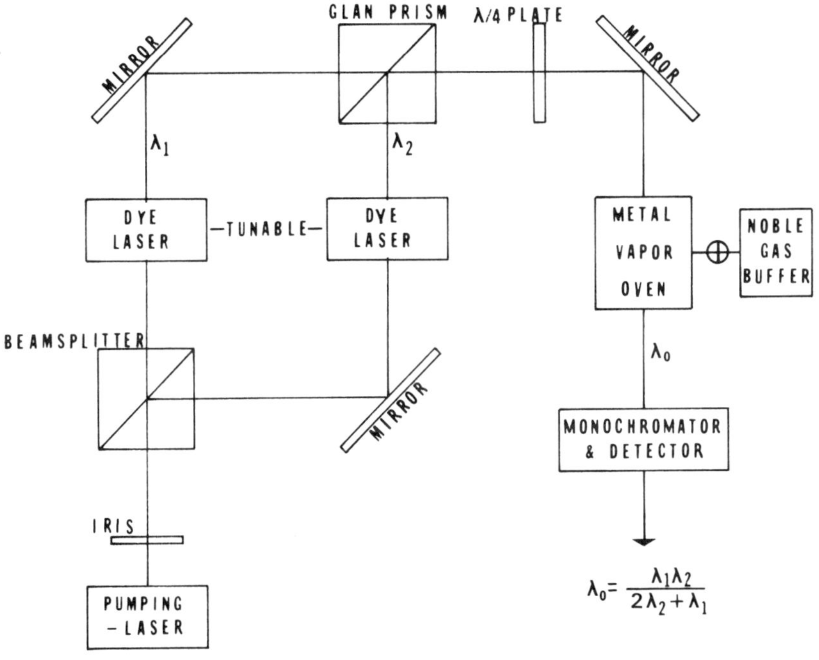

Fig. 14 Dual dye laser apparatus to synthesize a tunable single wavelength vacuum ultraviolet source. Source: P. P. Sorokin, J. J. Wynne, J. A. Armstrong, and R. T. Hudson, Resonantly enhanced, nonlinear generation of tunable, coherent, vacuum ultraviolet light in atomic vapors, Ann. N.Y. Acad. Sci., vol. 267, 1976.

Table 33 Typical Laser Q-Switches

Type	Spectral range (μm)	Minimum rise time	Contrast ratio
AO	0.30-2.10	70-200 ns	
EO	0.025-1.17	0.5-1 ns	600:1
KDP	0.25-1.30	0.09-2 ns	500 to 1500:1
Kerr	0.40-1.10	3-10 μs	
Laser trigger-spark gap	0.25-3.50	0.1-1 ns	
$LiNbO_3$	0.80-2.50	0.8-1 ns	100:1
Mechanical chopper	10.60	400 ns	
Pockels	0.60-1.10	1-15 ns	100:1
Pulsed chopper	0.40-1.21	0.8-1 ns	1000:1
Rotating	1.06	15 ns	100:1

Source: Product literature.

Electro-optical (EO) shutters can be used to obtain very fast Q-switching. The Pockels effect can be used with various crystalline materials such as KDP (potassium dihydrogen phosphate) using optically flat crystal surfaces connected to ring-shaped electrodes which can apply several kilovolts. A polarizer is used in combination with the Pockels cell. By controlling (switching) the cell bias, the transmitted light can be controlled to very fast speeds. The length of the Q-switch pulse can be controlled by adjusting the length of the laser-Pockels-cell-polarizer-mirror cavity. The combination of adjustable cavity length and rapid switching speeds permits the generation of large laser pulses at a specified frequency. Unlike the mechanical Q-switch, which can operate over a large range of laser emission wavelengths, the Pockels cell is essentially limited to one design wavelength. Another type of electro-optic shutter, the Kerr cell, can be designed to operate over a larger spectral region using the Raman effect in nitrobenzene. Here, instead of a crystalline material, nitrobenzene is used, switched at high potentials (15-35 kV) to the cell volume.

Table 34 lists some typical laser modulator characteristics for solid materials, including the ranges for use times, bandwidth, and percent transmission. Table 35 gives further details of the various types of coherent light modulators including expected beam diameters and insertion losses.

Table 34 Typical Laser Modulator Characteristics

Type	Crystal material	Range of rise times	Bandwidth (MHz)	Spectral range (μm)	Trans-mittance (%)
AO	Ge	100-120 ns	dc-5	3.3-10.6	50-60
	Glass		dc-7	0.44-0.64	85-90
	$LiNbO_3$	40-60 ns	90-200	0.30-1.06	60-70
	$PbMoO_4$	10-15 ns	dc-50	0.44-0.70	60-90
	Si	150-200 μs	0.3-0.5	0.30-2.10	
CO	ADP		200-450	0.25-1.18	95-99
	KDP	1-250 ns	dc-120	0.20-1.25	86-93
Pockel Long.	KDP	0.1 ns-5 ms	dc-4 GHz	0.25-1.49	70-97
Pockel Tranv.	ADP	0.8-3 ns	dc-120	0.35-1.30	90-02
	KDP	0.8-3 ns	dc-100	0.35-1.58	90-92
	$LiNbO_3$	0.8-2 ns	dc-150	0.49-2.01	80-85

Source: Product literature.

Laser Interferometers and Measurements

The laser is commonly used in the plano-Fizeau and Twyman-Green interferometers as shown in Figs. 15 and 16, respectively. These types of optical interferometers are used to determine the degree of optical flatness of various components and materials. Here, interference fringes between the reference flat and the test oomponent are observed or photographed. Each fringe is equivalent to an optical path length differential of $1/2\lambda$, where λ is the emission wavelength of the laser. The optical flatness of the test component is determined by measuring the straightness of the fringes on the photographed interference pattern or by direct measurement on the view screen. Interference fringes are also observed between the two surfaces of the test component, which can be defined by

$$N = \frac{2nD}{\lambda} \tan \theta$$

where

N = observed number of fringes
n = material index of refraction
D = diameter of the flat
θ = angle between the two near-flat surfaces

Table 35 Types of Coherent Light Modulators

Spectral range (μm)	Beam diameter (mm)	Device and materials	Bandwidth at 0.63 μm	Insertion loss at 0.63 μm	General remarks
0.22–1.65	50	EO (long and trans.); KDP and ADP	50 MHz	10% using (AR) anti-reflection coating	Bandwidth up to 3 GHz; requires polarized light
0.4–2.0	100	Faradayrotor, EO; dense optical glasses	1 MHz	~25%	Low-speed, large-aperture devices
0.4–2.2	10	AO (Bragg); $PbMoO_4$, TeO_2, dense glass	7 MHz (glass) 30 MHz ($PbMoO_4$)	10–20% using (AR)	Can modulate unpolarized light; can distort beam and cause frequency shifts
0.6–3.5	10	EO (total internal ref.) $LiNbO_3$, $LiTaO_3$	150 MHz	2–10%	Possible frequency response nonlinearities; resonances can occur
0.6–3.5	15	KTN, $LiNbO_3$, $LiTO_3$	150 MHz	~25%	Contract ratio ~100:1
3–15	15	CdTe, GaAs			Resonances can occur

Adapted from ADL World Electro-Optics Industry, 1975–1980, E-O Components Update, May, 1976.

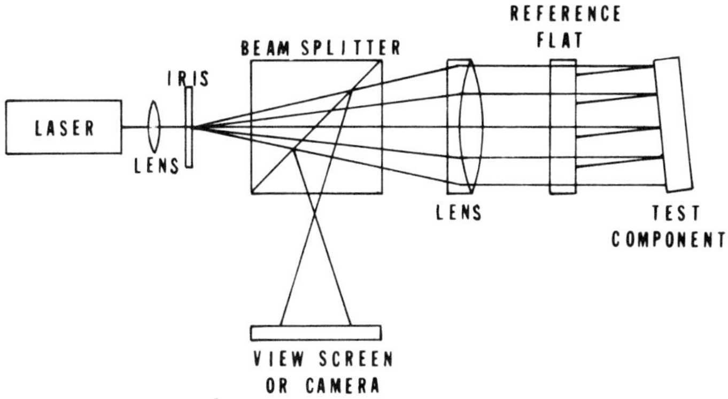

Fig. 15 Plano-Fizeau interferometer.

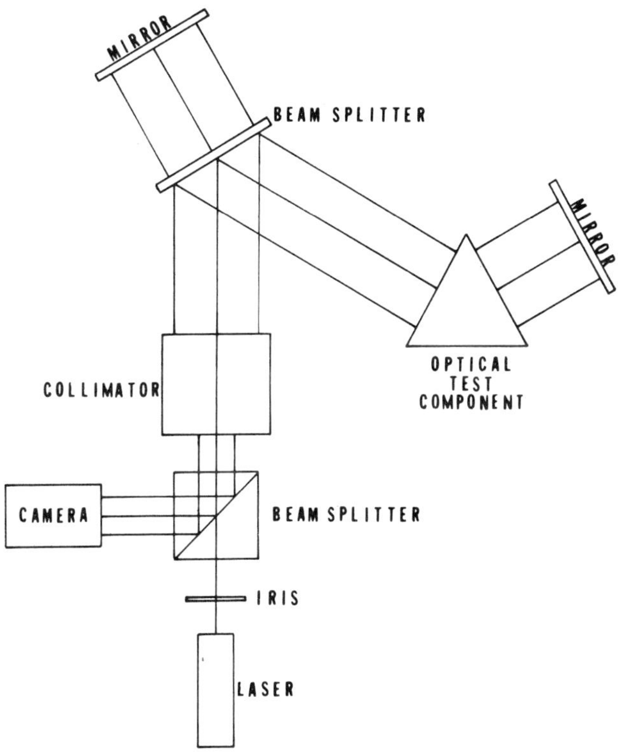

Fig. 16 Twyman-Green interferometer.

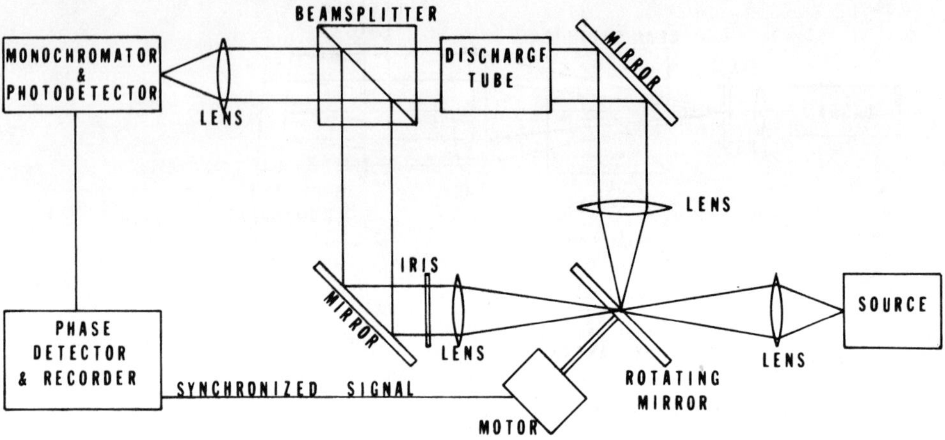

Fig. 17 Dual-beam spectrometer for laser spectral gain characteristics measurements. Source: Laser Technology And Applications, S. L. Marshall (Ed.), McGraw-Hill, New York, 1968, p. 205.

The internal characteristics of a laser can be measured using a dual-beam spectrometer as shown in Fig. 17, which can be used to measure the spectral-gain characteristics of various laser systems. In the single-beam mode the apparatus can measure the laser fluorescent lifetimes as well as the absolute gain. By varying the excitation the gain-vs.-pumping level function can be determined, and by varying the monochromator the gain-vs.-frequency function can also be established.

The laser is superior to conventional light sources for precision measurements, since it has greater spatial coherence and a narrow frequency range. A common type of interferometer used for measuring the various qualities of optical components is the Michelson interferometer. One configuration of this interferometer type is shown in Fig. 18. This apparatus can be used to measure the optical surface flatness of various types of components (filters, polarizers, etc.), the parallelism of mirrors, and the index of refraction variations in various plastic and glass components, and to measure distances in terms of the wavelength of light. The traditional light source was the cadmium red line. The preferred source now is a gas laser, which can be used to continuously measure distances less than 5×10^{-8} m. This type of interferometer can be used as a machining alignment measurement device for accurate cutting and tooling for automatic and manual industrial equipment.

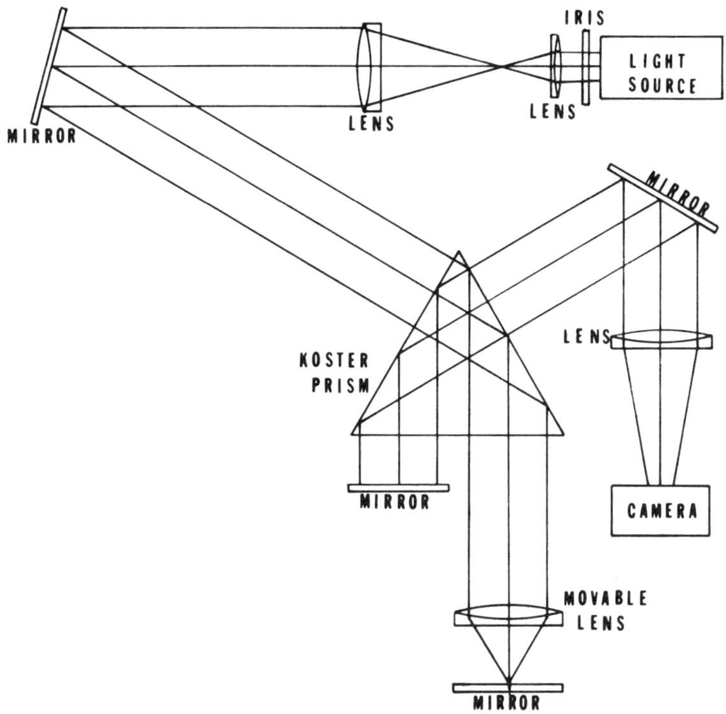

Fig. 18 Typical Michelson interferometer.

Laser Memory Systems

The use of lasers for optical memory storage and computing systems can be accomplished using holography. The basic components of the holographic memory system are a coherent light source (laser), beam deflectors, optical lenses and filters (beam control), a page composer (light modulator), a hologram array storage unit (memory storage), and a photodetector array (optical-to-electrical signal converter). The basic system is shown in Fig. 19. Some of the lasers and materials suitable for optical memory systems are listed in Table 36. The important laser characteristics to consider are output power, efficiency, and usable lifetime. The important material considerations are bit density, sensitivity, and erasability. The types of page formers that can be used are listed in Table 37. The practical considerations here are the page former mechanism, addressing mode, and material physical parameters.

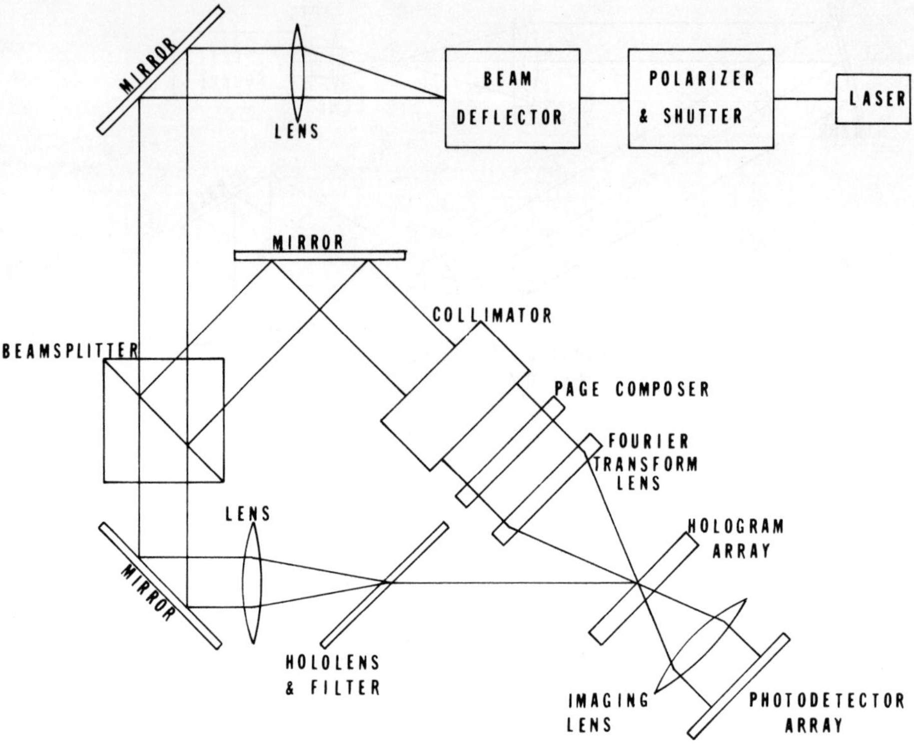

Fig. 19 Basic holographic memory system. Source: G. A. Bailey, Optical mass memories, Ann. N. Y. Acad. Sci., Third Conference on the Laser, vol. 267, 1976, p. 412.

The limitations on the storage capacity of optical memories are in part determined by the diffraction resolution limit and by the surface area of the storage material and the performance of the large-area modulator. Table 38 lists some types of large-area modulators suitable for laser-optical processing systems. The important factors in large-area modulations are the addressing method, modulation type, and resolution. The storage medium is considered to be the most important element in optical storage systems. Useful optical storage media must have high resolution, acceptable dark storage times and write-erase times, low noise, and reversibility. For high information densities most systems use binary coding on thick holograms for optimal performance. Various types of beam deflectors can be used successfully, including moving iron or coil galvanometers, acousto-optic types ($PbMoO_4$, TeO_2, As_2S_3, Tl_3AsS_4, or tellurite glass), and electro-optic types ($LiNbO_3$, nitrobenzene,

Table 36 Parameters of Lasers and Storage Materials Suitable for Optical Memory Systems

Lasers				
Type and wavelength (μm)	Output power (mW)	Input power (W)	Output efficiency (%)	Lifetime (hr X 10^{-3})
Ar (0.54)	4-300	200-6000	0.002-0.005	5-15
GaAs (0.93)	5	0.5	1.0	3
He-Ed	10	100	0.01	4
He-Ne (0.63)	1-10	20-100	0.005-0.010	15-20
Nd:YAG (0.53;1.06)	700-1000	667-700	0.020-0.150	8.15

Storage materials			
Material type	Bit density (bits/mm^2)	Sensitivity (J/mm^2)	Erasable
Compensation-point (GdIG)	2 X 10^3	10^{-3}	Yes
Curie-point MnBi	10^6	15 X 30 X 10^{-5}	Yes
Dichromated gel film	10^6	3 X 10^{-5}	No
Ferroelectric $LiNbO_3$:Fe	2 X 10^6	2 X 10^{-5}	Yes
Photochromic $SrTiO_3$:Ni:Mo	10^6	10^{-3}	Yes
Silver halide film	10^6	20 X 100 X 10^{-8}	No
Thermoplastic		10^{-3}	Yes
Vaporization of Bi	10^6	5 X 10^{-4}	No

Adapted from ADL World Electro-Optics Industry, 1975-1980, E-O Components Update, May, 1976.

potassium tantalium niobate, or strontium barium nitrate). For all optical components and elements used in optical storage and computing systems, radiation losses by absorption and reflection can be a significant factor in component choices and dictate the output power of the laser.

Light-Emitting Diodes

Light-emitting diodes (LEDs) generate light when carriers injected across a pn junction radiatively recombine, emitting radiation over a solid angle of 2π sr. The actual output power emitted into a unit solid angle per unit emission area is low for LEDs even though the total power output is reasonably high. The low output radiance results in high coupling losses to optical components.

Table 37 Types of Page Formers

Type	Mechanism	Addressing mode	Example materials[a]	Comments
Acousto-optic	Diffraction	Modulation of acoustic waves	Liquids, $PbMoO_4$, quartz	Excellent usage
Absorption	Optical density changes	Light or electron beams, electrodes for heating	As_2S_3, CdS, CdSe, photo and cathodo chromics	Materials suffer from fatiguing
Electro-optic	Polarization changes	Electrode matrix, electron beam	PLZT[a], KDP components, $Bi_4Ti_3O_{12}$	Bulky, poor performance
Electro-strictive	Optical pathlength changes	Electrode matrix	PLZT[a]	Slow, requires high voltage
Magneto-optic	Polarization changes	Conductor matrix thermomagnetic writing	MnBi, $FeBO_3$, FeF_3	Problems with absorption
Scattering	Electrically induced scattering	Electrode matrix	Liquid crystals, PLZT[a]	Materials suffer from fatiguing
Surface relief	Phase changes	Electron beam, electron matrix with changes, switched RF drive	Thermoplastics, mirrored pizzoelectrics, membranes	Used in large-screen display systems

[a]PLZT = lead zirconate litanate.

Adapted from ADL World Electro-Optics Industry, 1975-1980, E-O Components Update, May, 1976; and Appl. Opt., vol. II, February, 1972, p. 399.

Table 38 Large-Area Modulators Suitable for Coherent-Optical Processing

Type	Addressing method	Modulation type	Comments
Acousto-optic interactions	Transducers	Phase	Limited resolution, frequency response up to 500 MHz
Membrane deformation by electrostatic forces	Electrodes	Phase	Moderate resolution, low modulation factor
Oil film deformation by electrostatic forces	Electron beam	Phase	Moderate resolution, self-erasing
Thermoplastic deformation by electrostatic forces	Electron beam, light	Phase	Good resolution, slow speed, limited lifetime-in-use
Dynamic scattering + EO liquid crystal effects	Electrodes	Phase, amplitude	Poor resolution; slow speed, low contrast
EO effect on PLZT ($PbZrO_3/PbTiO_3$ + La)	Electrodes	Phase	Poor resolution, cross-talk, uniformity problems
Photodichroic effect	Light	Phase, amplitude	Excellent resolution, poor light sensitivity

Adapted from ADL World Electro-Optics Industry, 1975-1980, E-O Components Update, May, 1976.

The light intensity of typical LEDs as a function of angular distribution is shown in Fig. 20. For infrared-emitting diodes, Fig. 21 shows the typical relative photon intensity versus angular displacement.

The spectral bandwidth of LEDs ranges from 250 to 400° at room temperature operation. LEDs have a few advantages over semiconductor lasers, in that they are cheaper and have less of a temperature dependence on total emitted power. The three basic LED configurations are the edge, high-radiance, and surface emitters. The edge-emitting heterojunction structures use partial internal wave guiding to make the emitted light more directional. The lateral width of the emitting region can be adjusted to some extent to match the core diameter of the optical fiber.

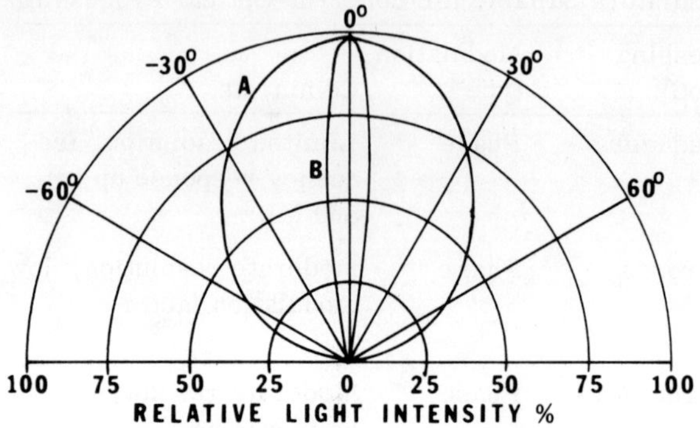

Fig. 20 Light intensity of typical LEDs as a function of angular distribution: A, with planar window; B, with lens attached.

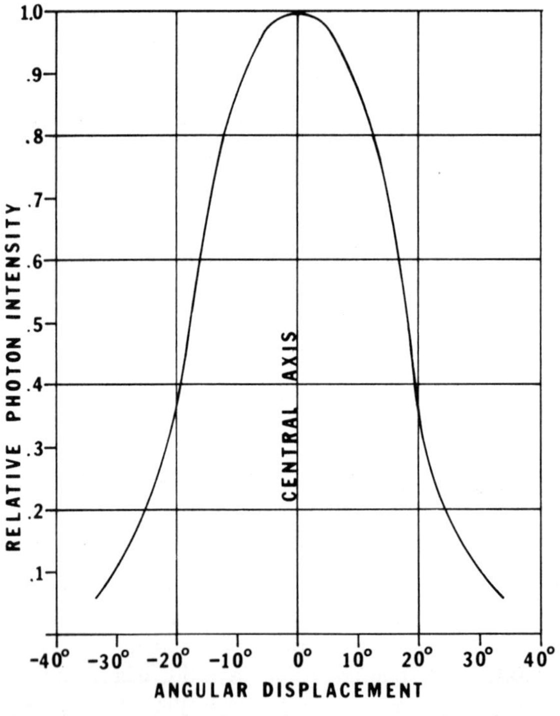

Fig. 21 Typical infrared-emitting diode relative photon intensity vs. angular displacement.

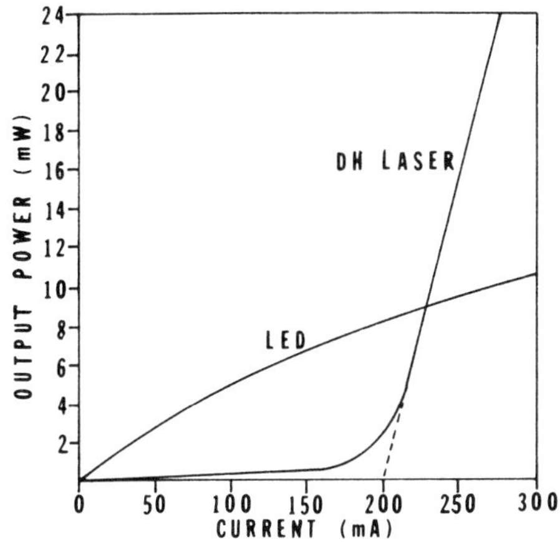

Fig. 22 Typical current-vs.-power curves for optical light sources.

A common high-radiance LED is the Burrus type, which is a small-area high-brightness double heterojunction (DH) LED with a brightness up to about 150 W/srcm^2 for current densities of 4-6 kA/cm^2 with quantum efficiencies of about 2-3%. This particular LED structure has efficient coupling into optical fibers. Modulation of LEDs is accomplished by varying the injected device current. The limitations of the speed of modulation are dependent upon the device response time determined electronically by the free-carrier generation and recombination time and the driving circuit capacitance. The typical power-vs.-drive response curves for an LED and semiconductor laser are as shown in Fig. 22.

Figure 23 shows the typical infrared-emitting-diode variations of wavelength change and relative power as a function of the LED temperature.

Two methods commonly used to prepare semiconductor materials are vapor- and liquid-phase epitaxy. Vapor-phase epitaxy is often used when there is a reasonably large lattice mismatch between the substrate and the epitaxial layers, since the composition can be graded to any desired degree, thus effectively reducing the defect density in the active region of the grown layer. Liquid-phase epitaxy is used especially for the introduction of Ge, Si, and Al into the semiconductor material layers. All dopants and materials used in the construction of LEDs and lasers should be of the highest purity possible to prevent the formation of nonradiative recombination sites caused by lattice defects and undesirable chemical precipitates.

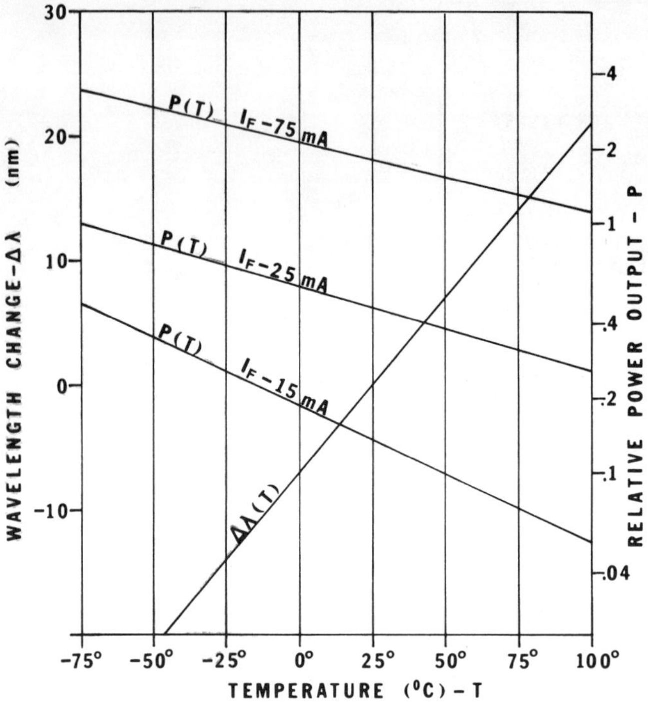

Fig. 23 Typical infrared-emitting diode variations of wavelength and relative power with device temperature.

Laser Diodes

Laser light sources are usually designed to have a high single-pass gain-to-loss ratio to maintain their physical structure as long as possible, thus sustaining their lifetime in use at room temperatures. Double-heterojunction laser structures can be made with relatively long lifetimes when the lattice parameters of the elemental compositions in the various layers and boundaries are carefully matched. The DH layers are made by careful epitaxial growth of materials above and below the active laser region where light is generated by the radiative recombination of carriers injected across the pn junction. The thin layers surrounding the active region have lower refractive indexes and high bandgap potentials, thus reducing optical losses and increasing gain by energy confinement. In the DH laser with the stripe geometry, the lasing region occurs only in the transverse direction under the width of the contact stripe. Figure 24 shows the typical layer structure of a DH laser.

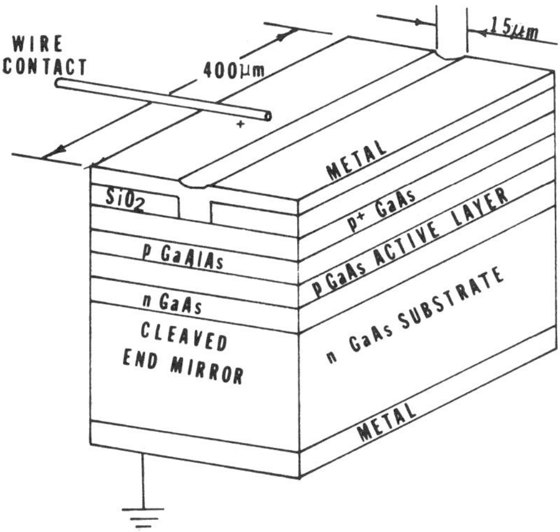

Fig. 24 Double-heterojunction injection laser.

The development of injection lasers has involved the examination of many materials. The basic semiconductor material for both LEDs and injection lasers is the GaAs combination with a characteristic emission wavelength of 905 nm. In optical communications fibers, one of the OH^- absorption bands covers this wavelength, requiring the wavelength to be shifted or the elimination of the absorption band to achieve the most efficient utilization of the light energy for low attenuation of signals. The means of achieving the wavelength shifting is to dope the basic GaAs material with other elements. The elements tested for these purposes must meet numerous physical and electrical criteria as well as simple limitation criteria such as having sufficiently high vapor pressures at room temperature operation to maintain material integrity. The elements commonly used in doping of the GaAs material are Al, As, In, P, and Sb, which must be available in ultrapurified forms to eliminate other metals from generating lattice imperfections in the epitaxial growth stages. The range of experimental wavelengths achieved by various doping techniques and elemental combinations for a variety of semiconductor materials is given in Table 39.

As in LEDs, lasers can be modulated by varying the injection current with modulation bandwidths of several hundred megahertz, or data rates of several hundred megabits per second. Compared to LEDs, injection lasers have a high differential quantum efficiency, in the sense of emitting substantial optical power for small changes in the driving current above the device threshold current. The threshold current for typical DH lasers

Table 39 Semiconductor Emission Wavelengths

Material composition	Wavelength (μm)
PbSnTe	28.0 → 6.5
PbSSe	8.5 → 4.3
PbSe	8.5
PbTe	6.5
InSb	5.3 → 5.1
In As_xSb_y	5.1 → 3.1
PbS	4.3
$InAsZn_x$	3.18 → 3.11
InAs	3.15 → 3.11
$InAs_xP_y$	3.15 → 0.91
AsGa In	3.15 → 0.84
Ge	1.8
$InPAs\text{-}Te_xZn_y$	1.6 → 0.90
GaSb	1.6
Si	1.1
InP	0.91 → 0.89
$GaAsTe_xZn_y$	0.90 → 0.84
$AsGa_xAl_y$	0.90 → 0.58
$GaAs_xP_y$	0.90 → 0.54
CdTe	0.855
GaAs	0.85 → 0.82
$TeZn_xCd_y$	0.85 → 0.53
$GaAsZn_x$	0.842 → 0.838
GaAsP	0.84 → 0.62
GaN	0.72 → 0.34
GaP	0.68 → 0.56
$Cu_2Se\text{-}ZnSe$	0.65 → 0.47
BP	0.64
$ZnTe_xSe_y$	0.627 → 0.47
ZnTe	0.62
ZnSe	0.47
SiC	0.456
ZnO	0.36
ZnS	0.33

ranges from 0.7 to 14 kA/cm^2 and the quantum efficiency from 10 to 50%, compared to 3% for LEDs. The lasing peak energy of heterojunction lasers decreases with increasing temperature at a rate of approximately 5×10^{-4} eV/K. The threshold current density increases with temperature. The typical power levels needed for optical communications systems are from 4 to 12 mW. The typical thermal resistance of stripe lasers is about

11-21 K/W, resulting in a temperature differential between the heat sink and junction of about 5-10 K.

The light output power (mW)-vs.-injection current (mA) characteristics of stripe lasers depend upon the width of the stripe region, among other factors. For stripe widths less than 10 μm the response curves are rather nonlinear due to intensity dependent losses. For wider stripes light intensity or current have little effect on the optical distribution of a given mode. For stripe widths greater than 30 μm the typical output power-vs.-injection current responses are very linear. Typical chip sizes use an average stripe width of 15 to 20 μm to keep the response curves reasonably linear by matching the width of the transverse optical distributions and gain regions. In all semiconductor lasers some noise is generated caused by quantum fluctuations. This noise or self-pulsing is greatest near the threshold current level. Wider stripes usually generate less noise except when higher order modes are generated at high current levels. The noise level generated by semiconductor lasers depends mostly upon the drive current and the quality of the laser construction. Below the threshold current the noise level is low, comparable to LEDs. The noise level increases rapidly at the threshold current and then decreases at higher current levels. As the current increases sufficiently beyond the threshold, additional lasing modes are usually created causing potentially higher noise levels. Noise is often generated at high frequencies (self-oscillation noise), above 650 MHz, which can be effectively eliminated or minimized by limiting the electronic bandwidth of the driving current or by using feedback on the transmitted optical signal.

Semiconductor Light Source Lifetime

Both LEDs and injection lasers have finite lifetime's-in-use. For LEDs the lifetimes are typically 10^5 h. For properly designed and prepared lasers, lifetimes of 10^5 h can also be achieved. The two types of failures associated with these communications light sources are gradual degradation and catastrophic degradation. Gradual degradation depends mostly upon the current density and duty cycle. Catastrophic degradation can be caused by many factors, including sudden system overload, causing surges in the current, and physical degradations, such as strain releases in the active lasing region. A common cause of catastrophic failure is to overheat a laser diode that does not have temperature feedback current control. The degradation process depends in part on the construction techniques and material purity used in the light source, and the initial concentration of flaws in the active region.

Lasers often exhibit two types of rapid decay mechanisms: facet damage and dark-line defects. Facet damage usually results from using an excessive energy density at the laser emitting surface, and can be minimized by using small operating power levels, and by dielectric facet coatings such as Al_2O_3. The minimal operational power level is determined by many factors, particularly fiber attenuation and the total length desired or required between end-devices or repeaters. Dark-line defects are caused by the migration of lattice vacancies in the active region, and can be minimized by matching the lattices of the surrounding regions and by minimizing externally applied stresses to the laser material. The common cause of externally applied stress to the laser material is soldering the metal heat sink or contact wire to the top conducting layer. By using indium solder, a soft, low-melting-temperature solder, residual strain introduced to the laser chip by post-production heating can be minimized. Internal stresses can be minimized by adding Al to the active region or P to the passive layers to partially compensate for the coefficient of expansion differences between the $Ga_xAl_{1-x}As$ and GaAs layers.

The procedure normally used to accelerate lifetime-in-use testing for LEDs is to increase the operational temperature and extrapolate to a projected failure time or degradation rate for room temperature operation, which assumes a uniform degradation process throughout the lifetime of the LED. For lasers this means of testing cannot be used since the operating current cannot be maintained constant, because of the temperature dependence and time dependence of the threshold current. The tests usually used to determine lifetime utilize rate measurements of the threshold current increase with operational time. These rates are typically 25% threshold current increase per 10,000 to 15,000 h for lasers with uncoated mirror facets. Coating the facets with dielectric films can effectively eliminate typical current degradation characteristics of the light output with time.

Semiconductor Laser Design

The requirements for semiconductor lasers include output power linearity, low temperature sensitivity, low current thresholds, and long lifetimes. The causes of nonlinearity behavior in the power output-vs.-current responses in lasers are similar to those in LEDs. Some of these causes are thermal nonlinearity, junction capacitance and uniformity, and undesirable radiationless carrier recombinations. Thermal nonlinearities can be controlled by using a feedback circuit with a photodetector that monitors the output power from the second laser face. Feedback circuitry also eliminates internal high-frequency modulations.

The design and construction of lasers must contend with the following physical and electrical requirements. First, there must be a close matching of the lattice parameters at the heterojunctions to minimize the expected interfacial recombination degradation effects. Second, each layer must have a minimum defect density and maintain low electrical and thermal resistance (10-20° C/W). Third, to keep threshold current densities below 3 kA/cm^2, the planar recombination regions that must be kept in the range of 0.1 to 0.5 μm wide and the bandgap energy step at the heterojunctions should be at least 0.2 eV. Fourth, the p layer should be doped high enough to prevent high-temperature electron leakage but not so high as to incur excessive current conduction away from the lasing area. Table 40 lists the major semiconductor properties to be considered in the design process, including lattice parameters, energy gaps, electron and hole mobility, and melting temperatures.

The degree of the interfacial lattice parameter mismatch (defined as the lattice parameter difference between the substrate and the epitaxial material divided by the substrate lattice parameter) must be kept small if quantum efficiency is to remain high. For quantum efficiencies greater than 40%, the lattice parameter mismatch should not exceed 0.15% for typical DH lasers. This is difficult to achieve for any binary material system, although quantum efficiencies higher than 40% can be attained. The matching of lattice parameters can be more precisely controlled when multicomponent materials are used in adjoining heterojunction layers.

Many combinations of group III-V compounds have been examined. One criterion used to select the semiconductor materials and elemental combinations in LEDs and laser diodes is the association of bandstructure and radiative lifetime. In indirect bandgap semiconductors, such as Si, Ge, GaP, and AlAs, the recombination process is relatively slow. In direct bandgap materials such as GaAs, InAs, and InP, the recombination process is very efficient resulting in short radiative lifetimes. The difference in the radiative lifetimes between direct and indirect bandgap materials is about four to five orders of magnitude. High internal quantum efficiency can thus be more easily accomplished in direct bandgap semiconductors.

The threshold current density can be minimized by restricting the width of the recombination region by placing a barrier at a distance less than the diffusion length from the pn junction. The degree of radiation confinement is determined in part by the refractive index differential between the heterojunction layers and the actual spacing of the layers. The refractive index difference between GaAs and $Ga_{1-x}Al_xAs$ varies as 0.45x. The emission wavelength decreases approximately by 6700x Å ($0 \leq x \leq 0.2$). In most lasers, the refractive index step changes at each heterojunction are symmetrical around the active region to optimize wave confinement. The degree of confinement affects the threshold current density and the radiation pattern by changing the effective source size. The threshold current density decreases approximately linearly with the heterojunction layer spacing depth.

Table 41 compares semiconductor lasers and LED light sources for a number of emission source parameters.

Table 40 Semiconductor Properties

Material	Lattice parameters (A at 300 K)	Energy gap E_g (eV at 300 K)	Mobilities (cm^2/V-s at 300 K) Electron u_e	Hole u_n	Melting point (K)
AlAs	5.6622	2.16	1,200	420	1870
AlP	5.451	2.5			1770
AlSb	6.1355	1.60	250	550	1330
BN	3.615	4.0			3000
BP	4.538	6.0	500	70	200
CdS	4.1348, 6,7490	2.42	400		2020
CdSb		0.45	250	500	
CdSe	6.052	1.74	600		1360
CdTe	6.477	1.44	1,000	70	1370
GaAs	5.6532	1.35	8,800	400	1570
GaN	3.180, 5.166	3.34			1500
GaP	5.4505	2.24	300	100	1750
GaSb	6.0954	0.67	4,000	1400	980
GaTe		1.66	14		824
Ge	5.6575	0.67	3,800	1820	1231
InAs	6.0584	0.36	33,000	460	1215
InN	3.533, 5.693	2.0			1200
InP	5.8688	1.27	4,600	150	1330
InSb	6.4788	0.16	78,000	750	798
InSe		1.8	900		890
PbS	5.9362	0.37	550	600	1390
PbSe	6.1243	0.26	1,020	930	1340
PbTe	6.454	0.25	1,620	750	1180
Se	4.3642	2.00	10		217
Si	5.4307	1.11	1,900	500	1685
Sn	6.4912	0.08	2,500	2400	503
SnTe	6.313	0.5			1080
Te	4.4570	0.33	1,700	1200	452
ZnO	3.2495, 5.2069	3.37	180		2250
ZnS	5.4093	3.54	180		1920
ZnSe	5.6676	2.58	540	30	1790
ZnSb		0.50	340	10	
ZnTe	6.101	2.26	340	100	1510

Sources: P. Aigram and M. Balkanski (Eds.), Selected Constants of Semiconductors, Pergamon Press, New York, 1961; E. D. Mitchell, et al., Proceedings of the International Conference on the Physics of Semiconductors, (M. Hulin, ed.), Dunod, Paris, 1964; E. H. Putley, The Hall Effect and Related Phenomena, Butterworth, London, 1960; R. K. Willardson and A. C. Beer (Eds.), Semiconductors and Semimetals, Vol. 1, Academic Press, New York, 1968.

Table 41 Comparison of Diode Light Sources

Light source types	Output power range (mw)	Bandwidth range (Mbps)	Spectral width (Å)	Lifetime average (h)	Emitter's area range (mm^2)	Axial radiance ($W/sr\text{-}cm^2$)	Source rise time (ns)
Large area LED	1-7	10–40	350	6×10^5	1–10	0.1–0.5	10–100
Small area LED	1–3	20–200	300	5×10^5	0.004–0.04	5–70	3–10
CW laser diode	3–40	30–1500	1–20	4×10^5	0.002–0.01	10–200	0.1–2
Pulsed laser diode	100–400	0.1	1–20	3×10^5	0.002–0.01	10–200	0.1–2

Optical Data Processing

The major advantages of using optical data processing (ODP) are that the processing of input data or information is performed at high speeds and that ODP can be used for parallel processing for Fourier analyses. The characteristics of the optical transform lens or filter can be changed at high speeds using real-time computers interacting with the optical system devices. The types of functions that can be used with ODP are listed as follows:

1. Addition: Accomplished by sequential addition of two or more signals on a single detector or by using sequential dichroic filters with inputs of differing wavelengths.
2. Analog to digital: This conversion is achieved using the electrical properties of EO components to attain the digital format.
3. Analog PDE solutions: By using the proper sequential format with differential and integrating filters the solutions to two- and three-dimensional linear and nonlinear partial differential equations can be attained.
4. Autocorrelation: Accomplished using a holographic filter of a scanned object placed in the Fourier filter plane.
5. Convolution: This function is obtained by placing two transparancies adjacent to each other and Fourier transforming the optically multiplied output.
6. Cross-correlation: Accomplished by using different or multiple autocorrelated scenes.
7. Differentiation: Obtained by placing a linear filter material in the proper filter plane where the optical transmission varies with distance.
8. Division: This function is effectively attained using negative transparencies.
9. Fourier transformation: The optical transformation between the front and back focal planes of a lens or lens system.
10. Fresnel transformation: Accomplished by Fresnel diffraction.
11. Geometric transformation: Various spatial or coordinate transformations can be accomplished by using computer-generated holograms with variable construct parameters.
12. Hilbert transformation: Attained using a 180° phase shifter coupled with the Fourier transform of a knife edge.
13. Integration: This function is accomplished by using filters that have transmissions inversely proportional to distance in the filter plane.
14. Laplace transformation: By using a Fourier transform coupled to a Laplace weighted filter of the general form exp(-ax).
15. Logarithmic filtering: By using the halftone method or various types of computer-generated holograms, noise can be successfully eliminated.

16. Matched filtering: Attained using a holographically matched filter in the filter plane.
17. Matrix multiplication: Accomplished using various combinations of optical fibers, modulators, and polarizers.
18. Melin transformations: Fourier transforms of f(exp x).
19. Multiplication: Attained by passing the object light output through a given transparency.
20. Partial differentiation: Optical differentiation along only one axis.
21. Rotational invariant transformations: Accomplished by switching to polar coordinates then using a combination of Melin and Fourier transforms.
22. Spatial filtering: By using a combination of two or more lenses to alter the input image.
23. Spectral analysis: This operation is performed when the Fourier spectrum of an object in the front focal plane is analyzed on the back optical focal plane.
24. Subtraction: This function is essentially the 180° phase shifting of objects using gratings or holograms.
25. Wiener filtering: By using various types of Wiener holographic filters.

By using various combinations of these optical data processing functions numerous practical applications are available. Some of these include optical memory systems, holographic and phase contrast microscopy, multiple image storage techniques, speckle interferometry, optical character recognition, radar multichannel correlation techniques, and various types of postdetection image restoration and amplification. By using on-line computation coupled with electro-optical modulators for opto-electrical (OE) conversion various hybrid ODP systems have evolved. Fraunhofer diffraction pattern analysis is one such example, and has been used for medical x-ray diagnosis, automated imaging grading, determining the MTF of lens systems, particle size analyses, and the screening of cloud-covered images. Another such application is the analysis of laser interference patterns of optical fibers to determine the radial refractive index profiles.

Chapter 6

DETECTORS

Detector Characteristics

There are numerous types of detectors available for both laboratory and commercial use. Detectors are usually characterized by type, material, spectral response, responsivity, rise time, detectivity, and noise voltage, or noise equivalent power. Table 42 lists various types of detectors with their corresponding characteristics from commercially available sources. This table shows the wide variety of detector types, materials, and response characteristics that can be used for a variety of applications. The noise generated in detectors is caused by 1/F noise, generation-recombination noise, or thermal (Johnson) noise (which is a function of the detector temperature). Figure 25 shows the general noise spectrum as a function of detector frequency for some of the basic detector types.

The conditions under which detector performance is measured must be completely specified. If the values are obtained by the use of a blackbody shource, the detectivity is specified by $D^*(T, f_c, B)$, where T = source temperature, f_c = chopping frequency, and B = bandwidth. If the values are measured at a given wavelength, then the detectivity is specified by $D^*(\lambda, f_c, B)$, where λ = wavelength. When the wavelength is at the peak (P) response, the detectivity is given as $D^*p(p, f_c, B)$. The detectivity of a detector device is given by noise equivalent power (NEP) given in units of W^{-1}. The higher the detectivity the lower the radiation level at which a photodetector can respond. The specific detectivity D^* is given by

$$D^* = A^{1/2}{}_s B^{1/2} (NEP)^{-1}$$

where A_s is the active detector area, and B is the bandwidth of the detector-amplifier given in units of $CmHz^{1/2}W$.

Table 42 Typical Detector Characteristics

Type	Material	Range of spectral response	Range of responsivity	Range of rise times
APD	Si	0.3-1.1 μm	0.14-4.8 X 10^6 V/W	7-70 ns
Calorimeter		0.25-100 μm	0.08-2.0 V/W	3-55 sec
PC	HgCdTe	8-16 μm	2.0-3.0 A/W	4-6 μs
PC	InSb	1-5.5 μm	2.0-2.4 A/W	1-2 μs
PC	PbS	0.5-4.5 μm	10^4-10^8 A/W	25 μs-5 ms
PC	PbSe	0.7-8.0 μm	10^3-10^7 A/W	3-50 μs
PC	Si	0.4-1.2 μm	0.4-0.5 A/W	6-50 ns
PIN	Si	0.35-1.13 μm	0.5-0.65 A/W	2-40 ns
PIN + amp	Si	0.17-1.15 μm	8-25 V/mW	20-85 ns
PIN + amp	GaAs	0.89-1.15 μm	3-15 V/mW	20-40 ns
PV	Ge	0.7-1.8 μm	0.3-0.4 A/W	8-10 ns
PV	HgCdTe	9.4-10.6 μm	3.0-3.8 A/W	3-3.2 ns
PV	InAs	1.0-3.6 μm	0.4-0.6 A/W	8-10 ns
PV	PbSnTe	5-13 μm	1.0-1.5 A/W	40-50 ns
PV	Si	0.20-1.13 μm	0.07-0.65 A/W	1 ns-50 μs
Pyroelectric	DTGS	0.2-30 μm	0.6-2.4 V/mW	100-150 ns
Pyroelectric	$LiTaO_3$	0.2-30 μm	0.4-0.6 V/mW	100-150 ns
Pyroelectric	PZT	0.2-30 μm	1.0-1.5 μA/W	100-150 ns
Pyroelectric	SBN	1.0-20 μm	10^{-3}-1.0 mA/W	12-15 ns
Thermocouple	Bimetallic	0.3-30 μm	0.4-1 mV/W	0.7-1.0 s
Thermopile	Bimetallic	0.3-30 μm	1-28 V/W	0.02-0.85 s

Source: Product literature.

The detector quantum efficiency is given by

$$n = 1.2395 \times \frac{R}{\lambda} * 10^6\%$$

where R is the responsivity in amperes per watt and λ is the measured or detecting wavelength in angstroms. For photomultipliers the quantum efficiency is limited by shot noise and reflection losses and theoretically the quantum efficiency should be able to reach 70%, but actual photocathodes rarely go beyond 30%.

Photodetectors can be designed to detect a wide range of frequencies from the ultraviolet to infrared regions. Most semiconductor photodetectors

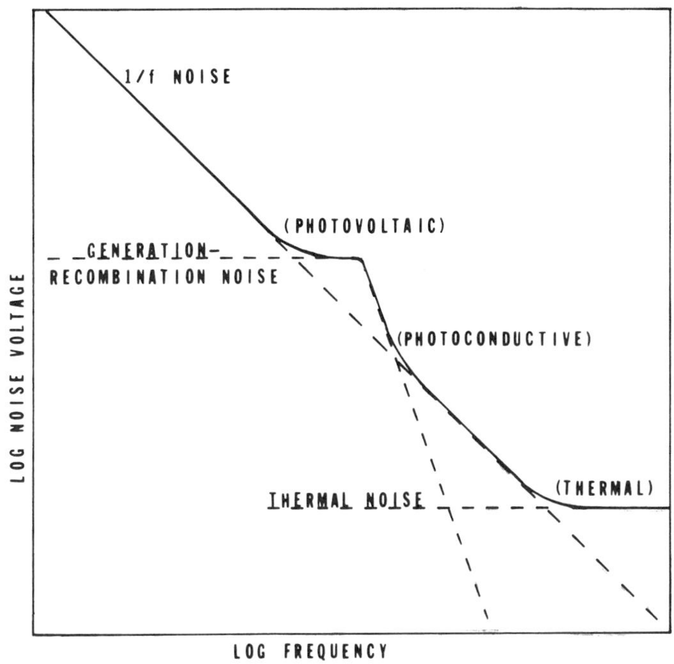

Fig. 25 Detector noise vs. frequency spectrum for various types of detectors and noise. Source: R. D. Hudson, Infrared System Engineering, Wiley, 1969.

are made to be responsive to wavelengths over a given spectral region with a reasonably large bandwidth. It is sometimes desirable to simultaneously detect and measure two or more wavelength regions that cannot be covered by a single detector. For some applications very narrow spectral bandwidths are desired for several wavelengths to continuously compare and compute incoming information. The multiwavelength detectors have several configurations, including rotating disks with different wavelength filters, beamsplitters with detector material of different spectral responses stacked together for simultaneous measurements. Examples of multispectral detectors include semiconductor systems such as PbSnTe, HgCdTe, InSb, and SiPbS. By using carefully chosen optical materials in the multiwavelength detectors, such as antireflection coatings and optical windows (absorbers and transmitters), sandwich designs can be made to have very narrow spectral bands to compare the relative intensities of two or more wavelengths. Wideband detectors can also be made to cover very large spectral ranges (uv to ir) for specialized broadband applications.

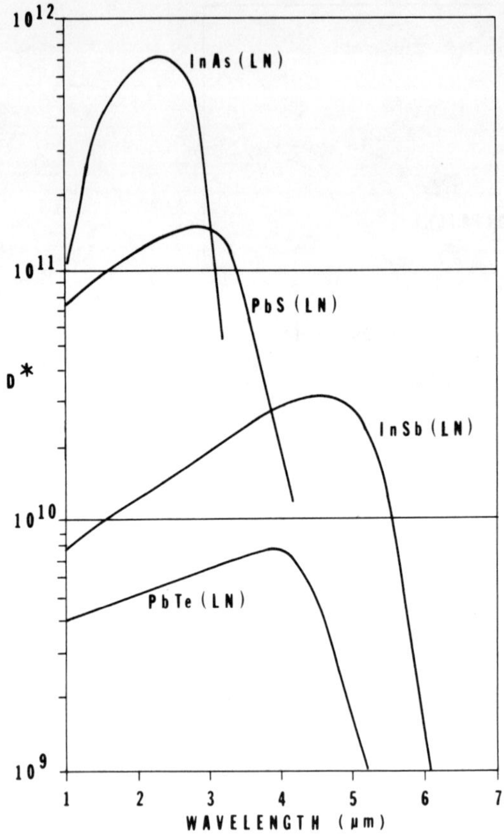

Fig. 26 Detectivity vs. wavelength for important detector materials.

Figures 26 to 31 show the detectivities of various detector materials for specified temperature ranges as a function of wavelength of the incident radiation. Note that these curves taken from typical manufacturers product literature show the approximate detectivity and spectral peak positions. However, by slight changes in doping compositions or manufacturing methods some of these materials can exhibit different forms of D^* as a function of λ.

Photodiodes

Semiconductor photodiodes all have a depleted region where electron-hole pairs are generated by the absorbed photons, with a high electric field

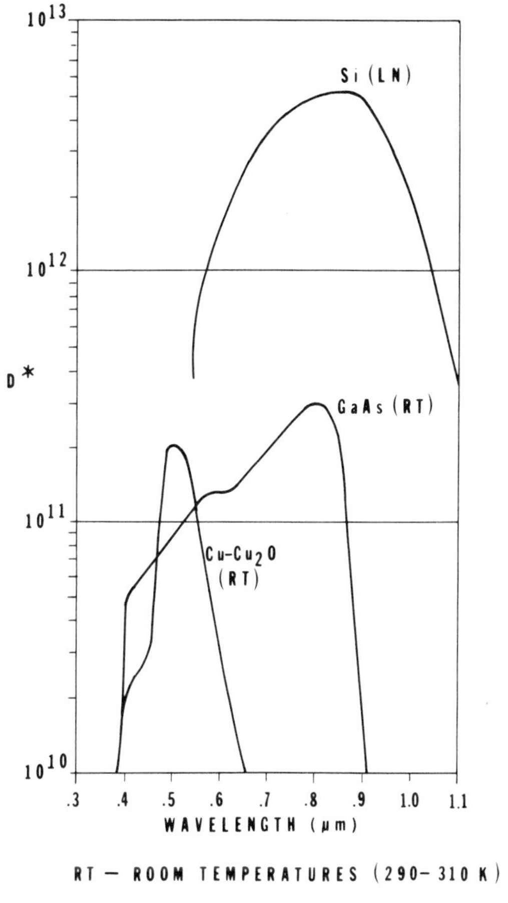

Fig. 27 Detectivity vs. wavelength for important detector materials.

between two semiconducting regions. To prevent the formation of local high-field regions are the edges of the junction (where the electron-hole pairs are collected), guard rings are often used in the diode structures. Fiber optic systems usually involve relatively low light levels; so the detectors are operated with a reverse bias such that the output current varies linearly with the intensity of the incident optical signal.

The probability of light absorption in the depletion region depends upon the wavelength of the incident signal, the type of semiconductor material used, and the thickness of the depletion region. For each semiconductor material, once a certain critical wavelength has been exceeded, the material becomes almost transparent to incident radiation. The thickness of the

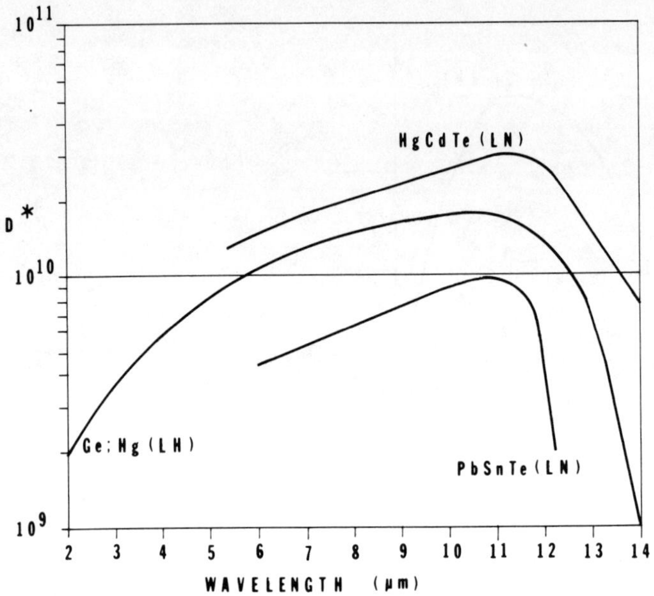

Fig. 28 Detectivity vs. wavelength for important detector materials.

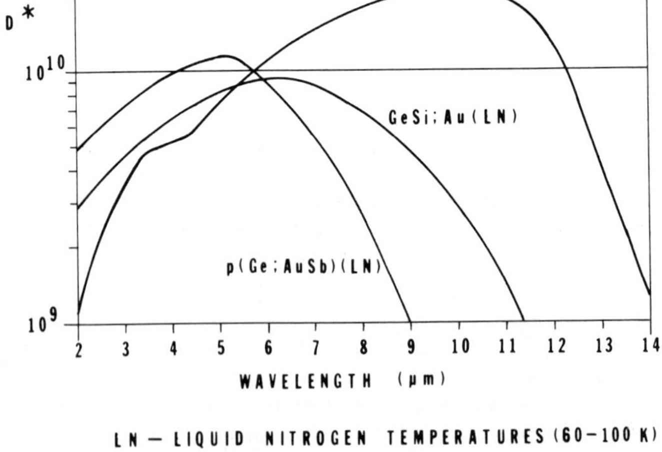

Fig. 29 Detectivity vs. wavelength for important detector materials.

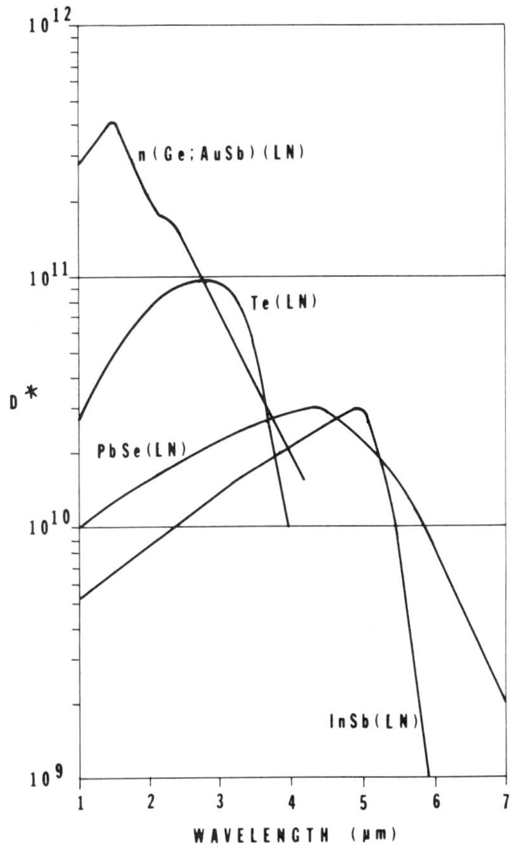

Fig. 30 Detectivity vs. wavelength for important detector materials.

depletion region can be increased to extend the wavelength before the transparency frequency is reached, but this extension is greatly limited by the necessary bandwidth (speed of response) of the detector.

Photodiodes are usually described by four basic quantities: response time, quantum efficiency, total noise equivalent power, and responsivity. The response time is the transit time for electrons in the photodiode to transverse from the cathode to anode, typically from 0.2 to 5 ns. The quantum efficiency is the percentage of incident photons that liberate photodetector electrons. The responsivity is the average emitted current divided by the average incident power. Figure 32 shows the quantum efficiency and responsivity vs. wavelength for typical PIN photodiodes. In avalanche photodiodes (APD) the two basic entities usually

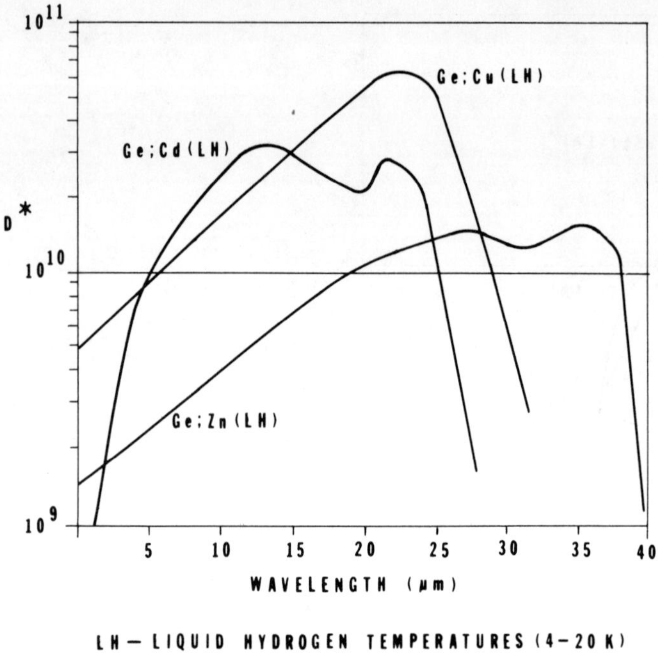

Fig. 31 Detectivity vs. wavelength for important detector materials.

considered are the response speed and the multiplication or gain characteristics. The physical structure of avalanche diodes usually includes a guard ring to prevent excessive leakage at the junction edges and low breakdown voltages. Many APD devices are silicon based with antireflection coatings to provide quantum efficiencies near 90% with gains of several hundred percent.

The responsivity of silicon decreases rapidly for wavelengths beyond 990 nm. Thus, for both PIN and APD devices other elemental systems must be used to attain high-speed detectors in the wavelength region extending beyond 990 nm. Typical PIN detectors are usually rated for 200 to 1100 nm with the responsivity peaks at 900 nm or less. The typical operating voltage range between 10 and 110 V with active surface areas ranging from 1 to 10 mm^2. Responsivities typically range from 0.2 to 0.7 A/W with minimum detectable light levels from 1 to 10 X 10^{-15} W and noise equivalent powers ranging from 1 to 10 X 10^{-14} $W/Hz^{1/2}$. Increasing the responsivity of existing silicon PIN detectors to wavelengths above 990 nm usually involves increased bias voltages and subsequent greater sensitivity to temperature changes. In addition, designs for longer wavelengths

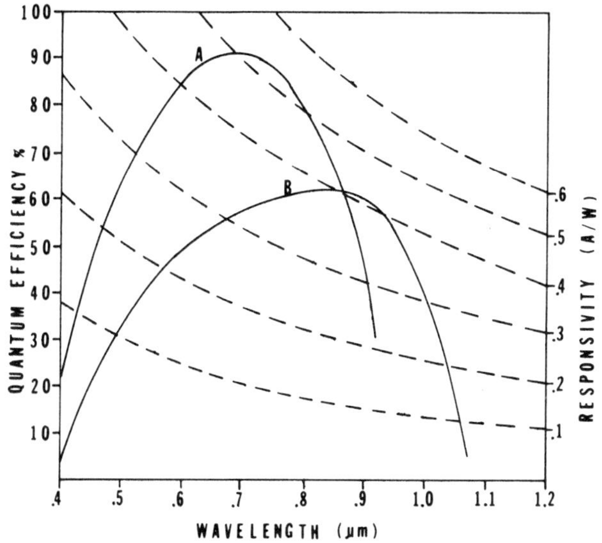

Fig. 32 Quantum efficiency vs. wavelength for typical silicon PIN photodiodes: A, antireflection coating used; B, typical response.

usually require greater junction capacitances, resulting in bandwidth reduction. For these reasons researchers and designers are attempting to develop other materials and device structures for use at wavelengths above 990 nm. Table 43 compares the typical properties of PIN and APD detectors.

The average current produced by a photodetector is given by

$$I(t) = G\left(\frac{2\pi Q\lambda_o}{hc}p(t) + I_d\right)$$

where

G	= average gain
Q	= quantum efficiency
λ_o	= nominal source wavelength
h	= Plank's constant
P(t)	= optical power incident on the photodetector (time dependent)
I_d	= dark current

Figure 33 shows the typical photodiode dark current as a function of the active area and ambient temperature.

Table 43 Comparison of Detector Types

Basic detector type	Sensing diameter area (mm^2)	Sensitivity at 1 MHz (dBm)	Responsivity at peak λ_0 (A/W)	Bias voltage (V)	Dynamic range (dB)	Maximum data rate (Mbps)	Lifetime range (h)	Rise time (ns)	Peak λ_0 response (nm)
PIN	0.8–8	-58	0.4–0.7	10–100	60	1–100	10^5–10^6	1–10	870
APD	0.8–8	-70	10–120	250–350	80	1–1500	10^5–10^6	2–5	850

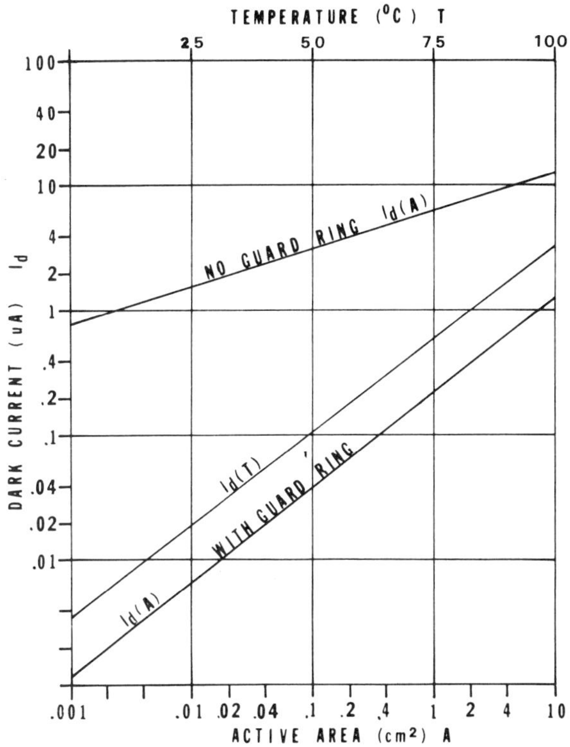

Fig. 33 Typical photodiode dark current vs. active area and ambient temperature.

The total noise equivalent power (TNEP) accounts for both the noise level production in the devices and its gain, to provide a measure of the minimum detectable signal. The TNEP is defined as the amount of light energy impinging on the active surface area which will produce an output signal equal in magnitude to the noise output of the photodetector device. The units typically used for TNEP are nanowatts. The NEP of a detector is the noise equivalent power which is simply the TNEP normalized to a 1-Hz bandwidth and is given in units of watts per hertz.

APD devices all have a large responsivity dependence on the bias voltage and temperature, as shown in Fig. 34. Since the detector breakdown voltage varies with temperature, the precise bias needed to maintain a fixed gain varies. Precise control is maintained by current feedback control or by monitoring the gain in an auxiliary APD matched to the active diode device.

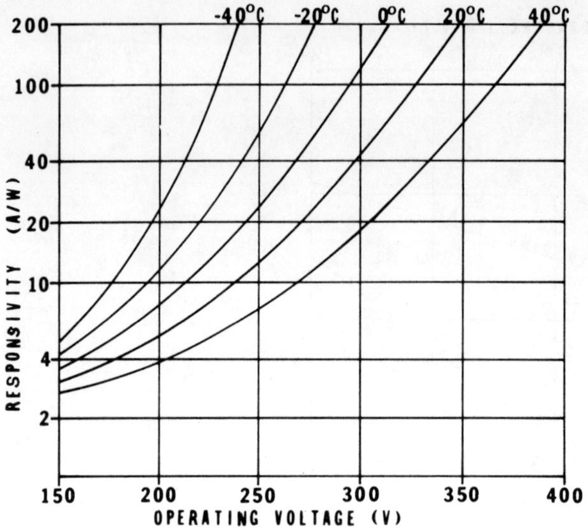

Fig. 34 APD responsivity vs. voltage for various device temperatures.

The formula used to calculate TNEP for PIN detectors is

$$\text{TNEP}_{\text{PIN}} = \frac{4hfB}{Qe}(2KTC)^{1/2}$$

where

h = Planck's constant
f = transmission optical frequency
B = bandwidth
Q = device quantum efficiency
e = electron charge
k = Boltzmann's constant
T = diode amplifier temperature
C = diode capacitance

For a given optical system, the TNEP can be optimized by reducing the detector capacitance and increasing the quantum efficiency. Detector capacitance depends mostly on the junction area and the width of the depletion region. For APD detectors, the formula to calculate the TNEP is given by

$$\text{TNEP}_{\text{APD}} = \frac{hfB}{QeG}(2KTC)^{1/2}$$

where G is the detector gain level where the shot noise and thermal noise at the temperature T are equal. The gain G typically ranges from 10 to 100. The value of G which minimizes the NEP is not necessarily the optimal value of G to use at high SNR levels.

Although the APD devices have a reduced TNEP, they must operate at higher voltage and require a greater degree of control on current and voltage. APD devices with 1-GHz bandwidth are available with response times less than 1 ns and a TNEP less than 10 μW. The NEP of typical detectors can be determined as a function of the gain for a specified bandwidth. The degree of gain in a given APD depends on the ratio of the hole collision ionization probability to the electron collision ionization probability given by K (not to be confused with the Boltzmann constant). For quality silicon detectors the value of k is usually about 0.025.

Figure 35 shows a typical silicon photodiode design monograph to determine the relationships between the resistivity, depletion depth, junction capacitance, and operating bias. For detectors without any temperature compensation circuitry coupled to the device the wavelength of the peak responsivity is shifted about 0.4 to 0.5 nm/°C. The emission wavelength of LEDs typically shifts 0.3 nm/°C when no temperature feedback circuitry is used. Since the emission wavelength of the light source and the peak responsivity of the detector can shift at different thermal rates, it is possible in some applications to incur significant efficiency losses.

Detectors are available as single devices or as a module unit coupled to a preamplifier. These modules are small, lightweight, reliable, have enhanced frequency response (stray inductances and capacitances eliminated), and have reduced stray noise (the detector-preamp unit is in a shielded environment). Typical detector modules have bandwidths ranging from 0.1 to 50 MHz and responsivities from 10 to 500 mA/W at peak wavelengths from 800 to 950 nm. The preamplifiers used in the modules are low-impedance operational amplifiers (op-amps) with fast response speeds and large linear dynamic ranges.

For many modules the actual usable bandwidth is only about 5 to 10 MHz, due to the large op-amp noise spectral density.

As with lasers and LEDs, PIN performance and linearity is effected by temperature effects which can be compensated for by an appropriate feedback loop signal circuitry. Even the best designed temperature-compensated modules have linearities less than those of the individual components, which can present problems for large-bandwidth, high-gain systems. APDs have high internal gains, high responsivity, fast response times, and large gain-bandwidth products, with small active areas. As the gain factor increases, the need for precise controls on voltage and

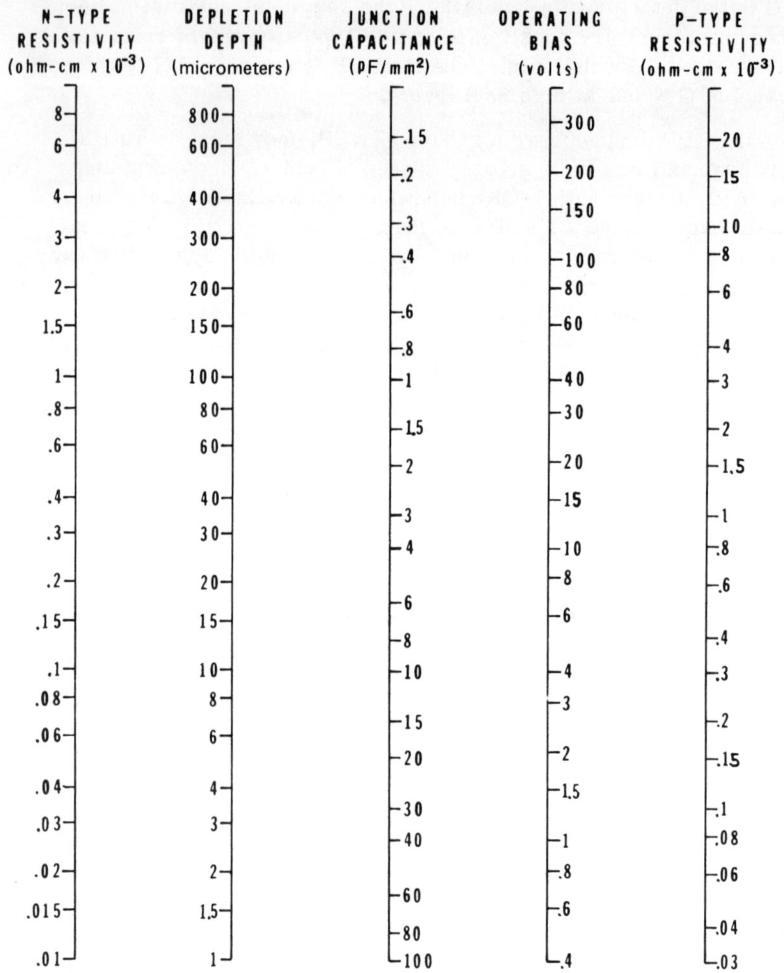

Fig. 35 Typical silicon photodiode nomograph for resistivity, bias voltage, depletion depth, and capacitance.

temperature becomes more critical if a constant gain is to be achieved. For most applications the temperature should be controlled to within 0.1°C and the voltage to within 0.1 V with appropriate feedback circuitry to avoid nonlinearities, excess noise, and drifting. Typical APDs have responsivity ratings at the spectral peak ranging from 10 to 100 with bias voltages ranging from 100 to 400 V. The bandwidth rating for APDs ranges from 100 to 1500 MHz.

Photocathodes

The materials used for photocathodes are classified by the S numbering system with associated values for the spectral response range using a given window material. Although there are many photosensitive materials available, only a few are in common usage: S-1, 4, 11, multialkali, and solar blind (Cs-Tc) photocathodes. The S-1 cathode (Ag-O-G) is often used, since its spectral response range is the highest of all the materials, rated for 0.4 to 1.2 μm. Although the S-1 photocathode normally has high noise and thermionic emission the signal-to-noise ratio (SNR) is often altered by cooling the cathode below 240 K. Semiconductor cathodes, on the other hand, have low thermal emission and noise but are fragile and difficult to manufacture. Table 44 lists the standardized photocathode characteristics and a few common unstandardized photocathode types. Figures 36 to 49 show the relative response vs. wavelength for the standard materials.

Phototubes

There are three basic types of phototubes, high vacuum, gas, and multipliers, all of which have photosensitive cathodes usually made from metallic sleeves coated with various types of photoemissive materials. The cathode is sensitive only to a defined wavelength region. Gas phototubes have a decreasing frequency especially above 10 kHz.

Photomultiplier tubes produce a current that is approximately proportional to the incident radiation. Linearity at low radiation levels is limited by the dark current. At higher light levels stability of the regulated power supply to the individual multipliers stages becomes important. Typical dynode materials can achieve a multiplication factor of 3 to 5.5 for a single-stage potential of 90 to 110 V, although semiconductor materials can sometimes achieve multiplication factors of 10 to 40. Most photomultiplier tubes are designed with equal gain factors from stage to stage. Optimal SNR is obtained by varying the multiplication factors with the highest gain at the first dynode and the lowest gain at the last stage. The root-mean-square (rms) shot-noise current $(i_n{}^2)^{1/2}$ at the output of a photomultiplier is given by

$$(i_n{}^2)^{1/2} = G_T \left[2e_I B \left(1 + \frac{1}{G_1} + \frac{1}{G_1 G_2} + \cdots + \frac{1}{G_1 G_2 \cdots G_n} \right) \right]^{1/2}$$

where

G_T = total gain of the photomultiplier $(G_1 G_2 G_3 \ldots G_n)$
G_n = individual dynode gain factors
e = electron charge

Table 44 Standardized Photocathode Characteristics

Spectral response designation	Photosensitive material	Sensor type[a]	Window material	Oper. mode[b]	λ_{max}[c] (nm)	Conv. factor[d] (lm/W)	Typical luminous R (μA/lm)	Typical radiant R (mA/w)	Typical quantum efficiency[e]	Typical photocathode dark emission range at 25°C (A/cm²)
S-1	Ag-O-Cs	Photoemitter	Lime glass	T, R	800	92.7	26	2.5	0.41	10^{-11}-10^{-12}
S-3	Ag-O-Rb	Photoemitter	Lime glass	R	420	285	6.5	1.8	0.54	10^{-11}-10^{-12}
S-4	Cs-Sb	Photoemitter	Lime glass	R	400	1044	40	41	12.5	10^{-14}-10^{-15}
S-5	Cs-Sb	Photoemitter	9741 glass	R	340	1262	40	50	18.1	10^{-14}-10^{-15}
S-8	Cs-Bi	Photoemitter	Lime glass	R	365	757	3	2.3	0.78	10^{-15}-10^{-16}
S-9	Cs-Sb	Photoemitter	Lime glass	T	480	683	30	20	5.3	10^{-14}-10^{-15}
S-10	Ag-Bi-O-Cs	Photoemitter	Lime glass	T	450	509	40	20	5.6	10^{-13}-10^{-14}
S-11	Cs-Sb	Photoemitter	Lime glass	T	440	808	65	50	14.8	10^{-14}-10^{-15}
S-13	Cs-Sb	Photoemitter	Fused silica	T	440	799	60	48	13.5	10^{-14}-10^{-15}
S-14	Ge	pn junction	Lime glass		1500	42	12,400[f]	520[f]	43.5	

S-16	CdSe	Photo-conductor	Lime glass		730	159				
S-17	Cs-Sb	Photo-emitter	Lime glass	R	490	667	125	78	21.0	10^{-14}-10^{-15}
S-19	Cs-Sb	Photo-emitter	Fused silica	R	330	1603	40	65	18.5	10^{-14}-10^{-15}
S-20	K-Na-Cs-Sb	Photo-emitter	Lime glass	T	420	428	150	64	18.4	10^{-15}-10^{-16}
S-21	Cs-Sb	Photo-emitter	9741 glass	T	440	783	30	23	6.6	10^{-14}-10^{-15}
S-23	Rb-Te	Photo-emitter	Fused silica	T	240			4	2.0	10^{-16}-10^{-17}
S-24	K-Na-Sb	Photo-emitter	7056 glass	T	380	1505	45	67	21.8	10^{-15}-10^{-16}
S-25	K-Na-Cs-Sb	Photo-emitter	Lime glass	T	420	215	200	43	12.7	10^{-15}-10^{-16}
g	K-Cs-Sb	Photo-emitter	7740	T	385	1117	77	86	28	10^{-16}-10^{-17}
g	CdS	Photo-conductor	Lime glass		510	643				
g	Cd-S-Se	Photo-emitter	Lime glass		615	276				
g	Si	np Voltaic	No window		860	75.9	7650[h]	580[h]	83.6	
g	Si	Photo-conductor	No window		1060			600[i]	71[i]	10^{-15}-10^{-16}

Table 44–Continued

g	Na-K-Cs-Sb	Photo-emitter	Lime glass	R	530	300	89	20.8	
g	Na-K-Cs-Sb	Photo-emitter	7740 pyrex	T	265	65	97	31	10^{-14}-10^{-15}
g	Cs-Te	Photo-emitter	Fused silica	T	250		15	7.4	
g	Cu-As	Photo-emitter	9741 glass	R	830	300	68	10	10^{-15}-10^{-16}
g	Cu-As-P	Photo-emitter	9741 glass	R	400	160	45	14	10^{-16}-10^{-17}
g	Ga-In-As	Photo-emitter	9741 glass	R	400	100	57	17.6	
g	Si	Photo-conductor	Lime glass		900	620[j]	620[j]	85[j]	

[a]Photoemitter with reflecting substrate; so emission is from side of incident radiation.
[b]Operation mode T is transmission mode; R is reflection mode.
[c]Wavelength of maximum response.
[d]Conversion factor is ratio of radiant responsivity at peak to luminous responsivity of tungsten source (T_c = 28).
[e]At the wavelength of maximum response.
[f]With 45-V polarizing voltage.
[g]Spectral response designation not yet standardized.
[h]Photovoltaic short-circuit responsivity.
[i]3.2-mm-thick depletion.
[j]150 μm wafer thickness.

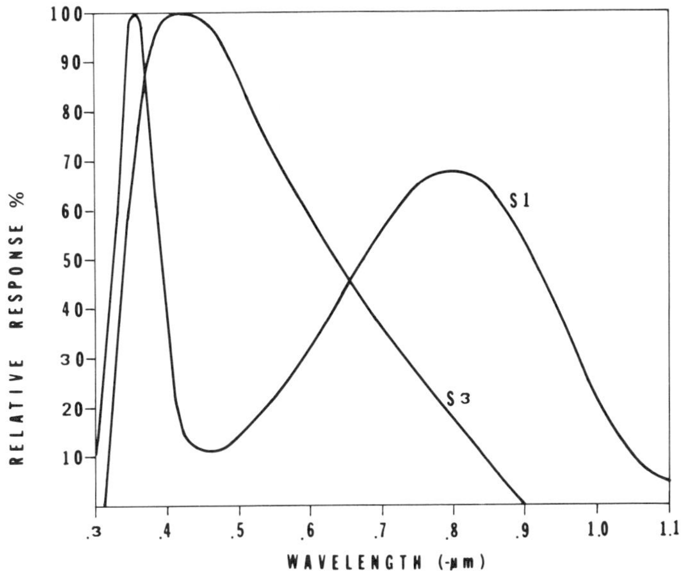

Fig. 36 Relative response vs. wavelength for standardized detector materials.

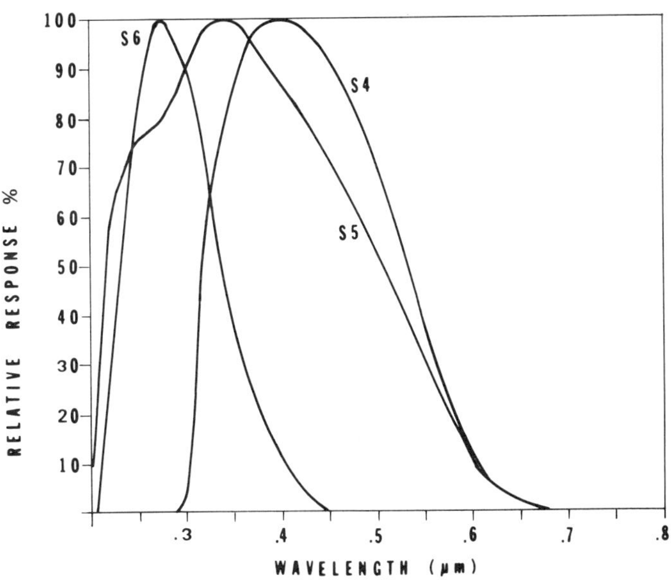

Fig. 37 Relative response vs. wavelength for standardized detector materials.

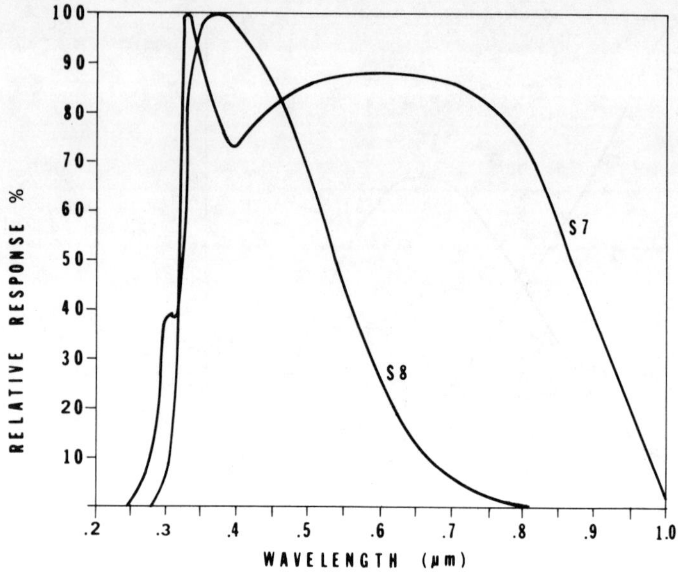

Fig. 38 Relative response vs. wavelength for standardized detector materials.

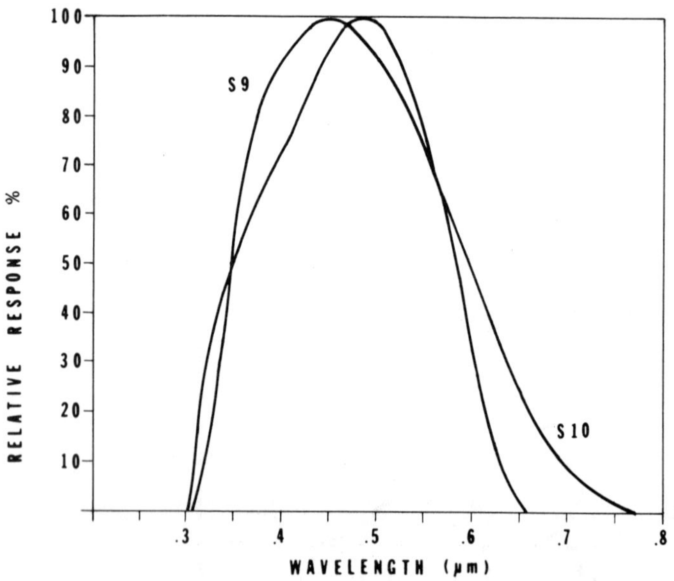

Fig. 39 Relative response vs. wavelength for standardized detector materials.

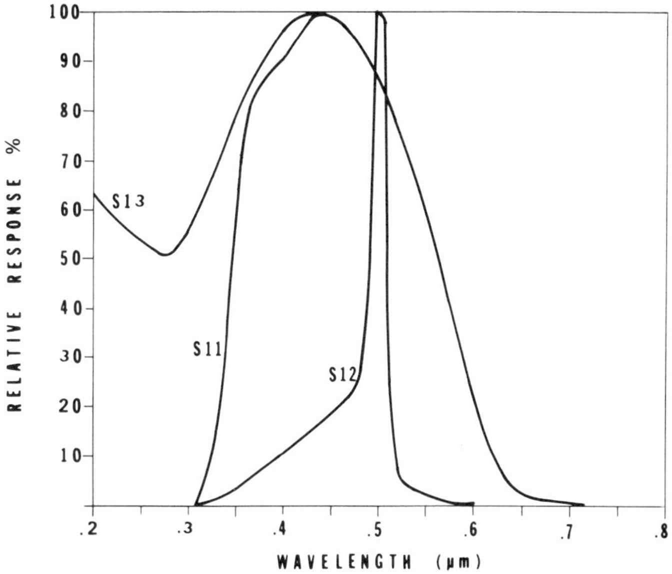

Fig. 40 Relative response vs. wavelength for standardized detector materials.

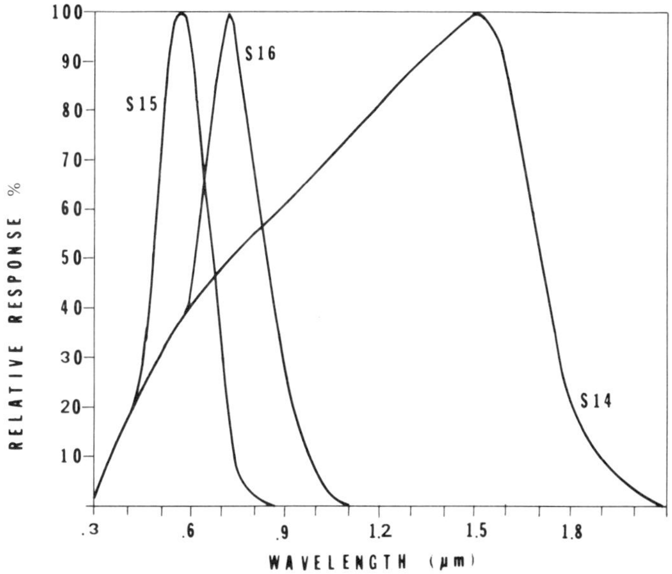

Fig. 41 Relative response vs. wavelength for standardized detector materials.

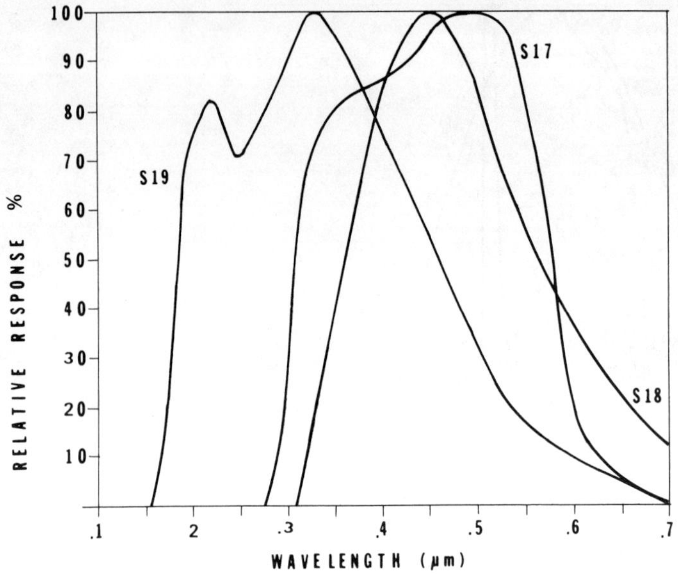

Fig. 42 Relative response vs. wavelength for standardized detector materials.

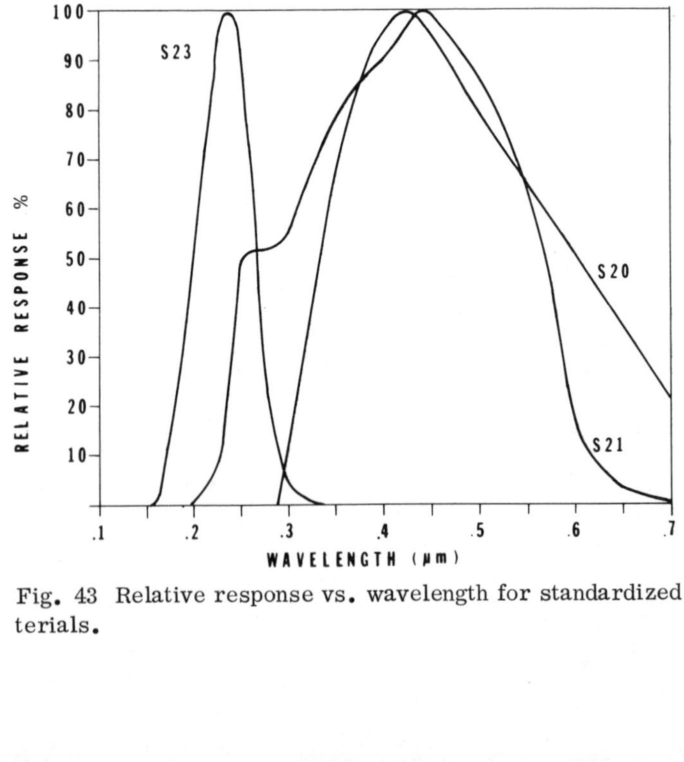

Fig. 43 Relative response vs. wavelength for standardized detector materials.

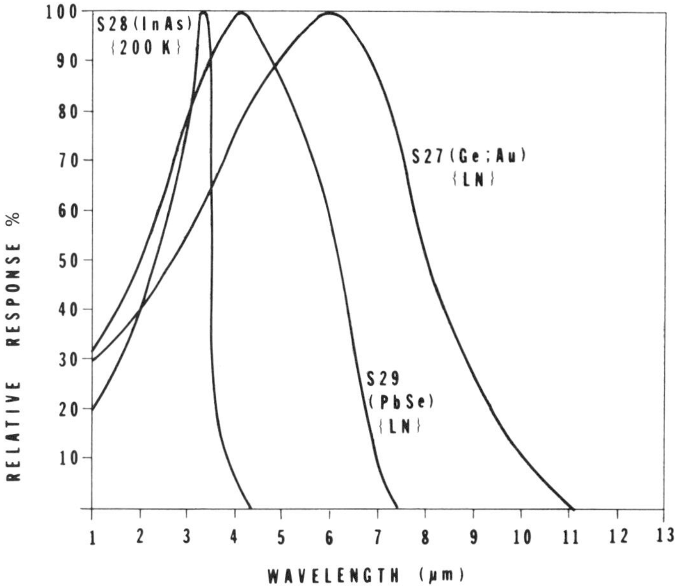

Fig. 44 Relative response vs. wavelength for standardized detector materials.

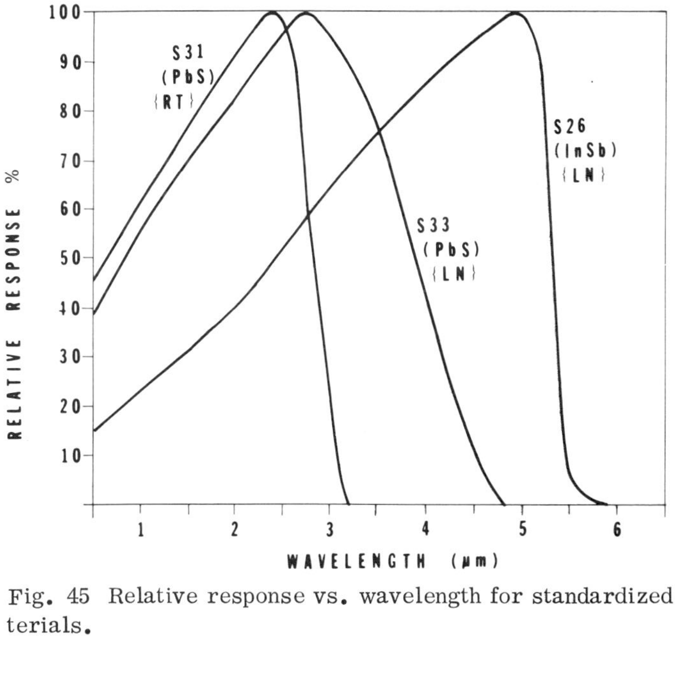

Fig. 45 Relative response vs. wavelength for standardized detector materials.

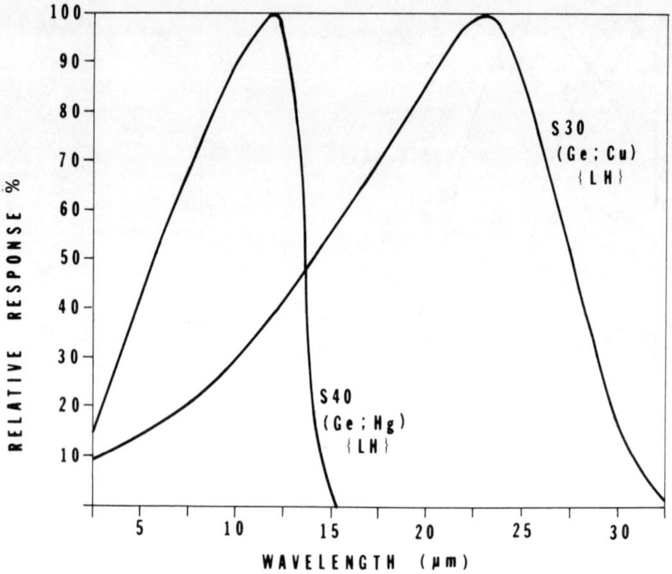

Fig. 46 Relative response vs. wavelength for standardized detector materials.

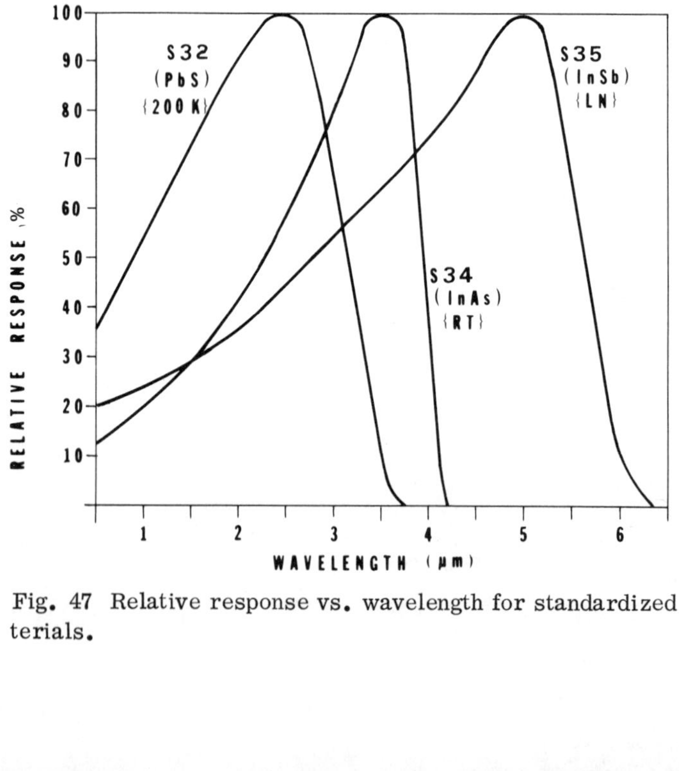

Fig. 47 Relative response vs. wavelength for standardized detector materials.

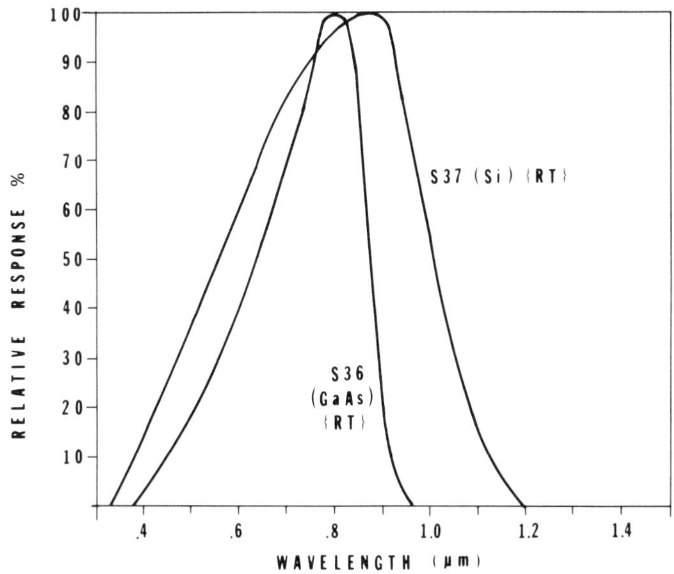

Fig. 48 Relative response vs. wavelength for standardized detector materials.

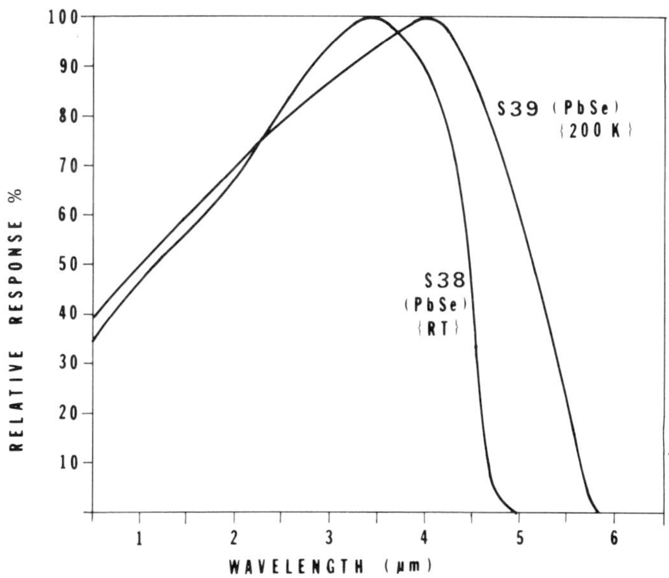

Fig. 49 Relative response vs. wavelength for standardized detector materials.

I = total average photocathode current
B = photomultiplier bandwidth

If $G_1 = G_2 = \cdots = G_n$ then the rms shot rise current is given by

$$(i_n^2)^{1/2} = G_T \left[2eIB\left(\frac{1}{1 - G_1 - 1}\right)\right]^{1/2}$$

Photovoltaic and Conductive Cells

Photovoltaic cells generate a potential of several millivolts in the presence of incident radiation, without the use of an external power source; however, the cells' large electrode capacitance causes poor frequency response. The semiconductor boundary layer response varies with time and material fatigue. The frequency response ranges from dc to 100 kHz. A photovoltaic device generates a current at a pn junction with zero applied voltage in the presence of incident light. The photocurrent generated divides between the load resistance and the internal diode junction resistance. When the load resistance is very small, the photovoltaic device is in the short-circuit operating mode. At higher load resistances and high levels of incident radiation, it is in the open-circuit (constant voltage) mode. At lower frequencies photovoltaic (pv) devices have better SNRs than photoconductive (pc) devices of equivalent size.

A photoconductive device operates with a reversed bias, which increases the junction field strength to reduce the junction capacity. Most of these devices are designed to detect high-frequency modulations or pulses operating over a dynamic range from dc to above 100 MHz, with rise times ranging from 2 to 20 ns. This type of radiation detector varies its resistivity with the incident radiant energy (photoresistive) in a semiconductor layer, where resistivity decreases with increasing incident radiation. Photoconductive cells have a wider spectral response than most gas phototubes.

Thermal Detectors

The basic types of thermal detectors are the balometer (resistive conductive element), thermistor balometer, radiation thermocouple, radiation thermopile (thermocouple pairs with fixed junction temperature), and pneumatic detector (thermal gas expansion and microradiometer, moving galvanometer, and thermopile combination). Thermopiles are classified into three basic types: the bulk, Schwartz, and thin-film devices. The thin-film devices are most commonly used due to their low noise and small size.

The radiation thermopile is a collection of thermocouples connected in series with the reference junctions all maintained at the detector ambient temperature. The temperature differential between the junction pairs generates a voltage determined by the thermoelectric power of the materials characterized by the Seeback coefficient α, given by

$$\alpha = \frac{dV}{dt}$$

Pyroelectric Detectors

The pyroelectric detector has a large spectral response range often extended by coating the detector surface with a thin black material to rapidly conduct heat to the sensor substrate. Pyroelectric detectors can measure radiation changes with rapid response times, unlike thermocouples and balometers, since the thermal rate of change is used instead of an absolute celebrated temperature value. The current (I) generated by a pyroelectric detector is given by

$$I = \beta(T)\, A_s \frac{dt}{df}$$

where $\beta(T)$ is the pyroelectric detector, which can be used to detect modulated radiation even when the modulation or impulse frequency extends the thermal relaxation frequency. For these applications the frequency-dependent output voltage V(f) is measured by

$$V(f) = P(f) \frac{\eta(\lambda)\beta(T) R_L}{Pc_p d\left[1 + (\alpha\pi f R_L Ci)^2\right]^{1/2}}$$

where

$P(f)$ = incident radiant power
$\eta(\lambda)$ = detector absorption efficiency at the peak incident wavelength λ
R_L = load resistance
P = mass density
C_p = detector material specific heat
d = detector thickness
ci = effective detector capacitance

For pyroelectric detectors used to detect high-frequency modulated signals, the load resistance must be kept low. Figure 50 shows the typical pyroelectric detector responsivity as a function of frequency and the load resistance.

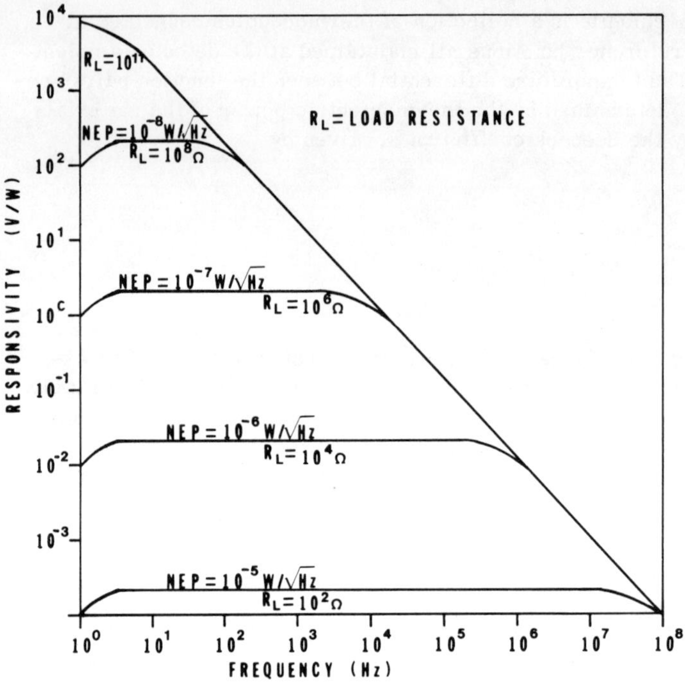

Fig. 50 Typical pyroelectric detector-preamp responsivity vs. frequency characteristics.

BIBLIOGRAPHY

Asthemer, R. W., and S. Weiner, Solid-backed evaporated thermopile radiation detectors, Appl. Opt., vol. 3, no. 4, 1964.

Baack, C., et al., GaAs MESFET: a high-speed optical detector, Electronics Lett., vol. 13, no. 7, 1977.

Barbe, D. F., Imaging devices using the charge-coupled concept, Proc. IEEE, vol. 63, no. 1, 1975.

Bear, A. C., and W. Vardsion, R.K. (Eds.), Semiconductors and Semimetals, Vol. 5, Academic Press, New York, 1970.

Biberman, L. M., and S. Wadelman, (Eds.), Photoelectric Imaging Devices, Plenum Press, New York, 1971.

Blevin, W. R., and J. Geist, Influence of black coatings on pyroelectric detectors, Appl. Opt., vol. 13, 1974.

Boisrobert, C. Y., et al., Low noise photodetection: preamplifier analysis, Second European Conference on Optical Fiber Communication, September, 1976, Paris, France.

Born, M., and E. Waofe, <u>Principles of Optics</u>, Pergamen Press, London, 1966.

Caprossi, D., et al., Design of impedance transformers for increasing the PIN photodiodes applications range, Second European Conference on Optical Fiber Communication, September, 1976, Paris, France.

Engstrom, R. W., Absolute spectral response characteristics of photosensitive devices, RCA Rev., vol. 21, no. 2, 1960

Fraser, M., and A. Laurral, Low-light and thermal imaging systems, IEE Conference Publication no. 124, 1975.

Holter, et al., <u>Fundamentals of Infrared Technology</u>, Macmillan, New York, 1962.

Hudson, R. D., <u>Infrared System Engineering</u>, Wiley-Interscience, New York, 1969,

Infrared detector product survey, Laser Focus, October, 1976, p. 66.

Ishii, K., et al., Double epitaxial silicon avalanche photodiodes for optical fiber communications, Electronics Lett., vol. 13, no. 10, 1977.

Jamieson, J. A., et al., Infrared Physics and Engineering, McGraw Hill, New York, 1963.

JEDEC Publication No. 50, Relative spectral response data for photosensitive devices, Electronic Industries Association, Washington, D.C., October, 1964.

JEDEC Publication No. 61, Typical characteristics of photosensitive surfaces, Electronic Industries Association, Washington, D.C., December, 1966.

JEDEC Publication No. 78, Relative spectral response curves for semiconductor infrared detectors, Electronic Industries Association, Washington, D.C., October, 1969.

Jenkins and White, <u>Fundamentals of Optics</u>, McGraw-Hill, New York, 1957.

Kondo, A., et al., A low-noise silicon avalanche photodiode, Electronics and Communications in Japan, vol. 58, no. 4, 1975.

Kondo, A., et al., Frequency response of avalanche photodiodes: Diffusion current dependence, Electronics and Communications in Japan, vol. 58, no. 12, 1975.

Logan, R. M., and T. P. Mclean, Analysis of thermal spread in a pyroelectric imaging system, Infrared Physics, vol. 13, 1973.

Misugi, T., et al., Detectors for optical fiber communications, International Conference on Integrated Optics and Optical Fiber Communications, July, 1977, Tokyo, Japan.

Moss, T. S., <u>Optical Properties of Semiconductors</u>, Butterworth, London, 1961.

Peterson, R. L., G. W. Day, P. M. Gruzeusky, and R. J. Phelan, Analysis of responses of pyroelectric optical detectors, J. Appl. Phy., vol. 45, 1974.

Proceedings of the SPK, Infrared Technology, vol. 62, 1975.

Putley, E. H., R. Watton, and J. H. Ludlow, Pyroelectric thermal imaging devices, Ferroelectrics, vol.,3, 1972.

Sommers, N. H., Photoemissive Materials, Wiley, New York, 1968.
Steedman, W., and C. Backer, Target size and visual recognition, Human Factors, vol. 2, 1960.
Zworykin and Ramsberg, Photoelectricity, Wiley, New York.

Chapter 7

IMAGE AND CAMERA TUBES

Image Tubes

The basic image tube consists of a photocathode where the radiant image is focused, an electron lens system, and a phosphor screen for image display. These tubes transmute visible images onto the display screen, intensify the actual screen brightness, or transmute nonvisible images and radiation into a visual format (display screen). The intensification or luminance gain is caused by the acceleration of electrons emitted from the photocathodes which can be up to 90 for one stage. Multitube gains up to 10^6 can be technically achieved, but at these high gains the SNR becomes rather small due to quantum noises. Figure 51 shows some typical image tube gains as a function of applied voltage.

There are various types of tubes including electrostatic, magnetic, and proximity focus. Figure 52 shows a basic electrostatic image tube. Microchannel plates can also be classified as image tubes, since their performance usually occurs in enclosed spaces under partial or high vacuum. All of these image tube types are additionally classified by the photocathode aperture, typically in the range of 18 to 60 mm. Table 45 lists some of the important parameters of the common types of image intensifiers. A single microchannel plate can attain a luminance gain of 10^3 to 10^4. When coupled together, gains of 10^7 to 10^8 are possible, again limited by quantum noise. The microchannel plate is made from a large parallel array of hollow glass cylinders about 10 mm in diameter, coated on the inside. A cascading secondary electron emission is obtained by applying a potential across the plate depth, usually 100 to 1500 V for a plate thickness of 1 to 3 mm. These image intensifiers are often used in field-ion microscopes in the field-emission and field-ionization modes to intensify the patterns of the projected electrons and ions to a visible image for direct visual observations or for photographic optics. The phosphor screens in all image tubes, particularly at high accelerating potential voltages, are susceptible to burn-out spots which can cause false or incomplete images.

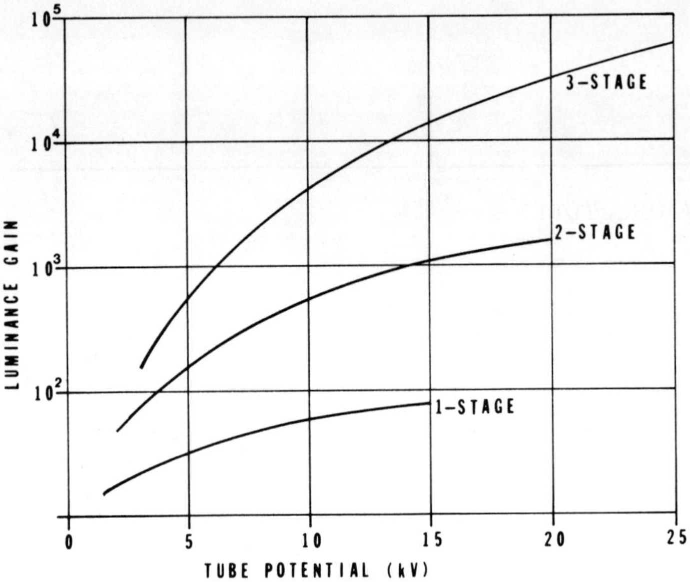

Fig. 51 Typical image tube gain vs. applied tube potential. Gain factor: 1-stage, 20-800; 2-stage, 60-1500; 3-stage, 200-10^5; orthicon, 300-5 X 10^4; SEC vidicon, 20-300.

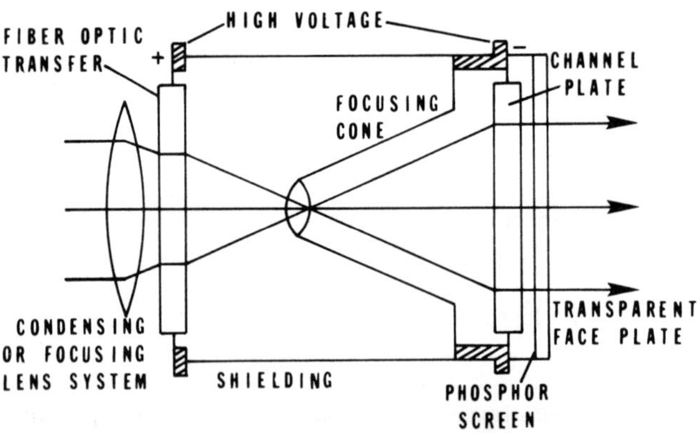

Fig. 52 Basic image intensifier or electrostatic image tube.

Table 45 Characteristics of Three Types of Image Intensifiers

	Electrostatically focused Cascade tube	Inverter tube	Microchemical plate
Size			
Optical length (mm)	190	40	30
Overall diameter (mm)	70	52	43
Image diameter (mm)	25	18	18
Performance			
Limiting resolution (spatial frequency avg.)			
at image center (cycles/mm)	25	38	25
at image edge (cycles/mm)	23	38	25
Modulation transfer function (MTF)			
contrast at 5 cycles/mm (%)	83	92	85
contrast at 20 cycles/mm (%)	30	43	18
Distortion of image at 80% of diameter (%)	22	14	None
Luminance gain	$2\text{-}5 \times 10^4$ (fixed gain)	10 to 10^5 (cont. variable)	10^2 to 1.5×10^4 (cont. variable)
Localized highlight suppression	No	Yes	Yes
Noise generated by intensification process	No	Yes	Yes
Sensitivity (minimum detectable illuminance at 5 cycles/mm (lux)	2.5×10^{-6}	10^{-5}	10^{-5}

Source: Electronics, vol. 46, no. 20, Sept. 27, 1973, p. 122.

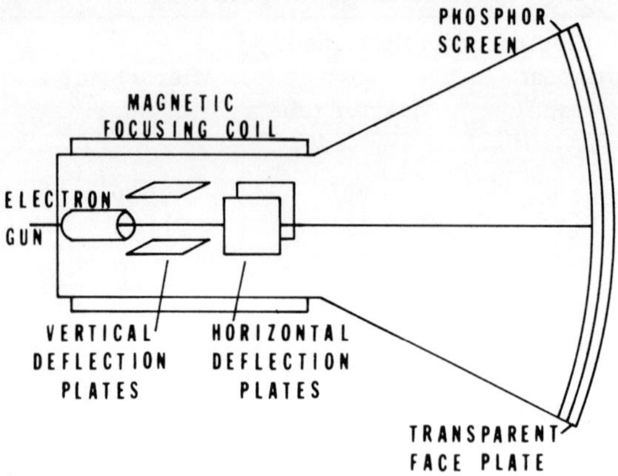

Fig. 53 Typical cathode-ray tube.

Cathode-Ray Tubes

A cathode-ray tube (CRT) basically consists of an electron gun, a magnetic focusing coil, vertical and horizontal deflection plates, and a phosphor screen coated into a face plate as shown in Fig. 53. The current density in CRTs is limited by the maximum angle of deflection of the electron beam, the cathode current density, and statistical variation in the electron transverse velocities.

At high scanning rates electrostatic deflection is usually used, and at lower rates magnetic deflection using external induction coils is commonly used. The ease of beam deflection is given in terms of the deflection sensitivity (S_D) defined by

$$S_D = Y_D\left(D_s V_D\right)^{-1}$$

where

Y_D = measured linear deflection distance on the CRT faceplate
D_S = spot diameter
V_D = deflection voltage

The ratio of the spot diameter to the scanning speed is the time resolution of the CRT. The spot-point-spread function for a typical high-resolution CRT is shown in Fig. 54.

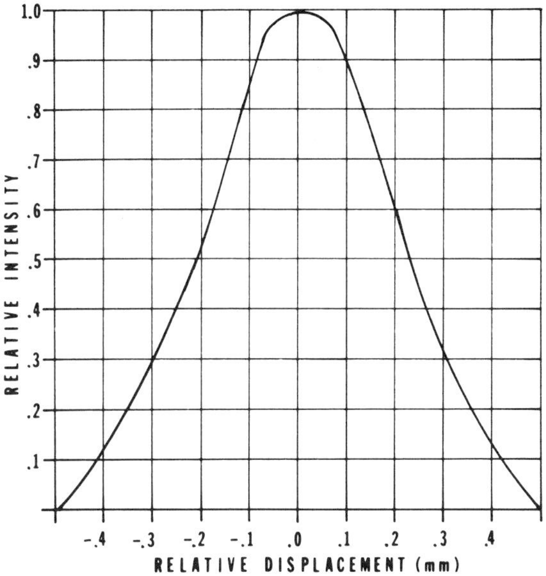

Fig. 54 Typical high-resolution CRT spot point-spread function. Source: Trans. IEEE, CS-12, 1964, and product literature.

In addition to accelerated electrons, the cathode in the CRT also emits negative ions from the materials of construction. These ions cause a screen darkening called ion-burn, which can be greatly reduced by using various types of ion traps to remove the ions in the space from the deflection plates to the phosphor screen, or by coating a thin metallic layer on the phosphor screen which permits electrons to be selectively transmitted. Aluminum is commonly used for this ion screen which increases the screen efficiency by removing accumulated charges on the screen and by limiting reflected light.

The raster (geometric pattern) luminance L_r for CRTs is given by

$$L_r = kf(V) \frac{P}{A_r}$$

where

k = efficiency parameter
$f(V)$ = screen voltage response
P = power delivered to the screen
A_r = raster area

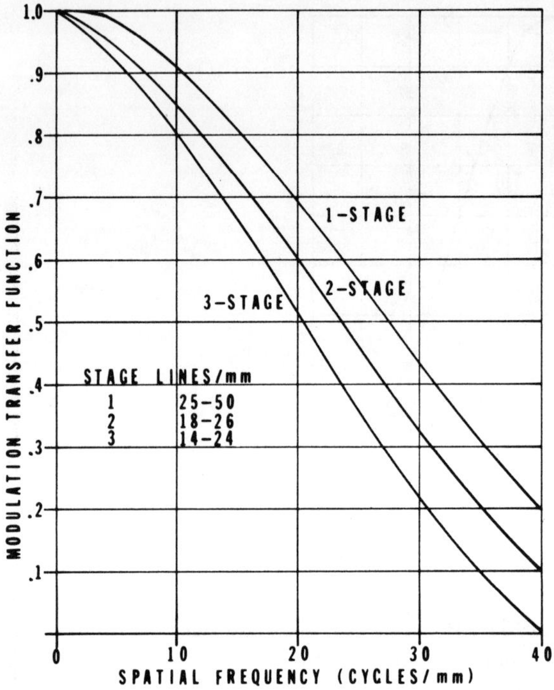

Fig. 55 Typical staged CRT modulation transfer function vs. spatial frequency.

The raster brightness is sometimes given in terms of the phosphor screen parameters (α and β) to define the efficiency and the exposure energy density E_p:

$$L_r = \frac{f(V)\beta P[1 - \exp(-E_p)]}{E_p A_r}$$

where

$$E_p = \frac{Pt_s}{A_s} = \frac{Pt_r}{nA_s} = \frac{PD_s}{A_s V}$$

t_s = spot exposure time
A_s = spot area
t_r = raster period

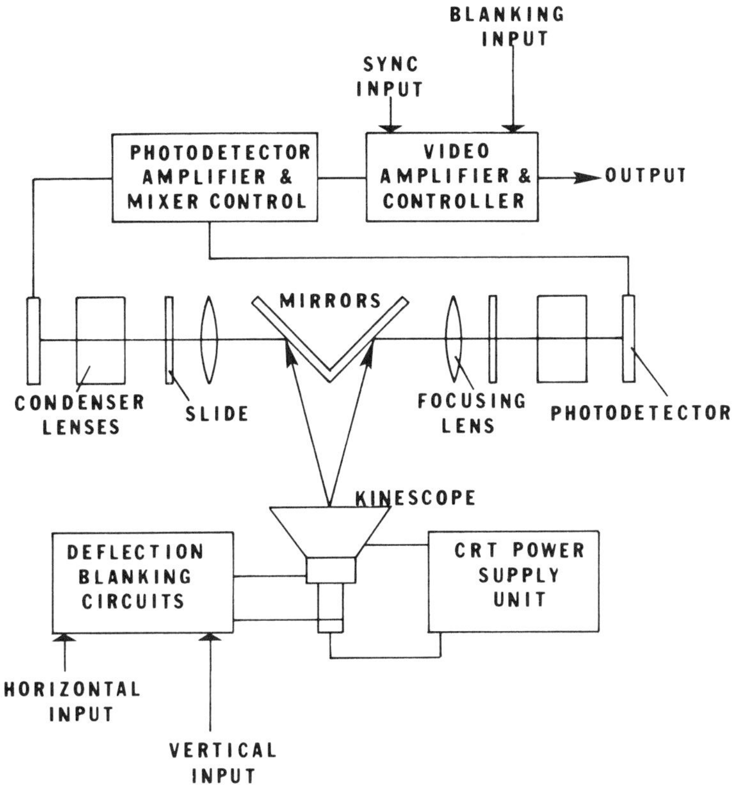

Fig. 56 Schematic diagram of a light-spot scanner.

n = number of spots of diameter D_s in the total raster trace length
V = writing speed

The modulation transfer function of staged CRTs decreases slightly with the addition of intensifier stages for a given spatial frequency. Figure 55 shows the modulation transfer function vs. spatial frequency for typical one-, two-, and three-staged CRTs. At higher gains (and higher stages) the variations in the modulation transfer function become more pronounced. The implementation of CRT tubes ranges from rather simple setups such as the display of slow-speed dc functions (medical applications) to rather complex implementations, such as the light-spot scanner shown in Figure 56.

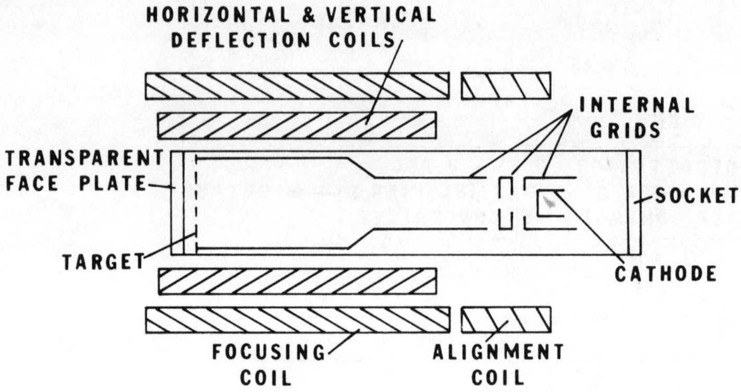

Fig. 57 Cross-sectional view of typical vidicon tube.

Television Camera Tubes

The various types of camera tubes include the vidicon (the most commonly used camera tube), image orthicon, image isocon, silicon intensifier target (SIT), secondary electron conduction (SEC), image dissector, and charge-coupled devices. Figure 57 shows a cross section of a typical vidicon tube. In addition to these basic types there are variations, modifications, and coupling of the different tube types to create such tubes as the ISEC (SEC tube coupled to an image tube), IV (vidicon coupled to an image tube), and the I^2V (vidicon coupled to a two-stage image tube). Other common variations mostly center around the scanning methods, materials of construction, and the target type. One important variation is the return-beam vidicon (RBV), which has electrical and optical characteristics vastly different from most other tube types with approximately 10 times the linear resolution of standard sensors.

A vidicon basically consists of a photoconductive target supported by a faceplate at one end and an electron gun at the other end of the tube. The electron gun produces a beam of electrons which scars the target surface. The camera optics images the desired scene through the faceplate onto the target with some transmission loss incurred in the lenses and faceplate material. The image at the front surface of the photoconductive layer is constantly being reproduced as a positive charge image on the rear surface of the photoconductive layer. The scanning electron beam discharges each picture element in turn. The greater the positive charge in an individual element the larger the number of electrons that must be removed

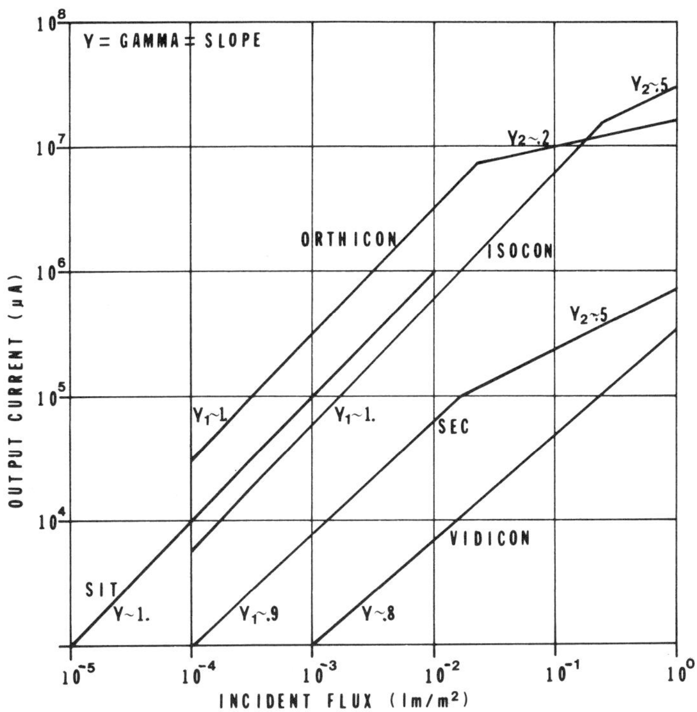

Fig. 58 Typical camera tube output current vs. incident flux.

from the beam to discharge the element. The discharged electrons flow into a load resistor circuit which produces the video signal.

The responsivity of camera tubes is defined as the ratio of the output current to the incident radiant flux. The slope of the log-log responsivity plot at a given faceplate illuminance (incident radiant flux) is defined as the gamma of the tube type. The gammas of vidicons and SITs usually have just one value over the practical usage range, meaning the log-log plot of the output current vs. incident radiant flux is linear. For SEC tubes at high illuminance values some nonlinearity in response is usually noted. Image orthocons and isocons characteristically have two gammas (or slope regions) with the lower gamma occurring at the higher values of incident radiant flux, and both have relatively high SNR values compared to vidicons. Figure 58 shows some typical camera tube output currents as a function of the incident flux with the associated gamma values.

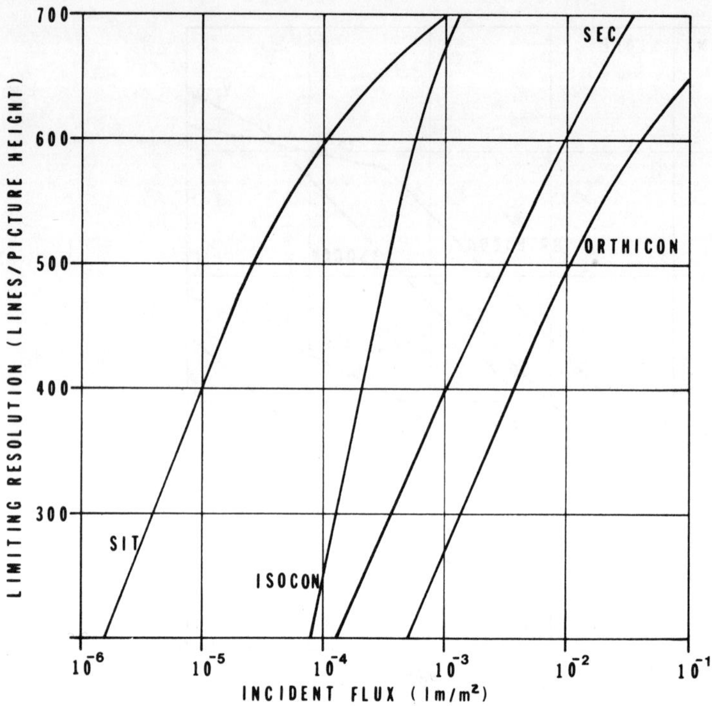

Fig. 59 Typical camera tube limiting resolution vs. incident flux at 100% scene contrast. Source: Product literature for lens transmission *85%.

All camera tubes have an associated log, which is the residual signal on the tube face after the original signal is removed and which is expressed as a percentage of the original signal. The two types of log normally considered are the photoconductive and capacitive logs, usually expressed as the ratio of the target capacitance per unit area to the gain ratio. Camera tubes have a limiting resolution given in terms of lines per picture height which is usually given as a function of the incident flux and the scene contrast. For a given incident flux, the higher the scene contrast, the higher the limiting resolution. Scene contrast values can range from below unity to near 100%, the ideal (unachievable) value. Figure 59 shows the 100% scene contrast values for limiting resolution for various common camera tubes. Real scene values are rarely ideal. To compute the various performance parameters of image intensifier systems, a nomogram such as that depicted in Fig. 60 is used. Here, the lens transmission value is taken to be 85%, which is an average value for most commercial systems.

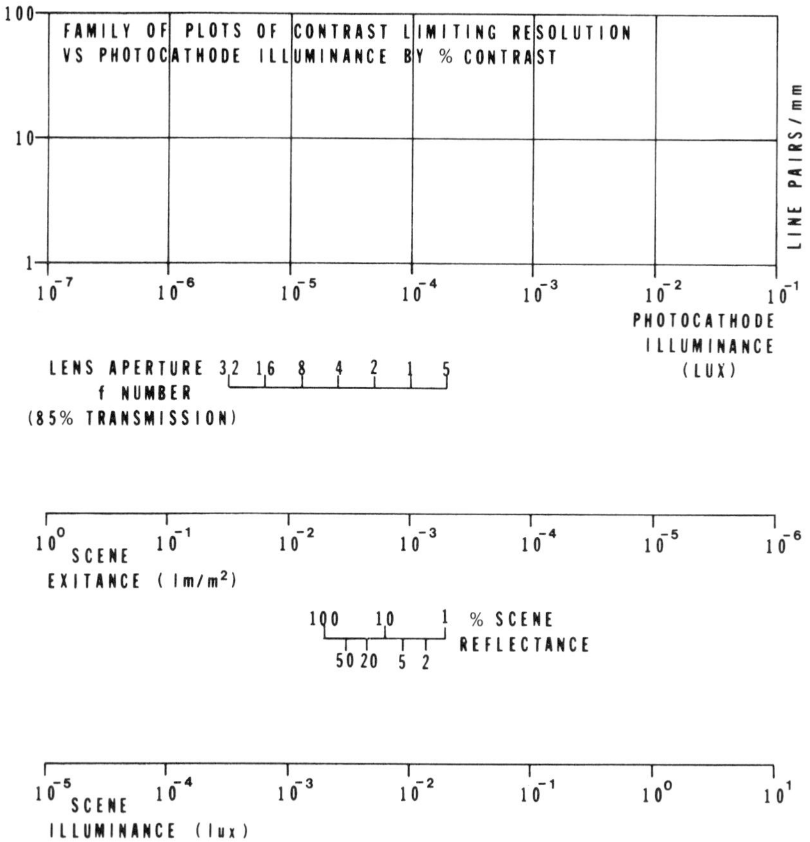

Fig. 60 Image intensifier system application nomogram for selecting performance parameters.

Image Orthicon

An image orthicon consists of three elements: the image, scanning, and multiplier sections. The image section is basically a semitransparent photocathode in the inside of the faceplate and a target disk which is usually a fine mesh screen and glass disk. Focusing is achieved by using an external coil and varying the photocathode voltage. The scanning section is basically an electron gun which sends a beam that scans the target in synchronism with the receiver tube. The multiplier section is basically a staged secondary electron-emission amplifier with a gain of about 10^3. Figure 61 shows the cross section of a typical image orthicon.

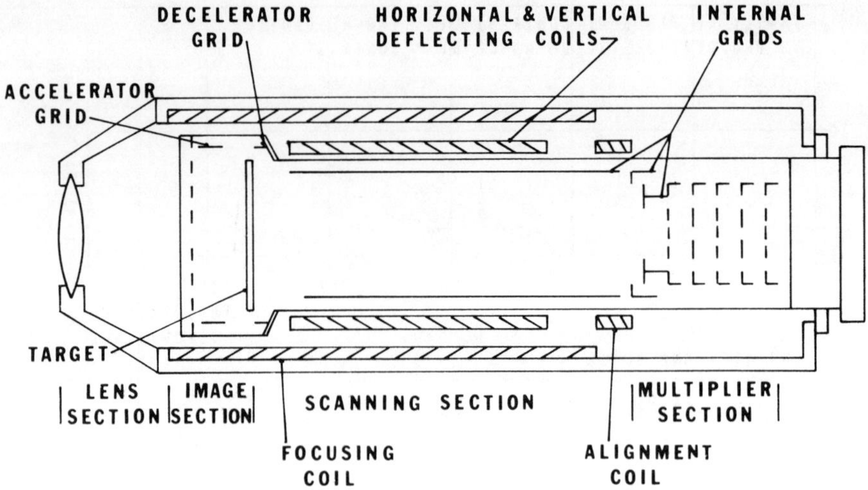

Fig. 61 Cross section of typical image orthicon.

The gun section of the image orthicon generates a scanning beam of electrons which slow to zero velocity as they approach the rear of the target, and then reverse direction and accelerate back beyond the electron gun section to the electron multiplier section. As the electron beam approaches the target, the individual picture elements attract and trap the electrons from the beam, in proportion to degree of surface charge. The beam which returns to the electron multiplier is modulated by the variations in the charge pattern previously stored at the target.

Pyroelectric Vidicon

In the pyroelectric vidicon the photoconductive retina in the standard vidicon is replaced by a pyroelectric disk which is polarized perpendicularly to the disk surface. The faceplate is chosen to be as transparent as possible to the given wavelength region. The incident radiant flux causes a charge pattern at the back of the disk which is neutralized by a scanning electron beam. The performance of the pyroelectric vidicon is limited by thermal diffusion onto the pyroelectric disk, discharge and thermal lag, and electrical and spatial noise.

BIBLIOGRAPHY

Biberman, L. M., and S. Nudelman (Eds.), Photoelectronic Imaging Devices, 1971.

Boot, H. A. H., and J. G. Castledine, Pyroelectric thermal image tube, Electronics Lett., vol. 10, 1974.

Cantella, M., Ultra-high resolution electronic imaging and storage with a return-beam vidicon, SPIE Proc., vol. 32, October, 1972.

Cantella, M., High-resolution non-destructive silicon target storage tube, Opt. Soc. Amer. Tech. Digest, vol. WB-3, March, 1975.

Cooper, H. G., Design of a high-resolution electrostatic CRT for the flying spot store, Bell System Tech. J., vol. 40, 1961.

Haines, J. H., Contrast in CRTs, Tele Tech., vol. 12, 1953.

Holeman, B. R., and W. M. Wrenthall, Thermal-imaging camera tubes with pyroelectric targets, J. Physical Devices, vol. 4, 1971.

Knoll, M., and B. Kazan, Storage Tubes and Their Basic Principles, Wiley, New York, 1952.

Levi, L., Applied Optics, Wiley, New York, 1967.

Martin, L. (Ed.), Advances in Electronics and Electron Physics, Vol. 16, Academic Press, New York, 1962.

Neuhauser, R. G., The silicon target vidicon, RCA Engineer, vol. 22, no. 4, January, 1977.

RCA Electro-Optics Handbook, Technical Series EOH-11, Harrison, N. J., 1974.

Richardson, C. R. W., The pyroelectric vidicon: camera design, system integration and applications, IEE Conf. Publ. 124, 1975.

Rodgers, R. L., Beam-scanned silicon targets for camera tubes, IEEE Intern. Proc., March, 1973.

Schade, D. H., Electron optics and signal readout of high-definition return-beam vidicon camera, RCA Rev., vol. 31, no. 1, 1970.

Soller, T., M. A. Sta, and G. E. Valley, Cathode Ray Tube Displays, McGraw-Hill, New York, 1948.

Woodley, W. A., and D. Rodgers, High-resolution and fiber-optic cathode ray tubes, Brit. Commun. Electronics, vol. 10, 1963.

Wotman, L., Closed-Circuit Television Handbook, 1974.

Chapter 8

OPTICAL FORMULAS, MATERIALS, AND COMPONENTS

Lenses

There are six basic types of optical lenses commonly used in electro-optics, as shown in Fig. 62. In addition to these basic lens elements, there are a few other types such as Fresnel lenses which are not usually encountered in modern electro-optic equipment and components. These basic lens types can be combined in a variety of ways to obtain different lens systems, each having a specialized purpose and optical function. A few of the common lens systems are shown in Fig. 63, including the retrofocus system, which is essentially a reversal telephoto lens used for wide-angle lenses, and the copy, photo, and projection basic lens schematics. Each lens system type has different ways of correcting optical aberrations by coupling different basic lenses and optical materials.

Some of the standard optical lens symbols and assignments are shown in Fig. 64. A more complete listing is given in Table 46. In defining these optical symbols and standards, light is assumed to be incident from the left of a lens or lens system. A given lens has two focal lengths f and f' which are equal only when the indexes of refraction of the material on each side of the lens are equal. With light incident from the left, f is to the right of a positive lens and to the left of a negative lens. Principal planes and nodal planes are perpendicular to the optical axis and through the principal points and nodal point, respectively.

Distances measured from a lens or mirror to real objects or images are defined by sign convention to be positive. Distances measured to virtual objects and images are defined as negative. A double convex lens is defined to have positive focal lengths with positive radii of curvature. A double concave lens has negative focal lengths and negative radii of curvature.

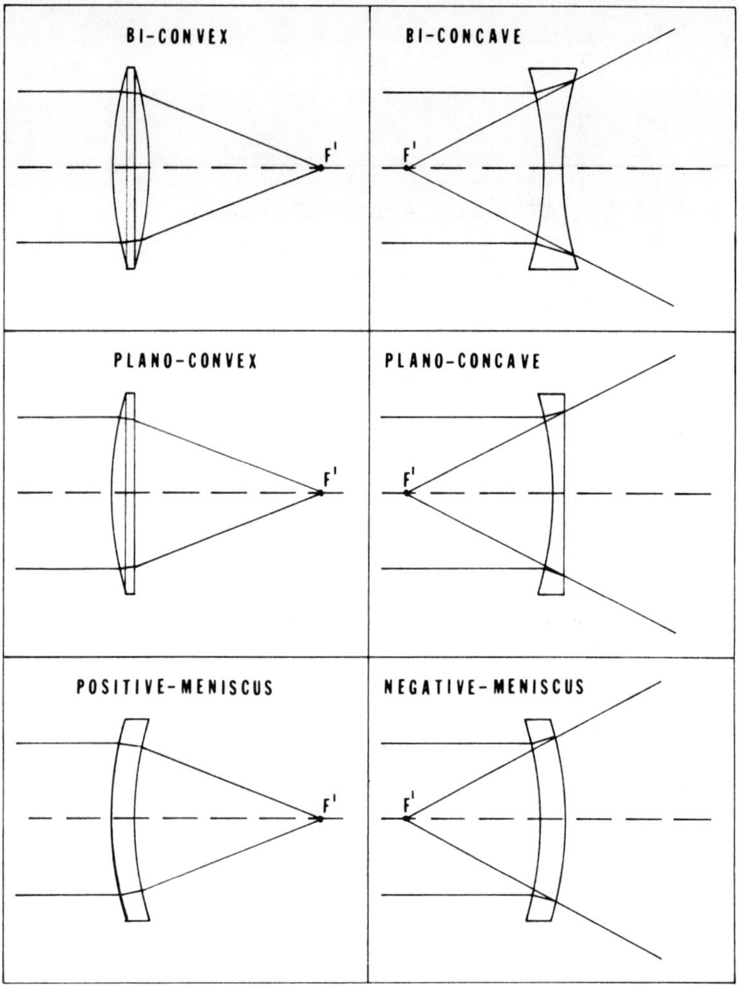

Fig. 62 Common types of optical lenses.

The circle of confusion c' in the image plane is the circle diameter of the image of a point source at infinity, measured in a plane normal to the optical axis at or near the principal focus. Optical system aberrations cause a point-source image to be larger than a point. At the principal focus of an aberration-free optical system, c' becomes the diameter of the diffraction image of a point source. The image of a point for an ideal lens occurs as a central circular bright region (airy disk) surrounded by a series

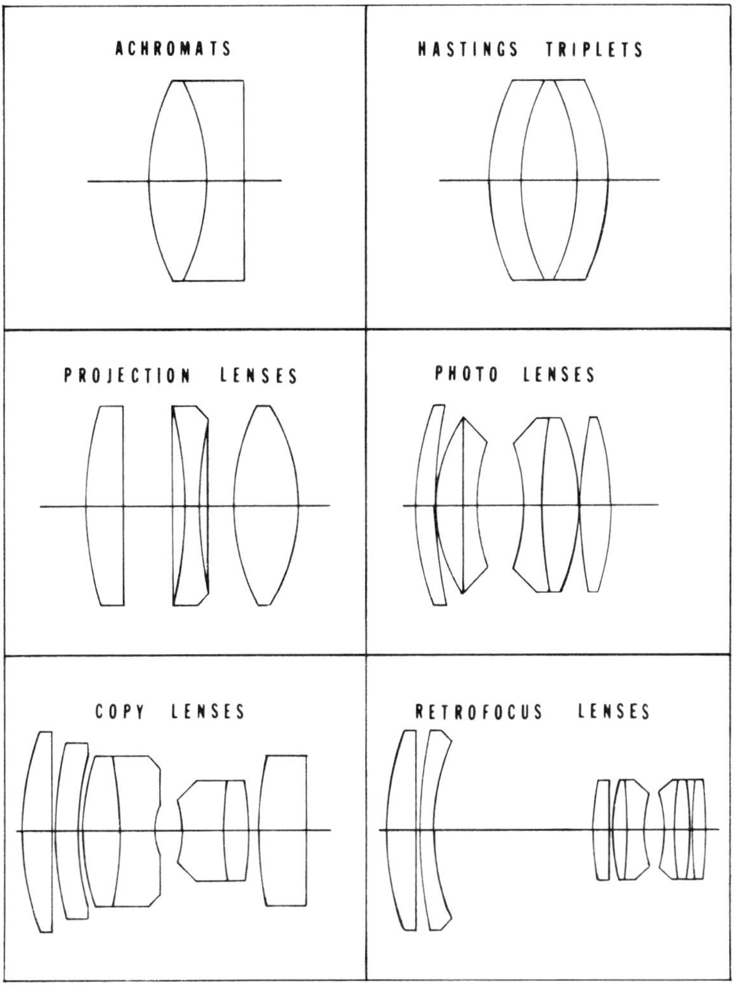

Fig. 63 Multiple lens systems.

of rings separated by dark rings. The central bright region theoretically contains 83.8% of the total radiant energy with the radius of the first null given by $r_1 = 1.22\lambda f/b$. By definition, two points can be distinguished as separate when their diffraction disk images do not overlap more than the radius of one of the disks.

A thin lens can be corrected for axial chromatic aberration by combining it with another thin lens such that the sum of the dispersions is as close

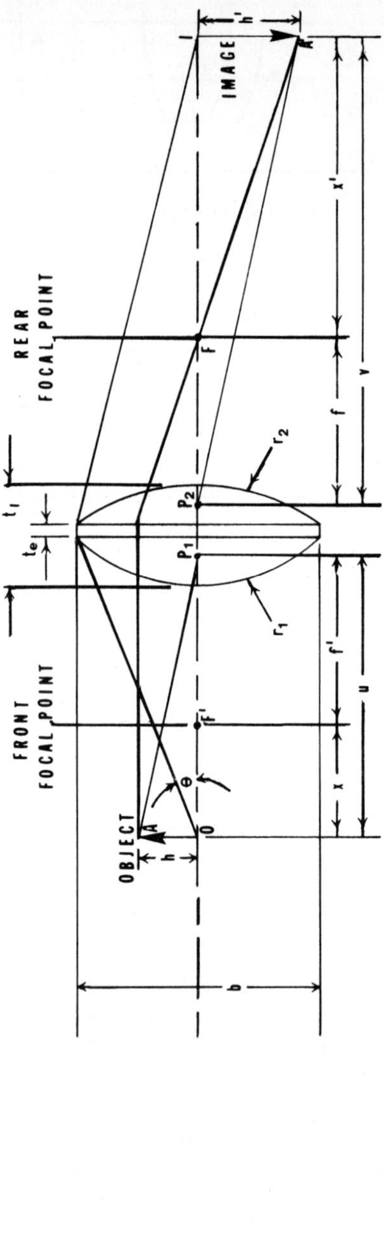

Fig. 64 Standard optical lens symbols and assignments.

to zero as possible. The blur circle diameters caused by the various optical aberrations of a lens are given as follows:

Spherical aberration:

$$c' = \frac{0.067f}{(f\#)^3}$$

Chromatic aberration:

$$c' = \frac{f}{130(f\#)}$$

Field of curvature:

$$c' = \frac{(h')^2}{3f(f\#)}$$

Coma:

$$c' = \frac{uf}{56(f\#)^2}$$

Astigmatism:

$$c' = \frac{u^2 f}{2(f\#)}$$

The modulation transfer function (MTF) of a lens is a function of the incident wavelength and the f# of the lens, given by

$$\text{MTF} = \frac{2}{\pi} \cos^{-1} y - y(1 - y^2)^{1/2}$$

where $y = \lambda f\#\nu$ and the spatial cutoff frequency is $\nu = 1/\lambda f\#$.

Selfoc Microlenses

A selfoc lens is a specialized optical lens with a radial parabolic refractive index distribution. These lenses typically have small diameters (less than 2 mm) and are capable of achieving very short focal lengths. The focal length can be altered by varying the length of the lens rod by cutting and polishing the end of the lens. These lenses have been used commonly to couple laser diodes and LEDs to optical fibers and in various combinations for use in wave-division multiplexing devices. They have also been used as objective lenses for endoscopes and microscopes, for optical reader units, and for optical duplicating and printing machines. Typical lens optical and physical parameters are shown in Table 47.

Table 46 Standard Optical Symbol Definitions

Symbol	Standard optical definition
a	Clear aperture of lens or mirror
A	Off-axis object point
A'	Off-axis image point
b	Diameter of lens or mirror
c	Diameter of the circle of confusion in the object plane
c'	Diameter of the circle of confusion in the image plane
C	Axial center of mirror radius
d	Distance between principle planes of two lenses
D	Object plane to image plane distance
$D_{L1}, D_{L2}, \ldots, D_{Ln}$	Diopter power of the first, second, etc., lens
D_{Lc}	Diopter power of a lens combination
D_1	Far limit of depth of field, measured from point focused upon
D_2	Near limit of depth of field, measured from point focused upon
f, f'	Focal lengths of a given lens
F, F'	Focal points of a given lens
f_b	Back focal length
f_c	Focal length of a lens combination
f_e	Effective focal length, distance between second principal point and the second focal point
f_f	Front focal length
f#	F-number or aperture ratio
f_e#	Effective f-number
$f_1, f_2, \ldots, f_n$	Focal length of the first, second, etc., lens
h	Object height perpendicular to the axis
h'	Image height perpendicular to the axis
H	Hyperfocal distance
I	Axial image point
$L_1, L_2, \ldots, L_n$	First, second, etc., lens numbered in order of incident light
L_1	Far limit of depth of field measured from lens
L	Near limit of depth of field measured from lens
M	Magnification of a lens or mirror
$M_1, M_2, \ldots, M_n$	First, second, etc., mirror numbered in order of incident light
n	Index of refraction with no subscript assumed to be n_D
N_1	First, frontal or entrant nodal point
N_2	Second, rear or emergent nodal point

Table 46-Continued

NA	Numerical aperture of a lens
O	Axial object point
PP_1, PP_2, ..., PP_n	First, second, etc., principal plane
r	Mirror radius
r_1, r_2, ..., r_n	Radius of first, second, etc., lens surface numbered in order of the striking incident light
R	Reduction of a lens or mirror
t_e	Axial edge thickness of a lens
t_1	Axial thickness of a lens
u	Axial object plane distance from thin lens or from the first nodal point of a thick lens or optical system
v	Axial image plane distance from thin lens or from the second nodal point of a thick lens or optical system
V	Relative reciprocal dispersion or constringence
x	Axial object distance minus focal length f'
x'	Axial image distance minus focal length f
α	Half the angle of view
θ	Angular size of circle of confusion
θ_i	Angle of incidence
θ_r	Angle of refraction or reflection
ω_a	Dispersive power for actinic achromatism

The parabolic refractive index profile is given by

$$N = N_0(1 - 0.5Ar^2)$$

where

N_0 = refractive index on the axis
N = refractive index at a distance r from the axis
A = positive constant determined by the lens pitch

The ray matrix of paraxial rays to determine exit ray position and angle is given by

$$\begin{bmatrix} X_1 \\ X_1' \end{bmatrix} = \begin{bmatrix} \cos A^{1/2} Z & N_0^{-1} A^{-1/2} \sin A^{1/2} Z \\ -N_0 A^{1/2} \sin A^{1/2} Z & \cos A^{1/2} Z \end{bmatrix} \begin{bmatrix} X_0 \\ X_0' \end{bmatrix}$$

where

X_0 = position of the incident ray
X_0' = angle of incidence

Table 47 Selfoc Microlens Specifications

Lens type	Diameter (mm)	Pitch Z_0 (mm)[a]	N_0 λ = 589 nm	N λ = 589 nm	NA (sin θ_c)	$2\theta_c$ (degrees)	$\frac{\Delta Z_0}{Z_0}$
SSL	1.0	16	1.553	0.025	0.30	36	0.07
SSL	1.5	24	1.553	0.025	0.30	36	0.07
SSL	2.0	32	1.553	0.025	0.30	36	0.07
SLW	0.35	3.4	1.616	0.080	0.50	60	0.05
SLW	0.50	5.2	1.616	0.080	0.50	60	0.05
SLW	1.0	10.4	1.616	0.080	0.50	60	0.05
SLW	1.5	15.6	1.616	0.080	0.50	60	0.05
SLW	1.8	18.4	1.616	0.080	0.50	60	0.05
SLW	2.0	20.8	1.616	0.080	0.50	60	0.05
SLL	1.0	45	1.525	0.004	0.10	12	0.004
SLL	1.0	57	1.525	0.0025	0.08	9.5	0.004
SLL	1.0	67	1.525	0.0017	0.07	8	0.004
SLAB	0.4	3.6	1.620	0.072	0.48	57	0.05
SLAB	0.5	4.5	1.620	0.072	0.48	57	0.05

[a] $Z_0 = 2\pi/A^{1/2}$.

[b] $\frac{\Delta Z_0}{Z_0} = \frac{[Z_0(\lambda_C) - Z_0(\lambda_F)]}{Z_0(\lambda_D)}$.

Source: Nippon Sheet Glass Company, Limited, Japan

X_1 = position of the ray at the outlet of the selfoc lens
X_1' = exit angle of the ray
Z = lens rod length

The lens focal length is given by

$$f = \frac{1}{(N_0 A^{1/2} \sin A^{1/2} Z)}$$

The length between the principal plane and the end of the lens is given by

$$h = N_0^{-1} A^{-1/2} \tan 0.5 A^{1/2} Z$$

The distance of the image plane to the end of the lens l_1 is related to the distance of the object plane to the end of the lens l_0 by

$$l_1 = \frac{1}{N_0 A^{1/2}} \frac{N_0 l_0 A^{1/2} \cos A^{1/2} Z + \sin A^{1/2} Z}{N_0 l_0 A^{1/2} \sin A^{1/2} Z - \cos A^{1/2} Z}$$

When l_1 is negative, the image is virtual, and when l_1 is positive, the image is real.

The lateral magnification is given by

$$M = \left(\frac{N_0 A l_1^2 + 1}{N_0 A l_0^2 + 1} \right)^{1/2} = -\left[N_0 A^{1/2} \sin A^{1/2} Z \left(l_0 - \frac{\cot A^{1/2} Z}{N_0 A^{1/2}} \right) \right]^{-1}$$

When M is negative, the image is inverted, and when M is positive, the image is erect.

The acceptance angle of the lens is given by

$$\sin \theta_{max} = 2A^{1/2} N_0 R = (2N_0 \Delta N)^{1/2}$$

where $\Delta N = N_0 - N_R$ and R is the outside lens radius.

The diffraction limitation of the lens is given by

$$D = \frac{1.22\lambda}{N_0 A^{1/2} R}$$

where D is the airy disk diameter and λ is the incident wavelength. This limit is particularly important to consider when attempting to use micro-lenses to couple into single-mode fibers with very small core diameters, since the airy disk diameter may easily exceed the fiber core diameter.

Laser Lenses

The angular radii Θ of the image spot for a laser lens is given by

$$\Theta = \frac{K(n)}{2f^3} b^3 + \frac{1.22\lambda}{b}$$

where K(n) is a function of the refractive index. The optimum diameter b(opt) and optimum focal length f(opt) where the angular radius of the blur circle is minimized are given by

$$b(opt) = \left(0.813 \frac{\lambda}{K(n)} \right)^{1/4} f^{3/4} \qquad f(opt) = \left(1.23 \frac{K(n) b^4}{\lambda} \right)^{1/3}$$

For the best lens form for minimum spherical aberration for specific laser lines,

$$K(n) = \frac{1}{32(n-1)^2}\left(n^2 - \frac{n(2n+1)^2}{4(n+2)}\right)$$

For plano-convex lenses,

$$K(n) = \frac{1}{32(n-1)^2}\left(n^2 - 2n + \frac{2}{n}\right)$$

For symmetric biconvex lenses,

$$K(n) = \frac{1}{32(n-1)^2}\left(n^2 - n - \frac{1}{2n} - \frac{1}{4}\right)$$

Magnification

The major factors in the selection of a magnifier are the power or magnification, working distance (the distance between the lens and the object), field of view, and depth of field. The power of a magnifying lens is given by 10/f. The diameter of the field of view of a magnifier is less than its focal length. As the power of a magnifier increases its working distance and lens diameter decrease. The curvature of the lens surface determines its power in that to increase magnification the lens radius must be decreased for a single lens system. The depth of field decreases with higher power magnifiers.

Corrected magnifiers are those in which various optical deviations or aberrations have been reduced or effectively eliminated by arrangements of simple lenses into multiple-lens systems, i.e., doublets, triplets, etc. The uncorrected lens can cause the image to have flares in localized zones, to contain false colors, to lack contrast, or to distort the true object curvature. When the image curves inward on the sides, it has pin-cushion distortion, and if it bulges, it has barrel distortion. The various types of lens aberrations usually do not occur independently. Spherical aberration is one of the causes of a dull or gray image. Coma, astigmatism, and curvative of field also contribute to image grayness and lack of detail in the outer areas of the image.

The smallest distance between two points that can be resolved by a light-lens system, due to the wave nature of light, is given by $\lambda/2(NA)$. To attain high resolution and high magnification the lower wavelength region (blue and ultraviolet) should be utilized. For an optical- visible-region microscope, the limit of magnification beyond which resolution is not improved is given by

$$M_{max} = \frac{R_e(2NA)}{\lambda_{min}}$$

where R_e is the limiting resolution of the eye. Typical practical values are λ_{min} = 400 nm, NA = 1.5 (oil immersion), and R_e = 0.2 mm. Thus,

$$M_{max} \simeq \frac{(0.2 \times 10^{-3} \text{ m})(2)(1.5)}{(400 \times 10^{-9} \text{ m})} \simeq 1500$$

Optical Formulas

Using the standard symbols as defined in Table 46, the optical formulas for various lens systems are shown in Table 48, listed in alphabetical order. By locating a desired quantity in the symbols table, the appropriate equation to relate the quantity to various other terms can easily be found in the optical formulas table.

In addition to these basic relationships there are some other commonly encountered equations in electro-optics listed with the appropriate symbol definitions.

Diffraction by a straight edge

$$X_b^{\,2} = \frac{V(u + v)}{u}(2n + 1) \qquad X_d^{\,2} = \frac{V(u + v)}{u}(2n\lambda)$$

When

$$u = x_b^{\,2} + (2n + 1) \qquad x_d^{\,2} = 2n\lambda$$

where

- u = distance from source to edge
- v = distance from slit to screen
- w = width of the slit
- n = band number from center reference point
- x_b = radius of bright band from center reference point
- x_d = radius of dark band from center reference point

Table 48 Optical Formulas

$a = f/f\#$	
$D = f(M + 1/M + 2)$	$D = f(M + 1)^2/M$
$D = f(R + 1/R + 2)$	$D = f(R + 1)^2/M$
$D_{Lc} = D_{L1} + D_{L2}$	$D_L = 1/f_e$
$D_1 = u^2 \tan \theta/(a - u \tan \theta)$	$D_2 = u^2 \tan \theta/a + u \tan \theta$
$D_L = (n - 1)(1/r_1 + 1/r_2) + (n - 1)^2 t_e/nr_1 r_2$	
$f = uv/(u + v)$	$f = uM/(1 + M)$
$f = u/(R + 1)$	$f = v/(1 + M)$
$f = uv/D$	$f = (u + v)M/(1 + M)^2$
$f = DR/(R + 1)^2$	$f = DM/(1 + M)^2$
$f\# = f_e/b$	$f\# = 1/2(NA)$
$f_e\# = vf\#/f$	$f_c = f_1 f_2/(f_1 + f_2 - d)$
$f_b = f - r_2 t_1/[M(r_2 - r_1) + t_1(n - 1)]$	$f_f = f + r_1 t_1/[M(r_2 - r_1) + t_1(n - 1)]$
$1/f = 1/u + 1/v$	$1/f_e = 1/f_1 + 1/f_2 + \cdots 1/f_n$
$1/f_c = 1/f_1 + 1/f_2 - d/f_1 f_2$	
$H = f^2/f\#c$	
$L_1 = Hu/(H - u + f)$	$L_2 = Hu/(H + u - f)$
$M = f/(u - f)$	$M = v/u$
$M = (v - f)/f$	$M = h'/h$
$NA = n \sin \theta$	
$R = (u - f)/f$	$R = f(v - f)$
$R = u/v$	
$u = vf/(v - f)$	$u = v/M$
$u = f(1/M + 1)$	$u = (u + v)/(1 + M)$
$u = (R + D)/(R + 1)$	$u = D/(1 + M)$
$v = uf/(u - f)$	$v = f(1 + M)$
$v = Mu$	$v = M(u + v)/(1 + M)$
$v = u/R$	$v = f/R + f$
$u + v = uv/f$	$u + v = v(M + 1)/M$

Table 48-Continued

$u + v = u(1 + M)$	$u + v = f(M + 1)^2/M$
$V = (n_D - 1)/(n_F - n_C$	
$x = f^2/(v - f)$	$x = u - f$
$x = f^2/x'$	$x = f/M$
$x' = f^2/(u - f)$	$x' = v - f$
$x' = f^2/x$	$x' = Mf$
$\theta = \arctan(b/2u)$	$\omega_a = (n_G - n_D)/(n_D - 1)$

Diffraction by a grating

$$n\lambda = d(\sin \theta_n - \sin \theta_i)$$

where

n = order of the spectrum
d = distance between adjacent lines
θ_n = angle between diffracted ray and the normal to the grating
θ_i = angle between incident ray and the normal to the grating

Newton's rings

$$2t = \frac{R_b^2}{r} = N + \frac{1}{2}\lambda \qquad 2t = \frac{R_d^2}{r} = N\lambda$$

where

R_b = radius of bright Newton's ring
t = thickness of air film
R_d = radius of dark Newton's ring
r = radius of lens surface
N = ring number from contact center

Interference from thin gas films

$$t_{max} = \frac{m\lambda}{n_g} \qquad t_{min} = \frac{m\lambda}{2n_g}$$

where

t = thickness of gas film
n_g = index of refraction of gaseous media
m = interference integer

Polarization by a stack of plates

$$P = M\left[M + \left(\frac{2n}{(1 - 2n)}\right)^2\right]^{-1}$$

where

P = degree of polarization of transmitted light from a beam of unpolarized light incident to a stack of parallel plates
M = number of plates
n = index of refraction

(for most types of optical glass $P \simeq M/M + 2.3$).

Polarizers

The transmittance T_S of a single polarizer element in a beam of linearly polarized light is given by

$$T_S = k_1 \cos^2\theta + K_2 \sin^2\theta$$

where θ is the angle between the polarization plane of the incident beam and the preferred transmission plane of the polarizer, and K and K are the principal transmittances of the polarizer and are functions of the incident wavelength ($K_1 \simeq 1$; $K_2 \simeq 0$).

For unpolarized light incident on a pair of polarizers, the transmittance of the pair T_p is given by

$$T_p = K_1K_2 \sin^2\theta + \frac{1}{2}(K_1^2 + K_2^2) \cos^2\theta$$

Illuminance and Irradiance Formulas

When an optical lens is used to transmit an image onto a planar surface the illuminance on the surface E_S is given by

$$E_s = \frac{E_0RT_r}{4f\#\,(1 + M)^2}$$

where

E_0 = real object illuminance
R = object area reflectance
T_r = lens transmission

The power of the radiant flux at the image surface Φ_S from a distant object is given by

$$\Phi_S = \frac{E_0 \pi b^2 T_r}{4}$$

where E_0 is now the irradiance of the real object and b is the actual lens aperture. When the object is a point source described in terms of radiant or luminus intensity

$$\Phi_S = \frac{I \pi b^2 T_r}{4u^2}$$

where I is the intensity and u is the lens to source distance.

Modulation Transfer Function

The modulation transfer function (MTF) is a means of measuring the resolution of an imaging system and is defined as the sine-wave spatial frequency amplitude response. The MTF applies to geometric optics (lens systems, etc.) as well as to more complex imaging systems such as camera and imaging tubes, image intensifiers, and display systems. At low values of the spatial frequency (N), MTF = 1.0. The MTF is used in simple and complex imaging systems to measure the overall resolution. For multicomponent or multielement systems the overall MTF is the product of the individual component or elemental MTFs.

The contrast frequency function (CTF) is defined as the square-wave spatial-frequency amplitude response. It can be related to the system MTF as follows:

$$\text{CTF} = C(N) = \frac{4}{\pi}\left(M(N) - \frac{M(3N)}{3} + \frac{M(5N)}{5} - \frac{M(7N)}{7} \cdots\right)$$

$$\text{MTF} = M(N) = \frac{\pi}{4}\left(C(N) - \frac{C(3N)}{3} + \frac{C(5N)}{5} - \frac{C(7N)}{7} \cdots\right)$$

Prisms

There are many types of prisms used in electro-optics systems. A few of the more common types are shown in Fig. 65. The major differences in optical prisms relate to their specific function, such as acting like mirrors, reversing an image direction or orientation, eliminating ordinary

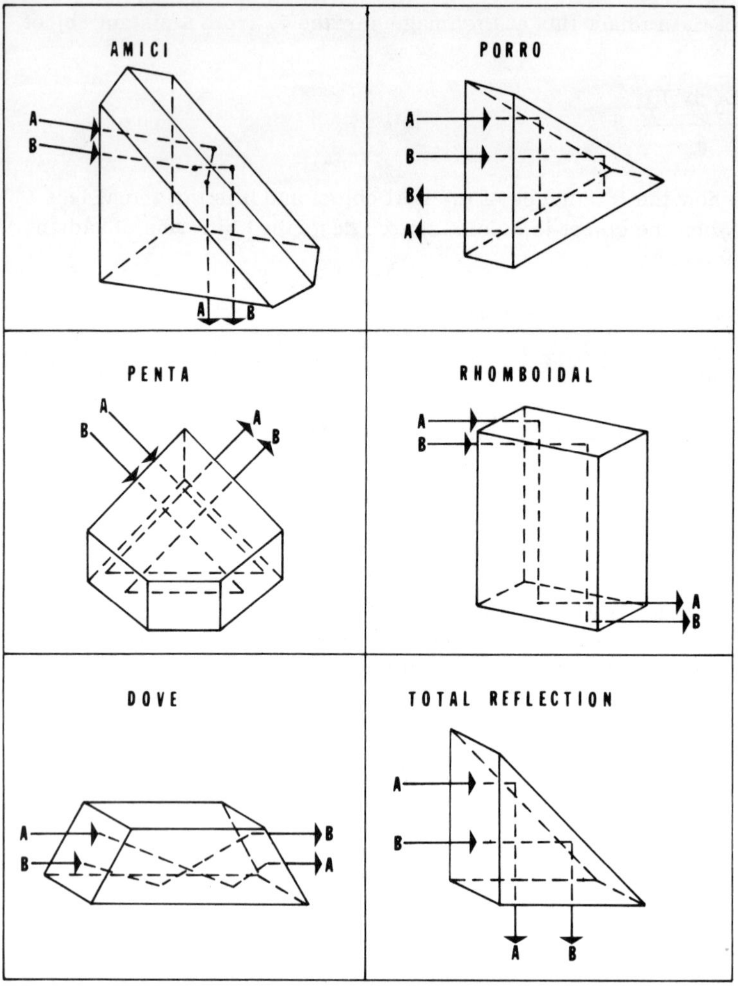

Fig. 65 Common types of optical prisms.

refraction rays, or eliminating extraordinary refraction rays. The dove prism allows on-axis inversion with a single reflecting surface and produces no net angular dispersion. Nichol and Glan-Fourcault prisms are made from two pieces of calcite cemented together with Canadian balsam (which has a refractive index between that of the extraordinary and ordinary ray), permitting only the extraordinary ray to pass and eliminating the ordinary ray by total reflection at the interface. Rochon and Wollaston prisms are similar except that the optical axes are shifted by 90°. They

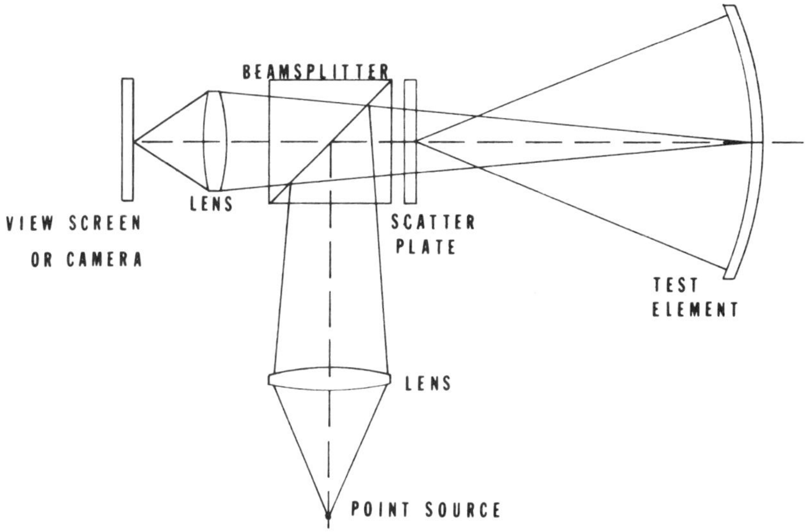

Fig. 66 Scatter-plate interferometer.

are made from two pieces of calcite or quartz and allow the ordinary beam to travel through without deviating from the entering path (achromatic) while the extraordinary ray is deviated (chromatic). The Cornu prism is made of quartz with its optical axis parallel to the base so that light can be transmitted without double refraction by setting the prism at a minimum deviation.

Interferometers

Various types of optical interferometers are used to measure the optical flatness, curvature, and homogeneity of optical components. These include the plano-Fizeau, Twyman-Green, Michelson, Mach-Zender, speckle, and scatter-plate interferometers (shown in Fig. 66). The various patterns seen by the observer or captured photographically depend upon the type of deviations existing in the optical component, such as having local high spots, axial and off-axis curvature, and other types. Figure 67 shows some of the interference patterns commonly encountered in these interferometers along with a brief physical interpretation of the defect depicted. Some interferometers use a simple beam-splitter configuration to superimpose the image of a measuring scale onto the interference pattern to give a direct indication as to the magnitude of the optical deviations at a specified wavelength.

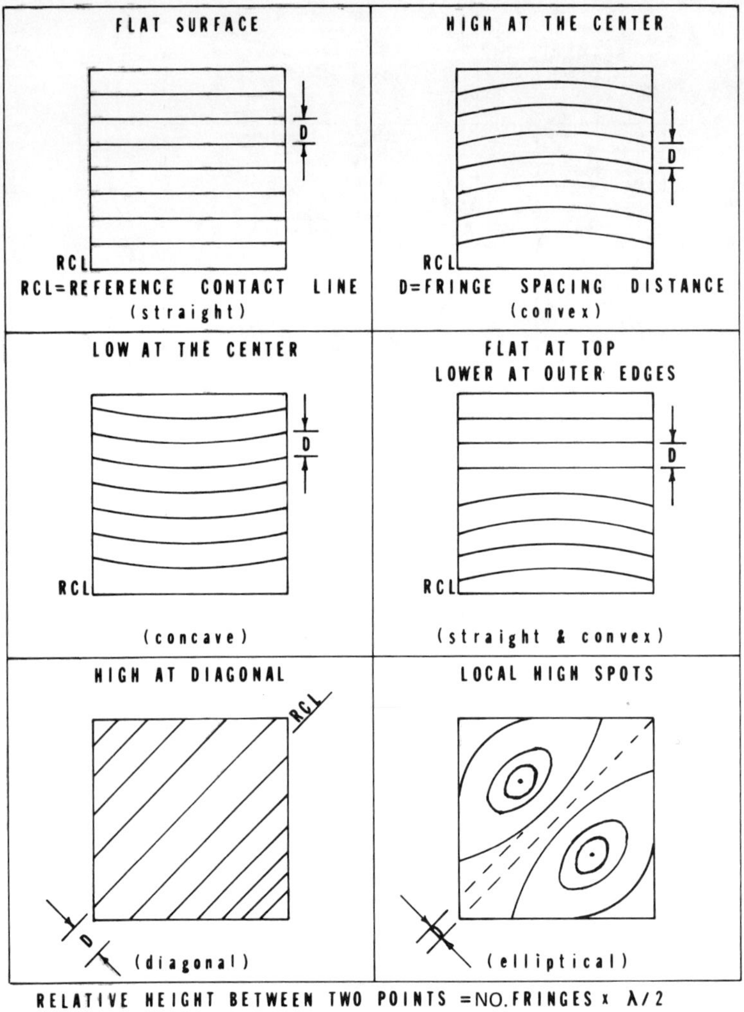

Fig. 67 Physical interpretation of interferometry patterns.

The speckle interferometer forms a speckle pattern on an optical component surface using the vibrational mode patterns formed by two laser beams. The interference generated by the expanded reference and object beams causes a pattern of light and dark speckles which fluctuate as the object moves. Vibrations at various frequencies will cause the generated speckle image to fade in vibration areas and to remain distinct in object modal areas. The relative coarseness, clarity, and image intensity of the speckle pattern are controlled using a combination of an iris and polarizer and collimating lens system.

Monochromaters

The two basic types of monochromaters are the reflection gratings (plano and concave) and the transmission gratings. The basic transmission grating monochromater consists of a narrow adjustable entrance slit and a low dispersion glass prism, followed by a finely ruled grating placed parallel to the entrance slit. To achieve accurate observation of the spectra transmitted through the grating, a corrective magnifying lens system is often used. Most transmission gratings have a useful spectral bandwidth from 350 to 800 nm. For low-intensity sources, the prism and lens system must be carefully selected and designed to optimize transmission at the desired wavelengths.

Concave reflection grating monochromators basically consist of an adjustable entrance slit and a concave grating which by various types of mechanical movements can swing in a circle. The spectra transmitted by the concave grating follow the equation

$$m\lambda = d(\sin \theta_i \pm \sin \theta_r)$$

where m is the order of the spectrum, d is the grating spacing, and θ_i and θ_r are the incident and reflected ray angles, respectively. The various types of mountings used for the concave grating monochromators are the Abbe, Beuther, Eagle, Paschen, Rowland, Sey, and Wadsworth mountings. Depending upon the optics and spacial configuration of these mountings varying degrees of astigmatism and linear dispersion can occur.

Plano reflection grating monochromators use planar diffraction gratings instead of curved (concave) gratings as the central dispersive element and concave spherical mirrors to reflect the light transmitted by the entrance slit and planar grating. The basic types of plano-grating systems are the Czerny-Turner monochromator, Fastie-Ekert grating monochromator, and the Littrow plano-grating and mirror spectrograph. These systems differ greatly by the extent of light transmitted to the object lens, detector, camera, or observation screen, by the degree of aberration correction, and by the spectral resolution achievable. Various types of optics can be used internally, such as cylindrical, magnifying, and anastigmatic lenses, to compensate for various aberrations or to achieve greater spectral resolution.

Thin-Film Coatings

For light incident upon a flat polished surface, the reflectance of light polarized in the plane of incidence ($P_{\parallel}$) is given by

$$P_{\parallel} = \frac{\sin(\theta_i - \theta_r)2}{\sin(\theta_i + \theta_r)}$$

and light polarized normal to the plane of incidence $P_{\perp}$ is given by

$$P_{\perp} = \frac{\tan(\theta_i - \theta_r)2}{\tan(\theta_i - \theta_r)}$$

For unpolarized light the reflectance $p(\theta)$ is given by

$$p(\theta) = \frac{1}{2} \frac{\sin^2(\theta_i - \theta_r)}{\sin^2(\theta_i + \theta_r)} + \frac{\tan^2(\theta_i - \theta_r)}{\tan^2(\theta_i + \theta_r)}$$

The significance of these equations is that light incident on a polished material surface can be expected to have a loss by reflection of 4 to 5% even at small incidence angles. For a large number of surfaces in a given optical system these losses can be rather significant.

To compensate for the anticipated losses by reflection or to meet other optical and electrical criteria, various thin-film coatings can be applied to lens, prisms, mirrors, filters, and other optical components. The various types of coating encountered are anti-reflection, high-reflection, partially reflective, low-reflection, filter-short pass, filter-long pass, bandpass, neutral, protective, conductive, metal, multilayer, and enhanced metal reflection coatings. A coating can be used to act as an electrical conductor such as the base for certain phosphor screens. Conductive coatings are usually made from copper, gold, silver, and tin. Protective coatings can be made from a variety of materials including quartz, sapphire, titanium dioxide, magnesium fluoride, and zirconium oxide. Optical thin-film coatings can also be used as a means to increase or decrease surface reflection as a frequency filter to selectively pass longer or shorter wavelengths or a narrow frequency band, as a fixed (neutral) or variable reduction of all wavelengths, or as a means of introducing varying degrees of polarization into the transmitted or reflected light beam.

The substrate for optical coatings can often present problems in matching the index of refraction, dispersion, surface flatness, and thermal properties. Substrate materials commonly used are quartz, sapphire, germanium, Pyrex, tempax, and silicon. Sometimes an isolating (neutral-transparent) layer is used between the substrate and desired coating to attain a more perfect match of the desired optical and electrical properties of the substrate and coating. Coatings can often be destroyed by physical contact with other objects, by oxidation, by thermal degradation (burning, sublimation, and melting), by various photochemical reactions, or by contact with water or other solvents.

To reduce the losses by reflection caused by the difference between the refractive indexes of two materials (n_1 and n_2), a material layer can be added with an index of refraction intermediate between the two given materials. When the antireflection coating has an index of refraction which is the average of n_1 and n_2, the reflectance at each interface is given by

$$p_c = \left(\frac{n_1^{1/2} - n_2^{1/2}}{n_1^{1/2} + n_2^{1/2}} \right)^2$$

Often, the index of refraction of the antireflection coating is given by $(n_g n_m)^{1/2}$, where n_g is the index of refraction of the external (gaseous) media and n_m is the index of refraction of the optical material (glass, plastic, etc.).

The thickness of the coating necessary to minimize losses is given by

$$t = \frac{(2m + 1)\lambda}{4}$$

where m is a positive integer or zero. By using several antireflection layers losses by reflectance can be greatly reduced, commonly by a factor of two or three over a wide wavelength region. Transmission losses at various wavelengths can become more significant than reflectance losses; so the coating material spectral characteristics must be carefully matched with the desired material usage.

High specular reflectance is usually obtained by coating metallic films or polished surfaces. Figure 68 shows the spectral reflectance of various metallic films that have been freshly evaporated onto a smooth glass surface. With aging the reflectance decreases as a result of oxidation of the metal surface. Protective coatings such as MgF_2 and SiO_2 are commonly used to protect mirrored surfaces. For mirror coatings on laser cavity mirrors, multilayer dielectrics are often used to attain very low reflection losses. When diffuse reflection is desired, magnesium oxide (MgO) and barium sulfate ($BaSO_4$) are commonly used. Within the spectral range of 400 to 1100 nm these materials have a reflectance greater than 98%. Magnesium carbonate ($MgCO_3$) also has a high reflectance over the visible and near-ir regions.

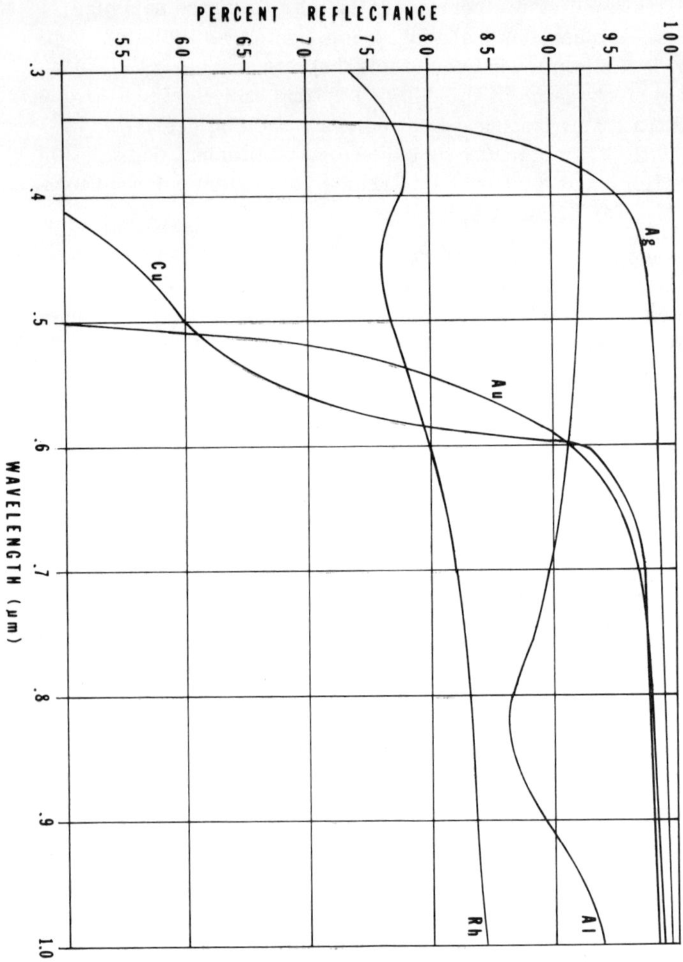

Fig. 68 Reflectance vs. wavelength for freshly evaporated mirror coatings.

Filters

Optical filters can be solid, liquid, plasma, or gaseous. The most common types are solid circular filters made by dissolving organic dyes in plastic or gelatin and coating transparent glasses or plastics, by dissolving metallic or organic dyes directly into glasses and plastics, or by cutting natural or synthetic crystals along various optical axes to obtain the desired filtering qualities.

The stability of optical filters is classified by the degradation rate of the filter color or spectral distribution. Class A is the most stable. Filters made by impregnating metallic compounds into glasses and natural and synthetic crystals almost all fall into this category. Classes B, C, and D (listed in order of increasing instability) cover the classification of most organic dyes impregnated in gelatin and various plastics. Some class D filters can have usable lifetimes in the order of seconds for certain high-intensity applications. Most optical filters follow the Bouger-Lambert law where the spectral transmittance of two or more optical filters (elements) is equivalent to the product of the spectral transmittance of each filter used, minus the losses incurred by multiple-lens (parallel surface) reflectance and air-layer interfaces.

Table 49 lists the spectral characteristics of some Kodak filters, listed by filter number. The wavelength vs. the percentage transmittance for each filter is given along with the various color coordinate information.

Optical transmittance (T) and optical density (D) are related by the equation

$$D = \log \frac{1}{T}$$

Table 50 lists the optical density vs. percent transmittance over a wide range of values. Optical densities greater than 4.0 are rarely encountered in electro-optical systems and are often difficult to measure, since the corresponding percent transmittance is less than 0.01%.

Many filters are listed by their mired shift value, using the mired system for light conversion. The mired value is defined by 10 $/T_c$, where T_c is the color temperature given in kelvin. The mired shift value is given by

$$\left(\frac{1}{T_{c2}} - \frac{1}{T_{c1}}\right) \times 10^6$$

where T_{c1} is the color temperature of the original light source and T_{c2} is the color temperature of the light through the homogeneous filter. Table 51 lists the mired value of various color temperatures in the range of 1100 to 9600 K.

Filters are often composed of multiple layers of materials of differing indexes of refraction using various layer thicknesses for each layer to control the desired filter characteristics. The common filter characteristics used in the design of multilayer filters include the transmission versus wavelength properties, the cutoff or cut-on peak wavelengths, material temperature stability, and the bandwidth of the central wavelength region. Typical construction involves the sequential disposition of high and low index layers (of $\lambda/4$ thickness) with an optical spacer (of $\lambda 2$ thickness) between each reflective layer unit. The number of layers in a given filter controls the bandwidth. The thickness of the spacer layer controls the

Table 49 Kodak Optical Filters[a, b]

KODAK COLOR-COMPENSATING FILTERS

	PERCENT TRANSMITTANCE						PERCENT TRANSMITTANCE					
WAVELENGTH	CC05R	CC10R	CC20R	CC30R	CC40R	CC50R	CC05B	CC10B	CC20B	CC30B	CC40B	CC50B
400	79.2	72.0	58.1	47.1	38.1	30.4	87.0	85.5	82.2	80.2	77.0	74.1
10	79.2	71.6	57.3	46.0	36.7	29.1	87.5	86.4	84.0	82.5	80.3	78.4
20	79.3	71.4	56.6	45.0	35.5	28.2	87.7	87.2	85.0	84.0	82.2	80.7
30	79.4	71.3	56.3	44.5	34.7	27.8	88.0	87.5	85.3	84.3	82.5	81.1
40	79.5	71.5	56.3	44.4	34.4	27.7	88.1	87.5	85.0	83.5	81.3	79.8
50	79.6	71.6	56.5	44.6	34.5	28.0	88.1	87.2	83.9	81.9	78.7	76.6
60	79.9	71.9	57.2	45.2	35.2	28.9	87.9	86.4	82.5	79.5	75.9	72.9
70	80.5	72.9	58.7	47.0	37.2	31.1	87.5	85.3	80.3	76.2	72.0	67.9
80	81.3	74.4	61.2	50.2	40.7	34.7	87.0	84.0	77.8	72.5	67.5	62.7
90	82.1	75.9	63.4	53.0	44.0	37.6	86.2	82.4	74.2	68.3	62.3	56.6
500	82.2	76.2	64.0	53.7	45.0	38.0	85.2	80.5	71.2	63.8	56.7	50.1
10	81.7	75.4	62.6	52.0	43.5	35.8	84.4	78.6	67.7	58.7	51.0	44.5
20	81.0	74.1	60.5	49.3	40.7	32.7	83.5	77.0	64.0	54.4	46.0	38.6
30	80.4	72.8	58.4	46.8	38.0	29.7	82.6	75.2	61.5	50.7	41.6	34.1
40	80.0	72.1	57.1	45.2	36.3	27.9	82.1	73.9	59.5	48.3	39.0	31.3
50	79.9	71.9	56.7	44.7	35.8	27.3	81.5	73.0	58.0	46.6	36.9	29.5
60	80.4	72.6	57.6	45.7	36.9	28.2	81.4	72.7	57.5	45.9	35.9	28.6
70	81.3	74.3	60.5	48.9	40.3	31.4	81.4	73.0	57.9	46.3	36.1	28.7
80	83.2	77.3	65.1	54.5	46.3	37.6	81.9	73.9	59.3	47.9	37.8	30.5
90	85.2	80.7	71.2	62.4	54.9	47.1	82.7	75.1	61.6	50.3	40.8	33.4
600	87.0	84.0	77.6	70.7	64.5	58.0	83.4	76.3	63.5	53.0	43.5	36.0
10	88.4	86.7	82.6	78.0	73.3	68.8	83.6	76.7	64.5	54.6	44.7	37.8
20	89.5	88.6	86.2	83.3	80.5	77.7	83.5	76.5	64.3	54.4	44.3	37.5
30	90.1	89.7	88.4	86.9	84.9	83.4	83.2	75.6	63.1	53.2	42.5	35.6
40	90.4	90.4	89.7	88.9	87.7	86.8	82.8	74.5	61.6	51.5	40.2	33.7
50	90.7	90.6	90.4	89.9	89.2	88.7	82.5	74.0	60.6	50.3	39.0	32.4
60	90.8	90.7	90.6	90.4	89.9	89.7	82.5	73.8	60.1	49.5	38.4	31.8
70	90.9	90.9	90.7	90.6	90.3	90.2	82.3	73.3	59.6	49.0	37.7	31.0
80	91.0	91.0	90.9	90.8	90.5	90.4	82.0	72.8	58.6	48.2	36.5	30.0
90	91.1	91.0	91.0	90.9	90.7	90.6	81.9	72.5	58.1	47.2	35.4	29.0
700	91.1	91.1	91.0	91.0	90.9	90.8	82.2	73.0	58.5	47.5	35.6	29.0
Dominant Wavelength (A)	610.0	607.0	611.0	613.5	613.5	615.5	478.9	460.0	443.0	448.0	463.0	461.0
Excitation Purity (%) (A)	4.0	8.0	15.0	21.5	29.0	34.5	1.0	2.5	5.0	8.5	12.0	16.0
Luminous Transmit. (%) (A)	84.6	78.7	68.1	59.6	51.7	46.1	81.1	75.0	61.5	50.9	41.3	34.1
x_A	0.4562	0.4649	0.4813	0.4965	0.5140	0.5290	0.4449	0.4404	0.4328	0.4235	0.4100	0.3995
y_A	0.4044	0.4021	0.3956	0.3895	0.3830	0.3764	0.4053	0.3985	0.3871	0.3751	0.3624	0.3481
Dominant Wavelength (C)	604.0	598.0	604.0	605.8	605.5	608.5	458.0	462.5	460.5	461.0	464.0	462.8
Excitation Purity (%) (C)	2.1	4.7	8.6	12.4	17.4	21.4	2.7	6.3	13.2	20.1	27.6	34.2
Luminous Transmit. (%) (C)	83.7	77.0	65.3	55.9	47.3	41.2	82.8	75.5	62.3	52.0	42.8	35.7
x_C	0.3170	0.3246	0.3385	0.3522	0.3695	0.3851	0.3057	0.2994	0.2881	0.2763	0.2626	0.2519
y_C	0.3171	0.3194	0.3199	0.3206	0.3222	0.3213	0.3087	0.2987	0.2789	0.2592	0.2393	0.2201
Dominant Wavelength (D_{65})	595.2	594.0	599.9	600.8	602.8	604.0	465.4	466.2	463.8	464.1	464.1	463.5
Excitation Purity (%) (D_{65})	2.4	5.3	10.0	13.4	19.6	23.8	2.2	5.6	12.8	19.5	26.7	33.2
Luminous Transmit. (%) (D_{65})	84.8	77.8	64.3	57.4	47.2	41.8	83.0	75.5	61.7	52.5	42.4	35.8
x_{D65}	0.3198	0.3277	0.3441	0.3554	0.3767	0.3915	0.3087	0.3025	0.2902	0.2784	0.2658	0.2548
y_{D65}	0.3307	0.3329	0.3334	0.3344	0.3352	0.3353	0.3227	0.3129	0.2916	0.2721	0.2511	0.2319

Table 49–Continued

KODAK COLOR-COMPENSATING FILTERS

WAVELENGTH	PERCENT TRANSMITTANCE					
	CC05G	CC10G	CC20G	CC30G	CC40G	CC50G
400	79.6	71.3	57.4	45.2	36.1	29.4
10	79.7	71.5	56.4	44.8	35.5	28.9
20	79.9	71.8	55.9	45.0	35.5	29.0
30	80.1	72.2	55.6	45.6	35.7	29.6
40	80.6	72.8	55.9	46.6	36.6	30.8
50	81.1	73.7	57.1	48.3	38.1	32.5
60	81.8	74.9	59.9	51.0	40.8	35.5
70	83.1	77.1	64.3	55.8	46.2	41.1
80	84.8	80.1	69.2	62.9	54.5	49.9
90	86.6	83.1	74.8	70.3	64.0	60.0
500	87.7	85.4	80.6	76.2	71.5	68.1
10	88.4	86.6	83.7	79.2	75.6	72.7
20	88.7	87.0	84.8	80.4	77.1	74.5
30	88.6	87.0	84.4	80.2	77.1	74.2
40	88.5	86.7	83.4	79.1	75.8	72.4
50	88.2	86.1	82.2	77.6	73.8	70.2
60	87.9	85.4	80.6	75.3	71.3	66.9
70	87.4	84.4	78.8	72.4	68.3	63.0
80	86.8	83.2	76.5	69.4	64.6	58.6
90	86.2	81.9	74.0	66.2	61.1	54.6
600	85.6	80.6	71.7	63.2	57.3	50.4
10	84.8	79.3	69.1	59.8	53.7	46.2
20	83.8	77.7	66.2	55.9	49.6	41.5
30	82.9	76.1	63.4	52.3	45.7	37.4
40	82.3	74.1	61.1	49.8	42.5	34.3
50	82.0	73.8	59.5	48.2	40.5	32.6
60	81.7	73.4	58.8	47.3	39.5	31.7
70	81.5	73.0	58.3	46.6	38.8	30.8
80	81.1	72.5	57.5	45.5	37.7	29.7
90	81.1	72.1	56.7	44.7	36.6	28.8
700	81.4	72.5	56.9	45.6	36.5	29.3
Dominant Wavelength (A)	554.6	559.0	559.0	558.0	558.5	557.0
Excitation Purity (%) (A)	2.5	5.5	10.5	15.5	21.0	24.5
Luminous Transmit. (%) (A)	86.8	83.7	76.3	70.2	65.3	60.5
x_A	0.4451	0.4434	0.4386	0.4333	0.4292	0.4232
y_A	0.4131	0.4193	0.4319	0.4439	0.4556	0.4670
Dominant Wavelength (C)	553.0	555.5	555.0	554.0	554.4	553.3
Excitation Purity (%) (C)	2.3	5.2	10.8	15.8	21.2	25.9
Luminous Transmit. (%) (C)	87.2	84.5	77.8	72.2	67.7	63.3
x_C	0.3105	0.3117	0.3131	0.3134	0.3151	0.3143
y_C	0.3245	0.3340	0.3533	0.3716	0.3900	0.4078
Dominant Wavelength (D65)	556.1	554.5	554.9	553.0	552.8	552.2
Excitation Purity (%) (D65)	3.1	5.2	11.7	16.3	22.2	27.5
Luminous Transmit. (%) (D65)	87.8	84.9	78.5	72.7	67.6	62.3
x_{D65}	0.3137	0.3138	0.3155	0.3144	0.3147	0.3138
y_{D65}	0.3391	0.3463	0.3678	0.3851	0.4054	0.4250

WAVELENGTH	PERCENT TRANSMITTANCE					
	CC05Y	CC10Y	CC20Y	CC30Y	CC40Y	CC50Y
400	78.8	72.3	58.2	47.1	38.3	31.3
10	78.6	71.6	56.8	45.3	36.3	29.2
20	78.6	71.4	55.9	44.3	35.3	27.9
30	78.8	71.6	55.7	44.2	35.0	27.5
40	79.2	72.4	56.3	44.9	35.7	27.9
50	79.7	73.3	57.4	45.3	37.3	29.2
60	80.6	74.7	59.7	49.0	40.2	31.8
70	82.2	77.2	63.9	54.1	45.9	37.2
80	84.4	80.4	70.6	62.2	55.6	47.1
90	86.6	84.1	77.4	71.5	66.6	59.7
500	88.1	86.8	82.9	79.4	76.6	71.8
10	89.1	88.4	86.3	84.3	82.6	80.0
20	89.6	89.3	88.1	86.9	85.9	84.4
30	89.9	89.9	89.0	88.3	87.5	86.7
40	90.2	90.1	89.6	89.0	88.4	87.9
50	90.4	90.3	89.9	89.5	89.0	88.6
60	90.6	90.5	90.0	89.8	89.4	89.2
70	90.7	90.6	90.4	90.0	89.7	89.6
80	90.8	90.7	90.5	90.3	89.9	89.9
90	90.9	90.8	90.6	90.5	90.1	90.0
600	90.9	90.8	90.6	90.6	90.3	90.2
10	91.0	90.9	90.7	90.6	90.4	90.4
20	91.0	90.9	90.8	90.7	90.5	90.5
30	91.0	91.0	90.8	90.7	90.5	90.5
40	91.1	91.0	90.9	90.8	90.6	90.6
50	91.1	91.0	90.9	90.9	90.7	90.7
60	91.2	91.1	91.0	91.0	90.8	90.8
70	91.2	91.1	91.0	91.0	90.9	90.9
80	91.3	91.2	91.1	91.1	91.0	91.0
90	91.3	91.2	91.2	91.1	91.1	91.1
700	91.4	91.3	91.3	91.2	91.1	91.1
Dominant Wavelength (A)	581.6	581.4	581.0	580.8	580.7	580.7
Excitation Purity (%) (A)	7.5	13.0	23.5	34.5	42.0	48.0
Luminous Transmit. (%) (A)	90.9	90.7	90.3	89.8	89.4	89.1
x_A	0.4531	0.4567	0.4643	0.4716	0.4767	0.4808
y_A	0.4125	0.4166	0.4249	0.4332	0.4388	0.4432
Dominant Wavelength (C)	572.0	571.5	571.2	571.2	571.2	571.2
Excitation Purity (%) (C)	5.4	9.5	18.8	28.4	35.6	42.0
Luminous Transmit. (%) (C)	90.4	90.1	89.1	88.2	87.4	86.9
x_C	0.3181	0.3239	0.3369	0.3506	0.3610	0.3699
y_C	0.3283	0.3382	0.3597	0.3816	0.3985	0.4131
Dominant Wavelength (D65)	570.1	570.6	570.4	570.4	570.5	570.5
Excitation Purity (%) (D65)	5.9	11.0	20.6	29.2	37.5	46.6
Luminous Transmit. (%) (D65)	91.4	90.5	89.9	89.2	88.6	88.0
x_{D65}	0.3204	0.3276	0.3403	0.3518	0.3632	0.3757
y_{D65}	0.3423	0.3535	0.3749	0.3940	0.4125	0.4325

Table 49–Continued

KODAK COLOR-COMPENSATING FILTERS

WAVELENGTH	CYAN SERIES PERCENT TRANSMITTANCE						CYAN-2 SERIES PERCENT TRANSMITTANCE					
	CC05C	CC10C	CC20C	CC30C	CC40C	CC50C	CC05C-2	CC10C-2	CC20C-2	CC30C-2	CC40C-2	CC50C-2
400	87.3	86.0	83.9	82.3	80.4	78.8	86.0	85.3	84.9	83.7	82.5	81.3
10	88.2	87.5	85.2	84.5	83.4	82.7	87.1	86.6	86.2	85.3	84.3	83.5
20	88.7	88.1	86.5	86.0	85.3	84.8	87.8	87.4	87.2	86.4	85.7	85.1
30	89.0	88.6	87.5	87.0	86.3	85.9	88.3	88.0	87.8	87.3	86.6	86.1
40	89.3	89.0	87.7	87.3	86.6	86.1	88.6	88.3	88.2	87.7	87.1	86.6
50	89.5	89.1	87.8	87.5	86.6	86.0	88.9	88.7	88.4	87.9	87.3	86.8
60	89.6	89.1	87.7	87.3	86.4	85.7	89.1	88.8	88.5	87.9	87.3	86.7
70	89.7	89.0	87.5	87.0	85.8	85.2	89.3	88.9	88.5	87.8	87.2	86.5
80	89.7	89.0	87.2	86.5	85.3	84.3	89.4	89.0	88.4	87.7	87.0	86.1
90	89.7	89.0	87.0	86.0	84.4	83.4	89.4	89.0	88.3	87.5	86.6	85.6
500	89.6	89.0	86.5	85.2	83.5	82.3	89.5	89.0	88.2	87.3	86.3	85.1
10	89.6	88.7	86.0	84.4	82.4	80.8	89.5	89.0	88.1	87.0	85.9	84.6
20	89.5	88.5	85.2	83.5	81.1	79.2	89.5	89.0	87.9	86.7	85.4	84.0
30	89.4	88.0	84.3	82.4	79.6	77.3	89.5	89.0	87.7	86.3	85.0	83.4
40	89.2	87.5	83.4	81.0	77.7	75.0	89.5	88.9	87.4	85.9	84.4	82.7
50	88.9	87.0	82.3	79.0	75.3	72.2	89.4	88.8	87.0	85.3	83.6	81.6
60	88.5	86.1	80.5	76.7	72.7	69.0	89.2	88.5	86.3	84.3	82.3	79.9
70	88.0	85.0	78.5	74.0	69.3	65.0	88.9	88.0	85.3	82.9	80.4	77.8
80	87.5	83.8	76.1	70.9	65.4	60.5	88.6	87.4	84.2	81.2	78.3	75.2
90	87.0	82.5	73.9	67.5	61.6	55.8	88.2	86.8	82.8	79.3	75.8	72.2
600	86.4	81.0	71.2	64.1	57.6	51.3	87.7	85.9	81.2	76.8	72.7	68.5
10	85.5	79.5	68.5	60.4	53.4	46.2	87.0	84.8	79.1	73.9	68.8	64.1
20	84.5	77.9	65.5	56.7	49.2	42.3	86.1	83.6	76.8	70.7	64.9	59.3
30	83.8	76.3	62.7	53.1	45.0	38.0	85.2	82.2	74.4	67.3	60.9	54.8
40	83.3	75.1	60.8	50.4	42.0	34.9	84.3	80.8	71.8	64.0	56.8	50.2
50	82.8	74.4	59.5	48.8	40.2	32.9	83.2	79.2	69.0	60.2	52.5	45.3
60	82.5	74.0	58.5	48.0	39.4	32.0	81.9	77.3	65.7	55.7	47.5	39.8
70	82.4	73.6	57.9	47.2	38.6	31.0	80.6	75.3	62.3	51.3	42.7	34.8
80	82.0	73.0	57.5	46.0	37.5	29.9	79.5	73.8	59.7	48.1	39.1	31.2
90	82.0	72.8	57.4	45.5	36.7	29.1	78.9	72.9	58.2	46.3	37.2	29.3
700	82.5	74.0	58.5	46.4	37.3	29.8	78.7	72.4	57.4	45.6	36.4	28.5
10	83.5	75.3	60.5	48.5	39.9	33.2	78.0	72.2	56.3	44.0	35.2	27.1
20	84.4	76.3	61.5	49.6	41.0	34.3	77.4	71.7	55.1	41.8	33.2	25.2
30	86.7	80.8	70.0	61.3	55.0	50.1	76.7	70.9	53.7	39.2	30.8	23.0
40	88.2	83.8	75.9	69.4	64.8	61.2	76.0	69.5	52.0	37.4	29.1	21.4
50	89.4	86.5	81.3	77.1	74.1	71.7	75.7	68.3	51.0	36.4	28.3	20.5
60	90.2	88.4	85.2	82.5	80.6	79.1	76.3	69.7	52.6	38.0	29.8	21.7
70	90.8	89.7	87.8	86.2	85.0	84.1	78.7	72.5	56.5	44.1	35.4	24.8
80	91.3	90.7	89.7	88.8	88.2	87.7	81.7	77.2	65.2	54.1	46.3	34.2
90	91.5	91.3	90.7	90.3	90.0	89.8	84.7	82.1	74.0	64.1	58.3	50.0
800	91.6	91.5	91.3	91.1	91.0	90.9	87.3	85.6	80.0	74.1	70.1	64.6
Dominant Wavelength (A)	499.7	498.6	497.6	497.4	496.9	496.7	500.0	500.0	500.0	499.5	499.0	498.0
Excitation Purity (%) (A)	1.0	3.0	6.0	9.0	12.0	14.5	0.9	1.5	3.4	5.1	6.7	8.5
Luminous Transmit. (%) (A)	87.4	83.8	76.4	71.4	66.4	61.9	88.2	86.8	83.3	79.9	76.9	73.6
x_A	0.4422	0.4350	0.4211	0.4084	0.3970	0.3846	0.4436	0.4409	0.4327	0.4253	0.4182	0.4106
y_A	0.4090	0.4102	0.4117	0.4133	0.4137	0.4141	0.4094	0.4102	0.4120	0.4136	0.4151	0.4162
Dominant Wavelength (C)	489.0	487.5	486.4	486.3	485.8	485.6	494.0	492.0	489.0	488.5	488.0	488.0
Excitation Purity (%) (C)	1.7	4.1	8.8	12.8	16.4	20.2	1.0	1.8	4.3	6.6	8.8	11.0
Luminous Transmit. (%) (C)	88.0	85.1	78.9	74.8	70.5	66.7	88.6	87.6	84.8	82.2	79.7	77.1
x_C	0.3058	0.2997	0.2882	0.2786	0.2699	0.2609	0.3072	0.3051	0.2988	0.2930	0.2877	0.2820
y_C	0.3156	0.3135	0.3084	0.3044	0.2997	0.2950	0.3172	0.3166	0.3146	0.3130	0.3111	0.3092
Dominant Wavelength (D65)	494.0	490.4	488.6	487.0	487.2	486.6	495.9	491.2	489.8	489.4	488.7	488.7
Excitation Purity (%) (D65)	1.7	3.7	8.1	13.0	16.4	19.6	1.4	2.6	6.0	8.5	11.6	13.5
Luminous Transmit. (%) (D65)	88.0	85.3	79.5	74.7	69.9	67.4	88.4	87.0	83.5	79.5	76.2	73.7
x_{D65}	0.3079	0.3028	0.2914	0.2797	0.2709	0.2635	0.3085	0.3055	0.2965	0.2901	0.2824	0.2774
y_{D65}	0.3300	0.3281	0.3239	0.3169	0.3144	0.3093	0.3305	0.3289	0.3268	0.3251	0.3220	0.3209

Table 49–Continued

KODAK COLOR-COMPENSATING FILTERS

WAVELENGTH	CC05M	CC10M	CC20M	CC30M	CC40M	CC50M
	PERCENT TRANSMITTANCE					
400	87.6	86.6	85.6	84.2	82.3	80.9
10	88.2	87.7	86.6	85.7	84.6	83.6
20	88.6	88.0	87.0	85.9	85.2	84.4
30	88.7	88.0	86.9	85.6	84.4	83.6
40	88.7	87.9	86.0	84.7	82.5	81.4
50	88.6	87.5	84.9	82.8	80.0	78.1
60	88.4	86.5	83.1	80.0	76.1	73.7
70	87.8	85.2	80.8	76.4	71.3	68.0
80	87.0	83.6	77.9	72.1	65.8	61.7
90	86.0	81.8	74.4	67.0	60.0	55.0
500	85.0	79.7	70.5	61.7	53.7	48.1
10	83.8	77.5	66.7	56.5	47.7	41.6
20	82.7	75.3	63.4	52.0	42.8	36.3
30	81.8	73.7	60.5	48.6	39.0	31.9
40	81.3	72.5	58.6	46.6	36.7	29.8
50	81.2	72.2	58.0	46.0	36.0	29.1
60	81.5	72.8	58.3	46.5	36.7	29.7
70	82.5	74.6	60.5	49.8	40.2	32.3
80	84.0	77.3	64.9	55.6	46.2	39.0
90	85.8	80.8	70.6	63.3	54.9	48.7
600	88.0	84.5	77.1	71.6	64.9	59.9
10	89.3	87.0	82.2	79.2	74.9	70.7
20	90.2	88.9	86.1	84.1	81.4	79.2
30	90.6	90.0	88.7	87.4	86.0	84.5
40	90.8	90.5	90.0	89.3	88.7	87.6
50	91.0	90.7	90.5	90.2	90.0	89.7
60	91.1	91.0	90.8	90.8	90.4	90.4
70	91.2	91.2	91.0	91.0	90.7	90.7
80	91.3	91.3	91.2	91.1	91.0	91.0
90	91.4	91.4	91.4	91.3	91.3	91.3
700	91.5	91.5	91.5	91.5	91.5	91.5
Dominant Wavelength (A)	535.4c	543.0c	549.0c	547.5c	548.0c	549.0c
Excitation Purity (%) (A)	3.5	7.0	13.0	19.5	25.5	31.0
Luminous Transmit. (%) (A)	84.9	79.1	69.0	60.8	53.2	47.5
x_A	0.4524	0.4563	0.4628	0.4706	0.4779	0.4833
y_A	0.4004	0.3929	0.3787	0.3647	0.3509	0.3387
Dominant Wavelength (C)	541.0c	547.5c	551.2c	550.0c	550.3c	551.2c
Excitation Purity (%) (C)	3.5	7.4	14.5	21.4	28.2	33.9
Luminous Transmit. (%) (C)	84.2	77.9	67.1	58.0	49.9	44.0
x_C	0.3114	0.3114	0.3104	0.3112	0.3112	0.3101
y_C	0.3082	0.2986	0.2808	0.2634	0.2467	0.2320
Dominant Wavelength (D65)	541.5c	545.1c	549.0c	549.0c	548.4c	549.8c
Excitation Purity (%) (D65)	3.7	7.0	15.1	21.2	27.7	33.9
Luminous Transmit. (%) (D65)	84.2	78.0	66.4	58.0	49.8	43.9
$x_{(D65)}$	0.3142	0.3147	0.3145	0.3153	0.3169	0.3157
$y_{(D65)}$	0.3202	0.3118	0.2908	0.2752	0.2590	0.2425

Table 49–Continued

WAVELENGTH	PERCENT TRANSMITTANCE No. 0	No. 1A	No. 2A
400	88.0	59.0	—
10	88.5	76.0	4.0
20	88.9	82.0	42.0
30	89.3	84.6	74.0
40	89.6	86.0	82.7
50	89.8	86.8	85.6
60	89.9	87.2	87.0
70	90.1	87.5	88.1
80	90.3	87.3	88.8
90	90.4	86.8	89.4
500	90.5	86.3	89.7
10	90.6	85.5	90.0
20	90.7	84.8	90.2
30	90.7	84.3	90.3
40	90.8	84.0	90.4
50	90.8	83.9	90.5
60	90.9	84.1	90.6
70	90.9	84.8	90.6
80	90.9	86.0	90.7
90	91.0	87.4	90.7
600	91.0	88.5	90.8
10	91.0	89.5	90.8
20	91.0	90.2	90.9
30	91.0	90.6	90.9
40	91.1	90.8	90.9
50	91.1	91.0	91.0
60	91.1	91.1	91.0
70	91.1	91.1	91.0
80	91.1	91.1	91.1
90	91.1	91.1	91.1
700	91.1	91.1	91.1
Dominant Wavelength (A)	578.5	507.5c	579.5
Excitation Purity (%) (A)	1.5	2.0	7.0
Luminous Transmit. (%) (A)	90.9	86.5	90.5
x_A	0.4484	0.4524	0.4519
y_A	0.4080	0.4042	0.4131
Dominant Wavelength (C)	573.5	497.6c	568.6
Excitation Purity (%) (C)	0.7	1.2	6.0
Luminous Transmit. (%) (C)	90.8	85.9	90.3
x_C	0.3112	0.3132	0.3177
y_C	0.3177	0.3147	0.3310
Dominant Wavelength (D65)	571.4	498.3c	568.5
Excitation Purity (%) (D65)	1.0	1.6	5.1
Luminous Transmit. (%) (D65)	90.7	86.4	90.6
x_{D65}	0.3140	0.3167	0.3188
y_{D65}	0.3311	0.3268	0.3410

WAVELENGTH	PERCENT TRANSMITTANCE No. 2B	No. 2C	No. 2E
400	19.0	40.5	—
10	48.0	63.7	—
20	67.0	75.2	8.7
30	75.3	80.4	51.1
40	80.0	83.4	75.8
50	83.0	85.3	82.2
60	85.2	86.8	84.8
70	86.7	88.0	86.4
80	88.1	88.8	87.6
90	88.8	89.4	88.4
500	89.5	89.8	89.0
10	89.9	90.1	89.4
20	90.3	90.3	89.7
30	90.5	90.4	89.9
40	90.6	90.5	90.1
50	90.7	90.6	90.2
60	90.8	90.6	90.4
70	90.9	90.6	90.5
80	90.9	90.7	90.6
90	91.0	90.8	90.6
600	91.1	90.8	90.7
10	91.2	90.9	90.8
20	91.3	90.9	90.9
30	91.3	91.0	90.9
40	91.4	91.0	91.0
50	91.4	91.0	91.0
60	91.5	91.1	91.1
70	91.5	91.1	91.2
80	91.6	91.2	91.3
90	91.7	91.3	91.4
700	91.8	91.4	91.4
Dominant Wavelength (A)	581.0	581.0	580.2
Excitation Purity (%) (A)	7.0	5.0	12.0
Luminous Transmit. (%) (A)	90.8	90.8	90.3
x_A	0.4526	0.4512	0.4557
y_A	0.4126	0.4113	0.4169
Dominant Wavelength (C)	569.8	570.2	569.0
Excitation Purity (%) (C)	5.6	4.1	10.7
Luminous Transmit. (%) (C)	90.5	90.5	89.9
x_C	0.3178	0.3158	0.3238
y_C	0.3298	0.3259	0.3426
Dominant Wavelength (D65)	569.2	569.6	568.5
Excitation Purity (%) (D65)	5.9	4.2	9.3
Luminous Transmit. (%) (D65)	90.1	90.3	90.0
x_{D65}	0.3201	0.3181	0.3239
y_{D65}	0.3427	0.3387	0.3511

WAVELENGTH	PERCENT TRANSMITTANCE No. 3	No. 3N5	No. 4
400	—	—	—
10	—	—	—
20	—	—	—
30	—	—	—
40	1.3	—	—
50	11.1	1.59	—
60	39.3	9.40	6.9
70	69.0	18.5	42.0
80	83.2	23.5	74.0
90	87.6	25.5	84.7
500	88.8	26.3	87.5
10	89.4	26.7	88.5
20	89.6	27.0	89.1
30	89.8	27.2	89.4
40	90.0	27.5	89.6
50	90.1	27.8	89.8
60	90.3	27.9	90.0
70	90.4	28.0	90.2
80	90.5	28.4	90.4
90	90.5	29.0	90.6
600	90.5	29.5	90.8
10	90.6	29.5	90.9
20	90.6	29.3	91.0
30	90.7	29.1	91.1
40	90.7	29.0	91.2
50	90.8	29.4	91.3
60	90.8	29.6	91.4
70	90.9	29.8	91.5
80	90.9	30.0	91.5
90	91.0	30.2	91.6
700	91.0	31.0	91.6
Dominant Wavelength (A)	580.0	581.2	580.0
Excitation Purity (%) (A)	50.0	52.0	64.0
Luminous Transmit. (%) (A)	89.7	28.1	89.5
x_A	0.4803	0.4851	0.4887
y_A	0.4470	0.4447	0.4579
Dominant Wavelength (C)	569.4	570.5	569.6
Excitation Purity (%) (C)	50.3	56.3	65.1
Luminous Transmit. (%) (C)	88.3	27.4	87.7
x_C	0.3752	0.3873	0.3955
y_C	0.4382	0.4485	0.4734
Dominant Wavelength (D65)	569.0	570.0	569.1
Excitation Purity (%) (D65)	52.7	53.4	64.8
Luminous Transmit. (%) (D65)	87.1	29.1	87.4
x_{D65}	0.3781	0.3827	0.3935
y_{D65}	0.4516	0.4498	0.4795

Table 49–Continued

WAVELENGTH	PERCENT TRANSMITTANCE No. 6	No. 8	No. 8N5
400	7.40	–	–
10	8.32	–	–
20	10.4	–	–
30	13.5	–	–
40	18.9	–	–
50	27.6	–	–
60	39.0	0.25	0.16
70	52.3	5.50	2.0
80	65.8	19.0	6.3
90	76.8	41.0	13.2
500	83.5	63.5	20.0
10	87.0	78.0	24.3
20	88.4	84.1	26.7
30	89.0	86.5	28.0
40	89.4	87.7	28.6
50	89.7	88.4	29.0
60	89.9	88.8	29.3
70	90.1	89.2	29.5
80	90.3	89.5	29.6
90	90.5	89.8	29.8
600	90.6	90.1	29.9
10	90.7	90.3	29.6
20	90.8	90.5	29.4
30	90.9	90.7	29.1
40	91.0	90.9	28.8
50	91.1	91.0	28.9
60	91.2	91.1	29.2
70	91.2	91.2	29.4
80	91.3	91.3	29.5
90	91.4	91.4	29.7
700	91.5	91.5	30.2
Dominant Wavelength (A)	580.3	581.2	581.1
Excitation Purity (%) (A)	47.5	83.5	84.0
Luminous Transmit. (%) (A)	89.2	86.6	28.3
x_A	0.4796	0.5078	0.5077
y_A	0.4444	0.4670	0.4673
Dominant Wavelength (C)	570.2	571.8	572.0
Excitation Purity (%) (C)	44.7	85.2	85.2
Luminous Transmit. (%) (C)	87.5	82.7	27.0
x_C	0.3704	0.4353	0.4362
y_C	0.4225	0.5088	0.5079
Dominant Wavelength (D65)	569.4	571.1	571.2
Excitation Purity (%) (D65)	48.1	84.7	82.5
Luminous Transmit. (%) (D65)	88.3	82.6	28.0
x_{D65}	0.3739	0.4306	0.4280
y_{D65}	0.4396	0.5134	0.5083

WAVELENGTH	PERCENT TRANSMITTANCE No. 9	No. 11	No. 12
400	–	–	–
10	–	–	–
20	–	0.16	–
30	–	0.29	–
40	–	0.56	–
50	–	1.32	–
60	0.2	4.00	–
70	1.8	12.0	–
80	7.6	26.0	–
90	20.0	43.7	–
500	34.7	55.0	1.5
10	49.1	60.0	17.3
20	64.8	60.2	55.0
30	78.0	57.8	77.8
40	85.3	54.2	86.0
50	88.3	50.0	88.4
60	89.4	44.8	89.4
70	89.7	38.9	89.7
80	90.0	33.1	90.1
90	90.1	27.6	90.3
600	90.2	22.7	90.4
10	90.2	19.0	90.5
20	90.3	14.9	90.7
30	90.4	11.4	90.8
40	90.5	9.10	90.9
50	90.5	8.05	91.0
60	90.6	7.50	91.1
70	90.6	7.05	91.2
80	90.7	6.50	91.2
90	90.7	6.10	91.2
700	90.8	6.20	91.3
Dominant Wavelength (A)	583.0	552.5	583.6
Excitation Purity (%) (A)	90.4	50.0	96.5
Luminous Transmit. (%) (A)	88.4	36.3	81.2
x_A	0.5211	0.3833	0.5331
y_A	0.4632	0.5421	0.4610
Dominant Wavelength (C)	574.3	550.3	576.3
Excitation Purity (%) (C)	91.0	60.7	97.7
Luminous Transmit. (%) (C)	77.6	40.2	73.8
x_C	0.4590	0.3063	0.4825
y_C	0.5067	0.5435	0.5083
Dominant Wavelength (D65)	573.7	550.1	575.2
Excitation Purity (%) (D65)	91.1	58.4	97.0
Luminous Transmit. (%) (D65)	76.4	41.4	75.2
x_{D65}	0.4557	0.3067	0.4752
y_{D65}	0.5112	0.5406	0.5128

WAVELENGTH	PERCENT TRANSMITTANCE No. 13	No. 15	No. 16
400	–	–	–
10	–	–	–
20	–	–	–
30	0.18	–	–
40	0.50	–	–
50	1.35	–	–
60	4.08	–	–
70	11.0	–	–
80	23.5	–	–
90	39.0	–	–
500	50.8	–	–
10	55.2	1.0	–
20	56.5	19.4	3.0
30	55.0	56.2	22.0
40	51.0	77.6	48.0
50	46.0	85.6	69.5
60	39.2	88.2	79.5
70	32.0	89.3	84.0
80	25.1	89.8	86.3
90	18.2	90.1	87.8
600	13.5	90.4	89.0
10	9.60	90.5	89.6
20	6.40	90.6	90.0
30	3.66	90.7	90.2
40	2.20	90.8	90.3
50	1.58	90.9	90.4
60	1.74	91.0	90.5
70	2.62	91.1	90.6
80	3.55	91.1	90.7
90	4.48	91.1	90.8
700	5.25	91.1	90.8
Dominant Wavelength (A)	537.1	585.9	588.7
Excitation Purity (%) (A)	46.0	98.0	99.0
Luminous Transmit. (%) (A)	30.0	75.5	67.9
x_A	0.3381	0.5485	0.5667
y_A	0.5741	0.4481	0.4314
Dominant Wavelength (C)	542.0	579.2	582.7
Excitation Purity (%) (C)	57.6	99.0	99.4
Luminous Transmit. (%) (C)	34.5	66.6	57.7
x_C	0.2723	0.5048	0.5289
y_C	0.5618	0.4905	0.4684
Dominant Wavelength (D65)	543.7	578.5	581.7
Excitation Purity (%) (D65)	57.0	98.3	99.0
Luminous Transmit. (%) (D65)	37.8	66.7	59.2
x_{D65}	0.2806	0.4989	0.5214
y_{D65}	0.5587	0.4941	0.4741

Table 49–Continued

WAVELENGTH	PERCENT TRANSMITTANCE No. 18A	No. 21	No. 22
400	–	–	–
10	–	–	–
20	–	–	–
30	–	–	–
40	–	–	–
50	–	–	–
60	–	–	–
70	–	–	–
80	–	–	–
90	–	–	–
500	–	–	–
10	–	–	–
20	–	–	–
30	–	–	–
40	–	2.5	–
50	–	29.0	0.25
60	–	65.0	19.0
70	–	80.6	60.0
80	–	85.4	81.0
90	–	87.3	87.0
600	–	88.1	88.5
10	–	88.7	89.0
20	–	89.0	89.5
30	–	89.5	89.8
40	–	89.9	90.0
50	–	90.2	90.1
60	–	90.4	90.2
70	–	90.5	90.3
80	–	90.5	90.4
90	–	90.6	90.5
700	0.36	90.6	90.6
Dominant Wavelength (A)		593.7	598.8
Excitation Purity (%) (A)		100.0	100.0
Luminous Transmit. (%) (A)		57.4	48.2
x_A		0.5955	0.6214
y_A		0.4037	0.3780
Dominant Wavelength (C)		588.9	595.1
Excitation Purity (%) (C)		99.9	99.9
Luminous Transmit. (%) (C)		45.6	35.8
x_C		0.5686	0.6030
y_C		0.4304	0.3964
Dominant Wavelength (D65)		588.4	595.0
Excitation Purity (%) (D65)		98.9	98.6
Luminous Transmit. (%) (D65)		45.6	34.3
x_{D65}		0.5625	0.5988
y_{D65}		0.4327	0.3955

WAVELENGTH	PERCENT TRANSMITTANCE No. 23A	No. 24	No. 25
400	–	–	–
10	–	–	–
20	–	–	–
30	–	–	–
40	–	–	–
50	–	–	–
60	–	–	–
70	–	–	–
80	–	–	–
90	–	–	–
500	–	–	–
10	–	–	–
20	–	–	–
30	–	–	–
40	–	–	–
50	–	–	–
60	–	–	–
70	11.0	–	–
80	47.0	4.55	–
90	69.6	37.3	12.6
600	82.7	72.3	50.0
10	85.8	82.9	75.0
20	87.2	86.4	82.6
30	87.9	87.8	85.5
40	88.5	88.5	86.7
50	89.0	89.0	87.6
60	89.4	89.3	88.2
70	89.6	89.7	88.5
80	89.8	89.9	89.0
90	90.0	90.2	89.3
700	90.2	90.3	89.5
Dominant Wavelength (A)	605.5	612.5	617.2
Excitation Purity (%) (A)	100.0	100.0	100.0
Luminous Transmit. (%) (A)	36.3	27.5	22.5
x_A	0.6498	0.6735	0.6850
y_A	0.3498	0.3263	0.3147
Dominant Wavelength (C)	602.6	610.6	615.3
Excitation Purity (%) (C)	99.9	99.9	100.0
Luminous Transmit. (%) (C)	25.0	17.7	14.0
x_C	0.6386	0.6675	0.6808
y_C	0.3610	0.3322	0.3190
Dominant Wavelength (D65)	602.7	611.1	615.4
Excitation Purity (%) (D65)	99.5	98.8	98.7
Luminous Transmit. (%) (D65)	24.6	17.3	13.4
x_{D65}	0.6371	0.6650	0.6762
y_{D65}	0.3607	0.3305	0.3188

WAVELENGTH	PERCENT TRANSMITTANCE No. 26	No. 29	No. 30
400	–	–	50.5
10	–	–	48.8
20	–	–	49.2
30	–	–	50.6
40	–	–	49.7
50	–	–	45.2
60	–	–	26.8
70	–	–	12.6
80	–	–	4.73
90	–	–	0.62
500	–	–	–
10	–	–	–
20	–	–	–
30	–	–	–
40	–	–	–
50	–	–	–
60	–	–	–
70	–	–	15.0
80	–	–	53.8
90	2.9	–	79.5
600	30.0	–	87.2
10	63.2	10.5	89.2
20	78.9	45.0	89.9
30	84.0	73.5	90.3
40	86.1	84.2	90.5
50	87.2	87.8	90.6
60	88.1	89.2	90.8
70	88.5	89.8	90.9
80	88.9	90.3	91.0
90	89.2	90.4	91.0
700	89.5	90.5	91.1
Dominant Wavelength (A)	620.6	632.7	505.5c
Excitation Purity (%) (A)	100.0	100.0	58.5
Luminous Transmit. (%) (A)	19.1	11.0	37.8
x_A	0.6926	0.7116	0.6023
y_A	0.3073	0.2883	0.3175
Dominant Wavelength (C)	618.9	631.6	498.6c
Excitation Purity (%) (C)	100.0	100.0	62.5
Luminous Transmit. (%) (C)	11.6	6.3	26.6
x_C	0.6894	0.7102	0.4520
y_C	0.3105	0.2898	0.2270
Dominant Wavelength (D65)	618.1	628.7	498.3c
Excitation Purity (%) (D65)	98.4	96.0	56.1
Luminous Transmit. (%) (D65)	12.0	6.6	29.8
x_{D65}	0.6815	0.6904	0.4561
y_{D65}	0.3126	0.2951	0.2500

Table 49–Continued

WAVELENGTH	PERCENT TRANSMITTANCE No. 31	No. 32	No. 33
400	13.8	38.0	0.85
10	14.5	37.9	0.71
20	16.4	40.0	1.17
30	25.5	43.0	1.69
40	42.7	55.5	5.36
50	50.2	66.0	14.3
60	40.4	66.0	12.4
70	22.6	57.0	5.00
80	8.20	40.0	0.50
90	1.85	21.0	–
500	0.12	9.56	–
10	–	2.51	–
20	–	0.13	–
30	–	–	–
40	–	–	–
50	–	–	–
60	–	–	–
70	–	–	–
80	–	–	–
90	0.63	–	–
600	26.0	6.04	–
10	67.2	41.0	0.80
20	84.0	75.0	24.9
30	88.1	86.1	60.8
40	89.8	89.0	78.0
50	90.2	90.0	85.0
60	90.4	90.6	87.5
70	90.5	90.7	88.7
80	90.7	90.8	89.4
90	90.8	90.9	89.8
700	91.0	91.0	90.0
Dominant Wavelength (A)	512.6c	546.8c	506.6c
Excitation Purity (%) (A)	80.0	77.0	90.0
Luminous Transmit. (%) (A)	19.8	16.9	8.9
x_A	0.6095	0.5410	0.6733
y_A	0.2623	0.2281	0.2641
Dominant Wavelength (C)	513.2c	551.8c	498.0c
Excitation Purity (%) (C)	81.8	79.5	88.3
Luminous Transmit. (%) (C)	12.9	12.5	5.2
x_C	0.4044	0.3079	0.5198
y_C	0.1572	0.1179	0.1947
Dominant Wavelength (D65)	517.0c	549.6c	498.7c
Excitation Purity (%) (D65)	77.5	74.6	84.6
Luminous Transmit. (%) (D65)	13.7	14.4	5.4
x_{D65}	0.3977	0.3197	0.5211
y_{D65}	0.1674	0.1387	0.2061

WAVELENGTH	PERCENT TRANSMITTANCE No. 34	No. 34A	No. 35
400	64.0	–	48.0
10	70.1	4.5	57.0
20	72.0	40.0	57.6
30	68.4	69.7	47.5
40	58.2	68.7	29.5
50	42.3	56.2	12.3
60	25.2	40.5	3.5
70	12.1	23.8	0.25
80	2.7	9.2	–
90	0.2	2.3	–
500	–	0.33	–
10	–	–	–
20	–	–	–
30	–	–	–
40	–	–	–
50	–	–	–
60	–	–	–
70	–	–	–
80	–	–	–
90	–	–	–
600	–	–	–
10	–	0.13	–
20	–	1.0	–
30	–	6.3	–
40	0.4	22.0	–
50	4.0	45.0	0.1
60	20.7	65.0	3.0
70	45.2	77.3	19.0
80	66.5	85.0	43.5
90	78.8	88.2	66.0
700	85.0	89.8	77.7
Dominant Wavelength (A)	577.5	574.0c	577.2c
Excitation Purity (%) (A)	94.0	91.5	97.0
Luminous Transmit. (%) (A)	1.12	3.5	0.47
x_A	0.2764	0.3917	0.2863
y_A	0.0692	0.1286	0.0650
Dominant Wavelength (C)	427.0	565.6c	566.9c
Excitation Purity (%) (C)	94.7	90.6	96.6
Luminous Transmit. (%) (C)	1.24	2.9	0.43
x_C	0.1772	0.2060	0.1813
y_C	0.0230	0.0432	0.0168
Dominant Wavelength (D65)	421.1	565.2c	565.7c
Excitation Purity (%) (D65)	92.5	89.3	94.3
Luminous Transmit. (%) (D65)	1.6	3.0	0.6
x_{D65}	0.1817	0.2073	0.1938
y_{D65}	0.0293	0.0482	0.0291

WAVELENGTH	PERCENT TRANSMITTANCE No. 36	No. 38	No. 38A
400	36.5	60.5	33.9
10	45.5	66.5	43.5
20	45.5	72.5	53.1
30	32.7	75.3	57.5
40	15.2	76.2	58.1
50	3.7	75.9	56.7
60	0.35	74.8	54.2
70	–	73.4	51.0
80	–	71.6	47.5
90	–	69.5	43.8
500	–	66.7	39.7
10	–	63.9	35.7
20	–	60.8	31.8
30	–	57.0	27.7
40	–	52.6	22.8
50	–	48.0	18.2
60	–	42.8	13.7
70	–	37.0	9.6
80	–	30.6	6.3
90	–	25.5	3.9
600	–	20.9	2.6
10	–	16.8	1.6
20	–	12.9	1.0
30	–	10.0	0.5
40	–	7.79	0.2
50	–	6.68	–
60	0.21	6.20	–
70	7.5	5.91	–
80	29.0	5.41	–
90	55.0	4.90	–
700	71.3	5.00	–
Dominant Wavelength (A)	576.9c	494.3	489.3
Excitation Purity (%) (A)	98.0	33.5	63.5
Luminous Transmit. (%) (A)	0.29	36.1	12.6
x_A	0.2969	0.3062	0.1934
y_A	0.0673	0.4038	0.3263
Dominant Wavelength (C)	566.6c	483.5	478.8
Excitation Purity (%) (C)	97.6	41.7	69.7
Luminous Transmit. (%) (C)	0.23	42.5	17.3
x_C	0.1837	0.2122	0.1606
y_C	0.0154	0.2583	0.1796
Dominant Wavelength (D65)	565.9c	484.4	480.0
Excitation Purity (%) (D65)	94.3	41.8	67.7
Luminous Transmit. (%) (D65)	0.4	41.8	17.5
x_{D65}	0.1917	0.2119	0.1629
y_{D65}	0.0281	0.2717	0.1959

Table 49–Continued

WAVELENGTH	PERCENT TRANSMITTANCE No. 39	No. 40	No. 44	No. 44A	No. 45	No. 45A	No. 46	No. 47	No. 47A
400	85.2	–	0.43	1.55	–	–	1.20	9.7	26.4
10	78.2	–	0.27	1.41	–	–	0.60	21.8	40.3
20	70.5	–	0.40	3.39	–	–	0.80	37.8	54.5
30	63.3	–	3.24	13.5	–	1.00	5.98	47.8	62.1
40	53.6	–	13.1	30.7	4.20	8.81	19.0	50.3	63.5
50	42.5	–	24.7	43.0	18.2	17.4	30.1	48.2	61.8
60	28.5	3.16	35.7	49.5	30.2	20.9	33.8	42.8	58.2
70	17.3	21.6	45.5	52.3	35.9	21.6	32.1	35.7	53.0
80	10.2	44.7	53.3	53.3	37.2	20.5	27.0	27.1	46.4
90	4.00	61.4	56.7	52.3	35.1	18.0	20.2	18.2	38.6
500	1.33	70.2	55.4	49.1	28.8	14.4	11.1	10.2	29.5
10	0.35	72.4	49.9	44.3	18.8	10.1	4.39	4.3	20.3
20	–	70.5	41.0	37.6	8.9	5.60	1.66	1.2	11.9
30	–	64.8	29.3	28.9	2.35	2.52	0.35	–	5.44
40	–	55.5	17.0	19.1	0.33	0.64	–	–	1.95
50	–	44.2	7.58	10.7	–	0.10	–	–	0.58
60	–	32.5	2.40	4.68	–	–	–	–	0.18
70	–	20.3	0.36	1.38	–	–	–	–	–
80	–	9.56	–	0.19	–	–	–	–	–
90	–	3.20	–	–	–	–	–	–	–
600	–	1.10	–	–	–	–	–	–	–
10	–	0.32	–	–	–	–	–	–	–
20	–	–	–	–	–	–	–	–	–
30	–	–	–	–	–	–	–	–	–
40	–	–	–	–	–	–	–	–	–
50	–	–	–	–	–	–	–	–	–
60	–	–	–	–	–	–	–	–	–
70	–	–	–	–	–	–	–	–	0.16
80	0.50	0.80	–	–	–	–	–	–	0.33
90	4.06	6.99	0.32	0.21	–	0.20	0.25	–	0.59
700	17.8	23.5	2.48	0.71	0.49	2.24	0.85	–	0.96
Dominant Wavelength (A)	454.5	513.4	495.1	491.9	486.6	483.5	475.1	470.1	477.5
Excitation Purity (%) (A)	98.0	53.5	76.5	76.0	90.0	90.5	95.5	96.0	91.7
Luminous Transmit. (%) (A)	0.39	26.2	10.3	10.1	2.8	1.5	1.05	1.2	3.4
x_A	0.1570	0.2229	0.1236	0.1348	0.0998	0.1122	0.1245	0.1371	0.1322
y_A	0.0287	0.6165	0.4126	0.3549	0.2451	0.1996	0.1014	0.0724	0.1361
Dominant Wavelength (C)	450.8	516.2	488.7	484.2	481.5	477.6	470.4	463.8	469.7
Excitation Purity (%) (C)	98.9	48.6	73.4	75.2	88.6	89.6	94.9	95.7	91.8
Luminous Transmit. (%) (C)	1.15	33.6	15.6	15.2	5.2	2.8	2.4	2.8	6.6
x_C	0.1574	0.1823	0.1200	0.1314	0.1108	0.1222	0.1327	0.1451	0.1410
y_C	0.0216	0.5611	0.2807	0.2199	0.1692	0.1305	0.0727	0.0485	0.0790
Dominant Wavelength (D65)	449.8	516.7	489.1	484.7	480.6	477.1	470.6	463.2	466.8
Excitation Purity (%) (D65)	97.3	45.5	73.2	75.5	88.6	87.3	94.1	95.6	95.9
Luminous Transmit. (%) (D65)	1.2	37.2	15.8	15.1	5.0	2.6	2.6	2.6	2.0
x_{D65}	0.1609	0.1931	0.1201	0.1296	0.1144	0.1291	0.1336	0.1464	0.1390
y_{D65}	0.0258	0.5543	0.2905	0.2282	0.1609	0.1320	0.0765	0.0484	0.0568

Table 49–Continued

WAVELENGTH	PERCENT TRANSMITTANCE No. 47B	No. 48	No. 48A
400	16.0	0.96	4.51
10	29.5	3.16	9.32
20	43.6	8.25	16.9
30	50.0	15.0	22.4
40	47.2	22.6	25.9
50	36.0	30.3	27.1
60	25.0	33.2	23.2
70	13.2	29.6	16.6
80	4.5	22.4	9.78
90	1.3	14.1	4.57
500	0.17	7.30	1.62
10	–	2.64	0.26
20	–	0.50	–
30	–	–	–
40	–	–	–
50	–	–	–
60	–	–	–
70	–	–	–
80	–	–	–
90	–	–	–
600	–	–	–
10	–	–	–
20	–	–	–
30	–	–	–
40	–	–	–
50	–	–	–
60	–	–	–
70	–	–	–
80	–	–	–
90	–	–	–
700	–	–	–

WAVELENGTH	PERCENT TRANSMITTANCE No. 49	No. 49B	No. 50
400	3.30	1.5	–
10	4.28	2.5	–
20	6.93	4.1	–
30	11.2	7.8	1.6
40	18.9	14.1	8.2
50	25.6	18.2	13.7
60	24.0	14.8	10.5
70	15.7	7.2	4.2
80	6.93	2.2	0.9
90	2.14	0.23	–
500	0.46	–	–
10	–	–	–
20	–	–	–
30	–	–	–
40	–	–	–
50	–	–	–
60	–	–	–
70	–	–	–
80	–	–	–
90	–	–	–
600	–	–	–
10	–	–	–
20	–	–	–
30	–	–	–
40	–	–	–
50	–	–	–
60	–	–	–
70	–	–	–
80	–	–	–
90	–	–	–
700	–	–	–

WAVELENGTH	PERCENT TRANSMITTANCE No. 52	No. 53	No. 54
400	2.18	0.30	–
10	1.51	0.22	–
20	0.80	0.09	–
30	0.44	–	–
40	0.41	–	–
50	0.69	–	–
60	1.45	–	–
70	2.70	0.10	–
80	4.90	0.71	–
90	8.50	2.14	–
500	13.3	4.47	–
10	18.2	7.24	0.10
20	23.7	10.7	0.31
30	28.5	14.0	0.64
40	32.1	16.6	0.89
50	33.1	17.3	0.93
60	31.0	15.4	0.62
70	25.6	11.4	0.21
80	19.1	6.90	–
90	12.6	3.60	–
600	7.78	1.41	–
10	4.17	0.40	–
20	2.34	0.15	–
30	1.38	–	–
40	0.80	–	–
50	0.54	–	–
60	0.36	–	–
70	0.27	–	–
80	0.23	–	–
90	0.19	–	–
700	0.17	–	–

		No. 47B	No. 48	No. 48A
Dominant Wavelength	(A)	452.7	471.1	462.5
Excitation Purity (%)	(A)	99.0	96.5	98.0
Luminous Transmit. (%)	(A)	0.23	0.75	0.31
x_A		0.1554	0.1320	0.1453
y_A		0.0220	0.0743	0.0407
Dominant Wavelength	(C)	449.4	466.5	458.0
Excitation Purity (%)	(C)	99.4	96.1	98.2
Luminous Transmit. (%)	(C)	0.79	1.86	0.88
x_C		0.1579	0.1393	0.1499
y_C		0.0187	0.0547	0.0318
Dominant Wavelength	(D65)	451.0	465.6	460.5
Excitation Purity (%)	(D65)	97.5	95.4	96.0
Luminous Transmit. (%)	(D65)	1.0	1.9	1.2
x_{D65}		0.1595	0.1425	0.1499
y_{D65}		0.0264	0.0547	0.0424

		No. 49	No. 49B	No. 50
Dominant Wavelength	(A)	461.0	457.7	457.7
Excitation Purity (%)	(A)	98.5	99.0	99.5
Luminous Transmit. (%)	(A)	0.23	0.11	0.079
x_A		0.1457	0.1492	0.1488
y_A		0.0353	0.0285	0.0279
Dominant Wavelength	(C)	457.8	455.3	455.7
Excitation Purity (%)	(C)	98.8	99.2	99.4
Luminous Transmit. (%)	(C)	0.69	0.36	0.26
x_C		0.1491	0.1517	0.1509
y_C		0.0299	0.0252	0.0254
Dominant Wavelength	(D65)	458.1	453.1	455.3
Excitation Purity (%)	(D65)	97.5	97.7	99.4
Luminous Transmit. (%)	(D65)	0.8	0.4	0.2
x_{D65}		0.1510	0.1570	0.1515
y_{D65}		0.0342	0.0278	0.0248

		No. 52	No. 53	No. 54
Luminous Wavelength	(A)	554.9	552.0	553.3
Excitation Purity (%)	(A)	71.0	82.5	56.3
Luminous Transmit. (%)	(A)	18.3	8.1	0.4
x_A		0.3684	0.3382	0.3789
y_A		0.5875	0.6327	0.5552
Dominant Wavelength	(C)	553.2	551.0	548.7
Excitation Purity (%)	(C)	77.2	89.2	54.7
Luminous Transmit. (%)	(C)	20.1	9.0	0.4
x_C		0.3213	0.3090	0.3003
y_C		0.5902	0.6459	0.5268
Dominant Wavelength	(D65)	553.0	551.1	547.9
Excitation Purity (%)	(D65)	75.0	84.4	55.8
Luminous Transmit. (%)	(D65)	21.1	10.2	0.4
x_{D65}		0.3204	0.3102	0.2980
y_{D65}		0.5864	0.6293	0.5394

Table 49–Continued

WAVELENGTH	PERCENT TRANSMITTANCE No. 55	No. 56	No. 57
400	–	–	–
10	–	–	–
20	–	–	–
30	–	–	–
40	–	–	–
50	–	–	–
60	0.20	0.36	0.44
70	2.90	2.9	3.10
80	13.1	12.5	13.1
90	34.2	32.2	31.9
500	56.0	55.2	50.5
10	67.0	71.5	60.6
20	69.3	78.8	63.3
30	65.1	80.3	61.0
40	56.7	77.6	55.0
50	45.0	72.4	47.1
60	33.1	65.7	37.3
70	20.7	56.9	26.5
80	9.00	44.7	16.6
90	2.70	32.0	8.69
600	0.40	23.1	3.70
10	–	18.1	1.60
20	–	12.9	0.49
30	–	7.0	–
40	–	2.9	–
50	–	1.7	–
60	–	2.3	–
70	–	6.7	–
80	0.66	20.0	–
90	6.90	42.2	–
700	27.8	63.3	–

WAVELENGTH	PERCENT TRANSMITTANCE No. 57A	No. 58	No. 59
400	–	–	–
10	–	–	–
20	–	–	–
30	–	–	–
40	–	–	–
50	0.7	–	0.40
60	3.3	–	1.90
70	11.0	0.23	7.70
80	26.8	1.38	21.0
90	46.5	4.90	41.5
500	61.2	17.7	59.0
10	68.1	38.8	67.7
20	69.2	52.2	69.8
30	66.3	53.6	67.2
40	60.3	47.6	61.5
50	51.8	38.4	54.0
60	41.8	27.8	45.0
70	30.6	17.4	35.0
80	19.1	9.0	24.0
90	10.0	3.50	14.0
600	4.7	1.50	7.95
10	2.2	0.41	4.90
20	0.8	–	2.70
30	–	–	1.00
40	–	–	0.17
50	–	–	–
60	–	–	–
70	–	–	0.63
80	–	–	4.00
90	–	–	12.0
700	1.2	0.53	22.6

WAVELENGTH	PERCENT TRANSMITTANCE No. 59A	No. 60	No. 61
400	0.13	–	–
10	0.15	–	–
20	0.22	–	–
30	0.39	–	–
40	0.95	–	–
50	2.37	0.19	–
60	6.39	1.38	–
70	15.9	5.38	–
80	32.5	15.0	0.33
90	52.1	32.0	4.00
500	66.2	48.4	16.6
10	73.6	57.2	32.3
20	75.5	59.2	40.0
30	73.6	55.5	39.6
40	69.1	47.5	34.5
50	62.7	36.8	26.3
60	55.2	25.2	17.3
70	46.1	14.4	9.70
80	34.4	6.3	4.40
90	23.3	1.82	1.66
600	16.6	0.48	0.38
10	12.2	0.10	–
20	8.3	–	–
30	4.3	–	–
40	1.62	–	–
50	0.74	–	–
60	1.29	–	–
70	3.98	–	–
80	10.8	–	–
90	21.3	2.10	–
700	31.1	8.70	–

		No. 55	No. 56	No. 57
Dominant Wavelength	(A)	524.1	555.5	530.5
Excitation Purity (%)	(A)	61.5	68.5	59.0
Luminous Transmit. (%)	(A)	25.2	47.7	26.9
x_A		0.2379	0.3746	0.2772
y_A		0.6669	0.5777	0.6413
Dominant Wavelength	(C)	530.1	552.2	536.3
Excitation Purity (%)	(C)	68.4	78.3	69.2
Luminous Transmit. (%)	(C)	31.5	52.8	32.5
x_C		0.2043	0.3164	0.2355
y_C		0.6512	0.5988	0.6339
Dominant Wavelength	(D65)	534.0	554.1	540.3
Excitation Purity (%)	(D65)	70.2	77.2	66.6
Luminous Transmit. (%)	(D65)	36.2	57.3	39.8
x_{D65}		0.2234	0.3265	0.2587
y_{D65}		0.6503	0.5887	0.6110

		No. 57A	No. 58	No. 59
Dominant Wavelength	(A)	526.8	538.2	532.3
Excitation Purity (%)	(A)	53.5	77.0	52.5
Luminous Transmit. (%)	(A)	30.8	19.8	32.6
x_A		0.2774	0.2693	0.3028
y_A		0.6277	0.6831	0.6118
Dominant Wavelength	(C)	533.8	540.3	538.3
Excitation Purity (%)	(C)	61.8	86.2	63.9
Luminous Transmit. (%)	(C)	37.2	23.7	38.7
x_C		0.2325	0.2425	0.2507
y_C		0.6076	0.6923	0.6023
Dominant Wavelength	(D65)	537.5	541.4	539.4
Excitation Purity (%)	(D65)	61.7	85.1	64.6
Luminous Transmit. (%)	(D65)	40.4	27.3	41.3
x_{D65}		0.2500	0.2509	0.2562
y_{D65}		0.6003	0.6836	0.6059

		No. 59A	No. 60	No. 61
Dominant Wavelength	(A)	536.2	520.0	533.8
Excitation Purity (%)	(A)	48.0	59.5	76.5
Luminous Transmit. (%)	(A)	39.5	20.7	13.7
x_A		0.3300	0.2249	0.2457
y_A		0.5838	0.6616	0.6989
Dominant Wavelength	(C)	541.5	525.7	536.8
Excitation Purity (%)	(C)	59.5	62.4	85.3
Luminous Transmit. (%)	(C)	45.8	26.1	16.8
x_C		0.2685	0.1917	0.2213
y_C		0.5718	0.6328	0.7053
Dominant Wavelength	(D65)	541.9	527.4	538.0
Excitation Purity (%)	(D65)	54.9	60.3	82.9
Luminous Transmit. (%)	(D65)	49.8	30.8	20.5
x_{D65}		0.2747	0.2047	0.2317
y_{D65}		0.5564	0.6235	0.6910

Table 49–Continued

WAVELENGTH	PERCENT TRANSMITTANCE No. 64	No. 65	No. 65A
400	0.30	–	–
10	1.50	–	–
20	3.30	–	–
30	6.40	–	0.16
40	11.5	0.20	1.32
50	17.8	2.04	5.50
60	26.3	9.0	13.0
70	36.5	19.1	24.9
80	47.1	27.3	36.6
90	55.7	31.9	45.1
500	61.5	33.6	45.8
10	63.1	32.4	39.7
20	60.2	28.1	29.7
30	53.7	21.3	17.8
40	44.4	13.3	7.90
50	29.8	6.6	2.40
60	15.7	1.82	0.32
70	7.90	0.43	–
80	3.70	–	–
90	1.20	–	–
600	–	–	–
10	–	–	–
20	–	–	–
30	–	–	–
40	–	–	–
50	–	–	–
60	–	–	–
70	–	–	–
80	–	–	–
90	–	–	0.20
700	–	–	2.18

		No. 64	No. 65	No. 65A
Dominant Wavelength	(A)	503.0	501.3	497.3
Excitation Purity (%)	(A)	62.6	74.0	82.0
Luminous Transmit. (%)	(A)	18.0	6.6	6.2
x_A		0.1682	0.1215	0.0952
y_A		0.5344	0.5266	0.4584
Dominant Wavelength	(C)	497.2	496.6	492.7
Excitation Purity (%)	(C)	55.7	67.8	77.5
Luminous Transmit. (%)	(C)	24.6	9.6	9.8
x_C		0.1451	0.1116	0.0953
y_C		0.4041	0.4090	0.3483
Dominant Wavelength	(D65)	499.4	496.5	492.6
Excitation Purity (%)	(D65)	50.8	65.0	75.1
Luminous Transmit. (%)	(D65)	27.9	10.8	10.5
x_{D65}		0.1587	0.1211	0.1030
y_{D65}		0.4277	0.4080	0.3478

WAVELENGTH	PERCENT TRANSMITTANCE No. 66	No. 70	No. 72B
400	13.2	–	–
10	13.4	–	–
20	14.8	–	–
30	17.9	–	–
40	22.8	–	–
50	30.1	–	–
60	41.2	–	–
70	54.5	–	–
80	67.0	–	–
90	76.4	–	–
500	81.9	–	–
10	84.2	–	–
20	84.7	–	–
30	83.3	–	–
40	79.7	–	–
50	74.6	–	–
60	67.6	–	–
70	58.6	–	–
80	46.7	–	–
90	33.8	–	1.26
600	24.6	–	5.89
10	19.4	–	5.25
20	13.7	–	2.88
30	7.3	–	1.26
40	2.92	–	0.48
50	1.36	1.3	0.14
60	1.94	12.4	–
70	8.4	35.5	–
80	26.5	58.0	–
90	49.7	72.1	–
700	68.1	80.3	–

		No. 66	No. 70	No. 72B
Dominant Wavelength	(A)	511.5	678.0	605.7
Excitation Purity (%)	(A)	26.5	100.0	100.0
Luminous Transmit. (%)	(A)	50.6	0.68	1.09
x_A		0.3333	0.7331	0.6506
y_A		0.5050	0.2668	0.3490
Dominant Wavelength	(C)	512.3	676.0	605.0
Excitation Purity (%)	(C)	21.5	100.0	99.9
Luminous Transmit. (%)	(C)	58.3	0.31	0.74
x_C		0.2486	0.7328	0.6481
y_C		0.4169	0.2672	0.3515
Dominant Wavelength	(D65)	521.0	641.1	606.3
Excitation Purity (%)	(D65)	19.4	67.6	90.4
Luminous Transmit. (%)	(D65)	64.5	0.5	1.1
x_{D65}		0.2679	0.7320	0.6204
y_{D65}		0.4271	0.2680	0.3448

WAVELENGTH	PERCENT TRANSMITTANCE No. 73	No. 74	No. 75
400	–	–	–
10	–	–	–
20	–	–	–
30	–	–	–
40	–	–	–
50	–	–	–
60	–	–	1.97
70	–	–	10.0
80	–	–	17.4
90	–	–	18.0
500	–	–	13.0
10	–	0.96	7.35
20	–	7.95	3.20
30	–	14.6	0.83
40	–	12.9	0.14
50	–	7.60	–
60	2.24	3.06	–
70	5.97	0.83	–
80	4.56	0.12	–
90	2.00	–	–
600	0.56	–	–
10	0.10	–	–
20	–	–	–
30	–	–	–
40	–	–	–
50	–	–	–
60	–	–	–
70	–	–	–
80	–	–	–
90	–	–	–
700	–	–	0.14

		No. 73	No. 74	No. 75
Dominant Wavelength	(A)	576.0	538.0	490.5
Excitation Purity (%)	(A)	100.0	93.5	92.5
Luminous Transmit. (%)	(A)	1.41	3.34	1.01
x_A		0.4853	0.2301	0.0732
y_A		0.5134	0.7421	0.3139
Dominant Wavelength	(C)	574.9	538.6	487.7
Excitation Purity (%)	(C)	100.0	96.7	90.6
Luminous Transmit. (%)	(C)	1.3	4.0	1.9
x_C		0.4780	0.2227	0.0801
y_C		0.5208	0.7473	0.2542
Dominant Wavelength	(D65)	575.5	539.3	488.3
Excitation Purity (%)	(D65)	90.5	92.1	85.2
Luminous Transmit. (%)	(D65)	1.6	6.4	1.8
x_{D65}		0.4659	0.2315	0.0915
y_{D65}		0.4990	0.7244	0.2706

Table 49–Continued

WAVELENGTH	PERCENT TRANSMITTANCE No. 79
400	24.0
10	26.0
20	29.0
30	31.0
40	32.2
50	32.7
60	31.4
70	28.8
80	25.6
90	22.2
500	19.3
10	16.8
20	14.2
30	12.7
40	11.8
50	11.0
60	9.76
70	8.81
80	8.50
90	8.29
600	7.56
10	6.45
20	5.13
30	4.17
40	3.47
50	3.16
60	3.09
70	3.16
80	3.16
90	3.16
700	3.31
Dominant Wavelength (A)	485.9
Excitation Purity (%) (A)	36.5
Luminous Transmit. (%) (A)	9.6
x_A	0.3085
y_A	0.3378
Dominant Wavelength (C)	474.8
Excitation Purity (%) (C)	52.7
Luminous Transmit. (%) (C)	11.3
x_C	0.2047
y_C	0.1949
Dominant Wavelength (D65)	476.5
Excitation Purity (%) (D65)	48.5
Luminous Transmit. (%) (D65)	12.5
x_{D65}	0.2116
y_{D65}	0.2173

WAVELENGTH	PERCENT TRANSMITTANCE No. 96	No. 97	No. 98
400	4.28	–	6.1
10	4.91	–	17.2
20	5.50	–	31.7
30	6.17	–	40.0
40	6.92	–	39.7
50	7.50	–	32.5
60	7.81	–	21.7
70	8.15	0.22	11.2
80	8.47	0.43	3.8
90	8.60	0.39	0.7
500	8.73	0.15	–
10	8.85	–	–
20	8.90	–	–
30	9.01	–	–
40	9.07	–	–
50	9.20	–	–
60	9.30	–	–
70	9.20	–	–
80	9.19	–	–
90	9.54	–	–
600	9.64	–	–
10	9.73	–	–
20	9.56	–	–
30	9.27	–	–
40	9.10	–	–
50	9.07	–	–
60	9.00	–	–
70	9.13	–	–
80	9.08	0.44	–
90	9.21	5.02	–
700	9.52	18.7	–
Dominant Wavelength (A)	582.5	516.0c	452.0
Excitation Purity (%) (A)	15.5	83.0	99.4
Luminous Transmit. (%) (A)	9.2	0.06	0.19
x_A	0.4602	0.60	0.1553
y_A	0.4170	0.25	0.0216
Dominant Wavelength (C)	572.7	540.0c	449.7
Excitation Purity (%) (C)	12.1	46.0	99.3
Luminous Transmit. (%) (C)	9.1	0.04	0.66
x_C	0.3286	0.33	0.1577
y_C	0.3430	0.21	0.0187
Dominant Wavelength (D65)	570.1	462.5	450.4
Excitation Purity (%) (D65)	7.0	14.3	98.6
Luminous Transmit. (%) (D65)	12.0	0.1	0.7
x_{D65}	0.3219	0.2879	0.1584
y_{D65}	0.3448	0.2868	0.0225

WAVELENGTH	PERCENT TRANSMITTANCE No. 92	No. 93	No. 94
400	–	–	–
10	–	–	–
20	–	–	–
30	–	–	1.2
40	–	–	4.9
50	–	–	8.5
60	–	–	8.3
70	–	–	4.7
80	–	–	1.4
90	–	–	0.3
500	–	–	–
10	–	–	–
20	–	1.05	–
30	–	4.07	–
40	–	6.02	–
50	–	4.37	–
60	–	2.00	–
70	–	0.68	–
80	–	0.24	–
90	–	–	–
600	–	–	–
10	–	–	–
20	0.40	–	–
30	17.4	–	–
40	51.2	–	–
50	72.2	–	–
60	80.7	–	–
70	84.8	–	–
80	86.5	–	–
90	87.4	–	–
700	88.0	–	–
Dominant Wavelength (A)	646.2	544.8	460.2
Excitation Purity (%) (A)	100.0	95.5	99.3
Luminous Transmit. (%) (A)	4.8	1.4	0.06
x_A	0.7237	0.2729	0.1459
y_A	0.2763	0.7110	0.0324
Dominant Wavelength (C)	645.2	544.8	458.0
Excitation Purity (%) (C)	99.9	98.0	99.3
Luminous Transmit. (%) (C)	2.6	1.6	0.19
x_C	0.7231	0.2655	0.1483
y_C	0.2768	0.7173	0.0289
Dominant Wavelength (D65)	643.7	543.5	456.6
Excitation Purity (%) (D65)	94.1	86.0	94.5
Luminous Transmit. (%) (D65)	2.9	1.7	0.2
x_{D65}	0.6978	0.2633	0.1577
y_{D65}	0.2809	0.6767	0.0414

WAVELENGTH	PERCENT TRANSMITTANCE No. 99	No. 102	No. 106
400	–	2.0	–
10	–	1.6	–
20	–	1.3	0.10
30	–	1.4	0.20
40	–	1.7	0.35
50	–	2.5	0.58
60	–	4.3	0.98
70	–	7.6	1.5
80	–	12.6	2.3
90	–	19.5	3.5
500	–	28.7	5.2
10	–	39.4	7.7
20	1.12	50.3	10.7
30	7.95	59.2	15.1
40	18.6	64.8	20.2
50	19.9	67.2	25.7
60	15.1	66.8	31.0
70	8.7	63.6	35.6
80	3.80	58.2	43.2
90	1.12	51.6	53.8
600	0.22	43.9	65.6
10	–	36.5	77.0
20	–	29.8	82.8
30	–	24.3	86.0
40	–	19.8	87.6
50	–	16.7	88.7
60	–	14.3	89.5
70	–	13.0	90.0
80	–	12.0	90.5
90	–	11.4	90.8
700	–	11.0	91.0
Dominant Wavelength (A)	554.6	571.5	596.1
Excitation Purity (%) (A)	96.5	77.5	96.5
Luminous Transmit. (%) (A)	6.4	50.2	43.2
x_A	0.3384	0.4534	0.6019
y_A	0.6530	0.5132	0.3916
Dominant Wavelength (C)	553.6	564.9	589.3
Excitation Purity (%) (C)	98.7	79.9	95.1
Luminous Transmit. (%) (C)	7.0	50.8	34.4
x_C	0.3268	0.3882	0.5583
y_C	0.6637	0.5355	0.4228
Dominant Wavelength (D65)	553.6	564.3	589.1
Excitation Purity (%) (D65)	93.7	76.3	93.7
Luminous Transmit. (%) (D65)	8.4	52.4	33.8
x_{D65}	0.3267	0.3820	0.5537
y_{D65}	0.6466	0.5316	0.4229

Table 49–Continued

WAVELENGTH	PERCENT TRANSMITTANCE HF-3	HF-4	HF-5
400	19.0	69.3	55.5
10	48.0	68.5	53.7
20	67.0	68.1	52.7
30	75.3	68.1	52.5
40	80.0	68.5	52.9
50	83.0	69.5	54.2
60	85.2	71.5	56.8
70	86.7	74.2	60.8
80	88.1	77.1	65.0
90	88.8	80.0	69.5
500	89.5	83.0	73.9
10	89.9	84.5	75.8
20	90.3	84.5	74.8
30	90.5	84.3	73.4
40	90.6	84.4	73.2
50	90.7	84.5	73.5
60	90.8	84.5	72.9
70	90.9	84.6	72.9
80	90.9	85.8	76.2
90	91.0	87.7	82.0
600	91.1	89.2	87.3
10	91.2	90.0	89.6
20	91.3	90.3	90.4
30	91.3	90.5	90.7
40	91.4	90.5	90.9
50	91.4	90.5	91.0
60	91.5	90.5	91.0
70	91.5	90.6	91.0
80	91.6	90.6	91.1
90	91.7	90.6	91.1
700	91.8	90.7	91.2

	HF-3	HF-4	HF-5
Dominant Wavelength (A)	581.0	586.0	587.0
Excitation Purity (%) (A)	7.0	16.1	26.7
Luminous Transmit. (%) (A)	90.8	86.3	78.9
x_A	0.4526	0.4608	0.4765
y_A	0.4126	0.4143	0.4130
Dominant Wavelength (C)	569.8	574.5	577.0
Excitation Purity (%) (C)	5.6	10.3	17.9
Luminous Transmit. (%) (C)	90.5	85.2	77.5
x_C	0.3178	0.3271	0.3422
y_C	0.3298	0.3375	0.3508

KODAK POLYCONTRAST GELATIN FILTERS

WAVELENGTH	PERCENT TRANSMITTANCE PC1	PC1½	PC2	PC2½	PC3	PC3½	PC4
400	38.0	71.2	85.8	83.3	79.5	71.8	46.8
10	36.6	70.9	86.9	84.5	81.1	74.3	51.5
20	35.8	70.7	86.6	84.6	80.9	73.9	49.2
30	35.6	70.8	86.7	83.4	79.3	70.4	41.5
40	36.0	71.3	86.0	81.6	76.2	64.2	32.3
50	37.4	72.1	84.8	78.7	71.4	56.9	22.2
60	40.8	73.4	82.7	74.8	65.0	47.3	13.1
70	48.7	76.6	79.7	70.1	57.1	37.0	6.3
80	59.7	80.4	76.3	64.2	48.9	27.1	2.6
90	70.5	84.1	72.8	58.5	41.1	19.3	1.0
500	78.8	86.7	69.1	52.3	33.8	13.4	0.3
10	83.7	88.2	65.3	46.3	27.9	9.2	–
20	86.2	89.2	62.1	41.5	23.5	6.5	–
30	87.6	89.6	59.6	38.1	20.4	4.9	–
40	88.4	89.9	58.1	35.8	18.7	4.1	–
50	89.1	90.2	57.4	35.1	18.2	4.0	–
60	89.5	90.4	58.6	36.2	19.6	4.7	–
70	89.7	90.5	61.7	39.9	23.9	6.5	–
80	89.9	90.5	66.8	46.6	31.6	10.6	0.1
90	90.1	90.6	72.6	55.6	42.5	18.8	0.8
600	90.3	90.7	78.7	66.0	56.5	33.2	3.7
10	90.4	90.8	83.3	74.5	68.3	48.7	11.5
20	90.5	90.9	86.8	81.2	77.5	64.6	28.7
30	90.6	90.9	88.8	85.7	83.6	75.8	48.5
40	90.7	91.0	90.2	88.6	87.4	82.9	66.0
50	90.7	91.0	90.6	89.6	89.1	86.7	77.0
60	90.8	91.0	90.9	90.3	90.0	88.6	82.9
70	90.8	91.0	91.0	90.7	90.5	89.6	86.2
80	90.9	91.1	91.1	91.0	90.8	90.3	87.9
90	90.9	91.1	91.2	91.1	91.0	90.7	89.1
700	91.0	91.2	91.2	91.2	91.1	90.9	89.9

	PC1	PC1½	PC2	PC2½	PC3	PC3½	PC4
Dominant Wavelength (A)	581.0	580.0	545.0c	544.0c	537.0c	534.0c	517.0c
Excitation Purity (%) (A)	42.3	12.6	13.4	26.3	41.4	65.0	88.0
Luminous Transmit. (%) (A)	87.8	89.7	69.0	52.8	39.8	23.1	9.3
x_A	0.4779	0.4573	0.4647	0.4809	0.5054	0.5407	0.6068
y_A	0.4381	0.4175	0.3788	0.3506	0.3200	0.2726	0.2402
Dominant Wavelength (C)	571.5	570.7	546.5c	548.0c	544.5c	544.5c	531.7c
Excitation Purity (%) (C)	36.2	10.4	14.8	28.8	44.5	68.0	89.5
Luminous Transmit. (%) (C)	85.8	89.0	66.8	49.3	35.0	17.9	5.7
x_C	0.3623	0.3251	0.3127	0.3140	0.3220	0.3292	0.3680
y_C	0.3990	0.3403	0.2811	0.2468	0.2115	0.1566	0.1210

Table 49–Continued

KODAK LIGHT-BALANCING FILTERS

WAVELENGTH	PERCENT TRANSMITTANCE No. 81	No. 81A	No. 81B	No. 81C	No. 81D	No. 81EF	PERCENT TRANSMITTANCE No. 82	No. 82A	No. 82B	No. 82C
400	77.7	65.1	55.1	46.1	38.2	30.7	83.0	80.1	76.7	73.4
10	78.1	65.9	55.8	46.6	38.4	31.5	83.7	80.8	78.0	75.0
20	79.0	67.6	57.7	49.0	41.0	34.3	84.6	81.6	79.2	76.4
30	80.5	70.2	61.0	52.5	45.5	38.6	85.1	82.2	79.7	77.2
40	81.9	72.8	64.5	57.2	50.0	43.2	85.4	82.4	79.7	77.2
50	83.0	74.8	67.2	60.5	53.9	47.4	85.4	82.4	79.2	76.6
60	83.7	76.0	69.1	63.0	56.5	50.2	85.0	81.7	78.0	75.2
70	84.3	77.1	70.6	64.2	58.1	52.0	84.6	80.7	76.3	73.2
80	84.6	77.8	71.3	65.0	59.0	53.0	84.0	79.3	74.4	70.7
90	84.9	78.3	71.8	65.7	60.0	54.0	83.3	78.0	72.1	68.1
500	85.3	78.6	72.6	66.4	60.8	55.4	82.6	76.6	70.2	65.7
10	85.4	79.0	72.9	66.5	61.1	56.2	82.0	75.3	68.3	63.5
20	85.5	79.5	73.2	67.0	61.6	57.0	81.4	74.0	66.5	61.5
30	86.0	80.4	74.5	68.8	63.5	59.5	81.0	73.1	65.5	59.9
40	86.5	81.5	76.0	71.0	66.1	62.7	80.8	72.7	65.0	59.1
50	86.8	82.3	77.0	72.0	67.3	64.5	80.6	72.4	64.5	58.3
60	87.0	82.6	77.6	72.5	68.0	65.3	80.4	71.8	63.8	57.2
70	87.1	82.7	77.8	72.7	68.3	65.8	80.2	71.5	63.2	56.2
80	87.1	82.8	78.0	73.0	68.5	66.0	80.2	71.5	63.2	56.1
90	87.4	83.1	78.2	74.0	69.5	66.5	80.3	71.7	63.4	56.0
600	87.6	84.0	79.1	75.6	72.0	68.1	80.2	71.5	63.0	55.0
10	88.1	85.0	81.0	78.5	75.0	71.6	79.3	70.3	61.5	53.0
20	88.8	86.1	83.1	80.8	78.0	74.7	78.4	68.5	59.0	50.2
30	89.2	87.0	84.2	82.1	79.8	77.0	77.5	66.9	56.9	47.4
40	89.4	87.4	85.1	83.0	80.8	78.4	76.8	65.5	55.0	45.2
50	89.5	87.7	85.6	83.5	81.5	79.2	76.5	64.8	54.1	44.1
60	89.8	88.0	86.0	84.1	82.1	80.1	76.2	64.6	53.7	43.6
70	90.0	88.2	86.5	84.8	83.0	80.9	76.1	64.5	53.7	43.5
80	90.1	88.5	87.0	85.5	83.7	81.8	76.1	64.4	53.5	43.1
90	90.3	89.0	87.5	86.1	84.6	82.9	76.2	64.2	53.4	42.8
700	90.5	89.2	88.0	86.8	85.5	84.0	77.1	64.6	54.1	43.5
Dominant Wavelength (A)	587.2	587.5	588.0	589.0	589.4	588.4	490.8	489.6	488.6	489.4
Excitation Purity (%) (A)	4.0	9.0	12.0	16.0	20.5	26.0	1.5	3.0	5.0	7.5
Luminous Transmit. (%) (A)	87.2	82.9	78.1	73.6	69.4	66.1	80.1	71.5	63.3	56.2
x_A	0.4522	0.4573	0.4620	0.4675	0.4729	0.4778	0.4416	0.4350	0.4275	0.4171
y_A	0.4089	0.4102	0.4108	0.4110	0.4116	0.4142	0.4061	0.4035	0.4003	0.3981
Dominant Wavelength (C)	577.2	577.5	578.0	579.0	579.4	578.7	477.8	476.6	476.0	477.3
Excitation Purity (%) (C)	2.9	6.0	8.7	11.5	14.7	18.9	2.9	6.3	10.1	14.3
Luminous Transmit. (%) (C)	86.8	82.0	76.9	72.0	67.4	63.9	80.6	72.5	64.6	58.1
x_C	0.3154	0.3213	0.3266	0.3329	0.3393	0.3468	0.3041	0.2970	0.2894	0.2803
y_C	0.3218	0.3275	0.3322	0.3367	0.3418	0.3501	0.3107	0.3028	0.2941	0.2863
Dominant Wavelength (D65)	575.2	576.2	577.1	578.0	578.2	577.5	480.5	478.6	477.4	479.7
Excitation Purity (%) (D65)	3.2	6.9	8.8	11.8	15.3	19.4	2.8	6.1	9.5	13.4
Luminous Transmit. (%) (D65)	87.2	82.2	77.2	71.9	68.2	64.6	80.4	73.0	65.4	60.0
x_{D65}	0.3180	0.3247	0.3285	0.3347	0.3413	0.3483	0.3065	0.2994	0.2926	0.2830
y_{D65}	0.3351	0.3417	0.3445	0.3492	0.3549	0.3629	0.3238	0.3161	0.3080	0.3023

Table 49–Continued

CONVERSION FILTERS FOR COLOR FILMS

WAVELENGTH	PERCENT TRANSMITTANCE KODAK WRATTEN Gelatin Filters No. 80A	No. 80B	No. 80C	No. 80D	PERCENT TRANSMITTANCE KODAK WRATTEN Gelatin Filters No. 85	No. 85B	No. 85BN3	No. 85BN6	No. 85C	No. 85N3	No. 85N6	No. 85N9
400	63.6	65.6	71.0	77.8	6.0	1.59	1.3	0.7	18.0	1.8	1.2	0.7
10	68.9	70.2	74.6	79.9	18.0	9.32	4.7	2.3	35.5	8.9	3.9	2.0
20	72.6	73.7	77.1	81.2	28.4	15.5	7.7	3.7	44.1	13.7	6.4	3.3
30	73.5	74.5	77.7	81.6	33.4	19.0	9.4	4.5	47.3	16.7	7.8	3.9
40	72.4	73.6	76.9	81.2	36.2	20.8	10.3	5.0	49.2	18.3	8.7	4.4
50	70.2	71.2	75.2	80.2	38.1	22.1	11.1	5.5	50.6	19.4	9.3	4.8
60	66.9	68.0	72.7	78.8	40.4	24.3	12.5	6.2	52.5	20.7	10.0	5.1
70	62.8	64.1	69.6	77.0	43.0	27.5	14.6	7.2	55.3	22.0	10.6	5.5
80	58.0	59.2	65.7	74.8	45.3	30.9	16.6	8.2	57.0	23.1	11.2	5.7
90	53.0	54.0	61.7	72.3	47.2	34.3	18.8	9.4	58.6	24.0	11.6	5.8
500	47.7	48.8	57.6	69.7	48.9	38.3	21.2	10.5	60.5	24.7	12.0	6.0
10	42.5	43.7	53.4	66.9	49.2	40.7	22.2	11.0	60.6	24.8	12.0	5.9
20	38.2	39.4	50.2	64.6	48.2	40.6	22.2	11.0	60.0	24.5	11.9	5.9
30	34.6	36.3	47.7	63.0	48.3	40.7	22.6	11.2	60.6	24.6	12.1	6.0
40	31.3	33.5	45.7	61.5	49.2	41.6	23.3	11.5	61.6	25.3	12.5	6.2
50	28.4	31.2	43.6	60.0	51.0	43.2	24.5	12.1	63.0	26.3	13.1	6.6
60	26.5	29.9	42.6	59.2	55.8	47.1	27.4	13.7	67.0	29.0	14.7	7.4
70	25.2	29.4	42.4	59.3	64.5	56.0	33.3	16.7	74.1	33.8	17.5	8.9
80	23.7	28.9	42.0	59.0	75.0	68.1	40.3	20.5	81.6	39.5	20.7	10.6
90	22.2	27.9	40.6	58.0	83.0	78.1	45.2	23.1	86.4	43.4	22.9	11.7
600	20.8	26.8	38.8	56.7	87.2	85.0	47.7	24.5	88.8	45.4	23.8	12.2
10	20.1	25.9	37.4	55.5	88.9	88.0	48.5	24.6	89.9	46.0	23.9	12.3
20	19.3	24.8	35.8	54.4	90.0	89.6	48.5	24.3	90.3	45.9	23.6	12.0
30	18.5	23.5	34.5	53.6	90.5	90.3	48.3	23.7	90.6	45.6	23.0	11.7
40	17.9	22.5	33.4	53.2	90.7	90.7	48.2	23.5	90.8	45.5	22.7	11.5
50	17.9	22.2	33.1	53.2	90.9	90.9	48.3	23.5	90.9	45.5	22.7	11.4
60	18.3	22.4	33.1	53.5	91.0	91.0	48.4	23.5	91.0	45.6	22.8	11.5
70	18.6	22.5	33.2	53.3	91.0	91.2	48.3	23.5	91.1	45.6	22.7	11.5
80	18.2	21.9	32.7	52.7	91.0	91.3	48.2	23.4	91.1	45.5	22.6	11.5
90	17.5	21.2	31.9	52.6	91.0	91.3	48.4	23.5	91.1	45.7	22.8	11.6
700	17.7	21.4	32.1	53.8	91.0	91.3	49.4	24.3	91.1	46.6	23.8	12.1
Dominant Wavelength (A)	486.0	483.0	486.0	487.0	595.0	593.8	592.0	592.0	593.2	592.0	591.0	595.0
Excitation Purity (%) (A)	27.4	21.3	12.9	6.7	41.5	58.5	61.1	60.6	30.0	47.2	52.9	48.7
Luminous Transmit. (%) (A)	26.9	30.4	43.6	59.7	68.0	62.3	34.9	17.8	75.3	33.4	18.0	8.9
x_A	0.3472	0.3690	0.3971	0.4215	0.5120	0.5353	0.5333	0.5322	0.4913	0.5115	0.5161	0.5183
y_A	0.3554	0.3605	0.3829	0.3957	0.4030	0.4047	0.4096	0.4100	0.4070	0.4043	0.4054	0.4031
Dominant Wavelength (C)	475.0	474.0	475.0	476.0	587.3	585.7	584.4	584.3	585.3	586.6	586.4	587.7
Excitation Purity (%) (C)	43.3	37.2	25.2	13.8	30.5	48.0	50.4	50.2	21.0	31.0	35.0	34.5
Luminous Transmit. (%) (C)	30.2	33.1	45.8	61.5	62.5	55.5	31.2	15.9	71.1	32.0	16.4	8.1
x_C	0.2237	0.2367	0.2608	0.2820	0.3860	0.4247	0.4271	0.4258	0.3600	0.3866	0.3951	0.3968
y_C	0.2170	0.2259	0.2586	0.2857	0.3542	0.3805	0.3881	0.3876	0.3450	0.3558	0.3614	0.3583
Dominant Wavelength (D65)	471.6	480.3	476.1	477.2	586.8	584.1	584.4	584.6	585.0	582.1	580.6	589.4
Excitation Purity (%) (D65)	45.4	26.7	31.7	14.4	31.7	47.3	51.3	54.9	20.2	36.7	37.3	32.6
Luminous Transmit. (%) (D65)	27.2	51.1	42.2	59.9	63.5	59.4	30.1	13.1	71.7	33.4	18.4	8.7
x_{D65}	0.2251	0.2400	0.2470	0.2823	0.3896	0.4197	0.4297	0.4386	0.3594	0.3910	0.3887	0.3970
y_{D65}	0.2095	0.2400	0.2549	0.2967	0.3653	0.3911	0.3954	0.3994	0.3543	0.3817	0.3862	0.3611

Table 49–Continued

KODAK WRATTEN PHOTOMETRIC FILTERS

WAVELENGTH	PERCENT TRANSMITTANCE No. 78	No. 78AA	No. 78A	No. 78B	No. 78C	PERCENT TRANSMITTANCE No. 86	No. 86A	No. 86B	No. 86C
400	36.5	42.7	57.5	68.9	79.1	0.50	8.00	20.0	44.0
10	40.4	46.1	60.2	70.9	80.2	0.81	12.2	26.1	55.0
20	43.6	49.4	62.6	72.5	81.2	1.55	16.7	31.6	62.0
30	44.1	50.0	63.3	72.9	81.6	2.88	21.5	37.5	66.6
40	43.2	49.4	63.1	72.8	81.8	5.50	27.8	44.0	70.8
50	40.9	47.4	61.5	71.8	81.6	9.10	34.2	50.1	74.3
60	36.7	43.6	58.5	69.8	80.9	13.5	40.4	55.4	76.8
70	31.6	38.5	54.6	67.0	79.7	17.8	45.0	59.5	78.7
80	26.4	33.1	50.5	63.9	78.5	21.3	48.7	62.5	80.2
90	21.7	28.4	46.1	60.8	76.9	24.5	51.2	64.6	81.2
500	18.0	24.4	42.4	57.7	75.4	26.8	52.8	66.0	81.7
10	14.9	21.1	38.6	54.9	73.9	27.9	53.4	66.4	81.9
20	12.6	18.2	35.4	52.2	72.6	28.6	53.7	66.6	82.0
30	11.0	16.2	33.3	50.2	71.7	30.4	55.0	67.6	82.4
40	10.0	15.1	31.9	49.0	71.1	32.5	56.5	69.0	83.0
50	9.1	13.8	30.5	47.7	70.5	35.0	58.5	70.2	83.5
60	8.1	12.6	28.9	46.1	69.7	41.2	63.0	73.0	84.6
70	7.4	11.6	27.6	44.9	69.0	53.0	70.9	78.1	86.8
80	7.2	11.1	27.3	44.4	68.8	67.5	79.0	84.0	88.9
90	6.9	10.8	27.0	44.1	68.9	76.5	85.2	87.5	89.9
600	6.2	10.0	26.0	43.0	68.2	85.0	88.1	89.3	90.6
10	5.2	8.8	24.2	41.2	67.0	88.1	89.8	90.3	91.0
20	4.2	7.4	22.0	38.9	65.5	89.6	90.5	90.7	91.1
30	3.4	6.0	19.9	36.7	64.0	90.4	90.8	90.9	91.2
40	2.9	5.1	18.4	35.2	63.1	90.7	91.1	91.1	91.3
50	2.6	4.8	17.7	34.6	62.8	91.0	91.2	91.2	91.4
60	2.6	4.9	17.8	34.8	62.8	91.1	91.3	91.3	91.5
70	2.7	5.0	18.1	35.1	63.0	91.2	91.4	91.4	91.6
80	2.7	4.9	18.2	35.2	63.0	91.3	91.4	91.5	91.6
90	2.6	4.8	18.0	35.4	63.2	91.3	91.5	91.6	91.6
700	2.8	5.0	18.4	36.4	64.1	91.3	91.5	91.6	91.6
Dominant Wavelength (A)	482.0	484.6	486.8	488.5	491.1	593.6	590.7	589.0	587.7
Excitation Purity (%) (A)	45.0	37.0	19.5	11.5	3.5	76.0	46.0	31.5	13.0
Luminous Transmit. (%) (A)	8.8	13.5	29.1	44.4	69.2	57.8	72.2	78.9	86.8
x_A	0.2836	0.3083	0.3715	0.4029	0.4330	0.5607	0.5083	0.4859	0.4618
y_A	0.2953	0.3288	0.3736	0.3916	0.4044	0.4045	0.4132	0.4145	0.4112
Dominant Wavelength (C)	471.0	473.6	476.4	477.3	479.8	585.8	580.7	579.5	577.7
Excitation Purity (%) (C)	63.1	54.4	33.7	20.7	6.7	69.6	36.0	24.2	9.0
Luminous Transmit. (%) (C)	10.7	15.8	31.6	46.7	70.4	49.6	67.1	75.5	85.4
x_C	0.1911	0.2035	0.2418	0.2670	0.2954	0.4773	0.3849	0.3584	0.327
y_C	0.1563	0.1857	0.2406	0.2725	0.3039	0.4093	0.3757	0.3582	0.333
Dominant Wavelength (D65)	471.8	473.8	476.1	477.6	479.6	585.1	581.0	578.6	576.1
Excitation Purity (%) (D65)	62.9	52.4	32.0	19.3	7.5	68.3	36.7	25.9	9.8
Luminous Transmit. (%) (D65)	9.4	15.2	31.0	48.2	69.5	51.2	68.7	76.2	86.2
x_{D65}	0.1909	0.2082	0.2465	0.2716	0.2963	0.4718	0.3883	0.3619	0.3297
y_{D65}	0.1640	0.1981	0.2541	0.2865	0.3141	0.4141	0.3844	0.3722	0.3471

Table 49–Continued

FILTERS FOR INFRARED PHOTOGRAPHY

WAVELENGTH	PERCENT TRANSMITTANCE No. 87	No. 87A	No. 87B	No. 87C	No. 88A	No. 89B
700	–	–	–	–	–	11.2
10	–	–	–	–	–	32.4
20	–	–	–	–	1.1	57.6
30	–	–	–	–	14.0	69.1
40	0.8	–	–	–	33.7	77.6
50	3.5	–	–	–	54.0	83.1
60	12.3	–	–	–	70.5	85.0
70	22.1	–	–	–	77.3	86.1
80	33.2	–	–	–	81.8	87.0
90	44.9	–	–	0.55	85.2	87.7
800	56.9	–	–	3.0	87.0	88.1
10	63.2	–	–	8.3	88.0	88.4
20	68.0	–	0.10	16.2	88.8	88.6
30	73.3	–	0.60	25.7	89.0	88.8
40	76.3	–	1.86	38.0	89.2	89.0
50	78.5	–	4.07	48.4	89.2	89.2
60	79.5	–	7.58	57.5	89.2	89.4
70	80.4	–	12.7	65.3	89.1	89.6
80	80.7	0.10	18.0	72.5	89.1	89.8
90	81.4	0.28	24.5	78.5	89.1	89.9
900	81.9	0.73	31.2	80.6	89.1	90.0
10	82.1	1.55	38.0	82.2	89.1	90.1
20	82.7	2.95	44.1	83.3	89.1	90.2
30	83.1	4.89	49.5	84.7	89.1	90.3
40	83.4	7.33	53.7	85.6	89.1	90.4
50	83.6	10.2	58.2	86.4	89.3	90.5
60	84.0	14.0	61.7	87.2	89.3	90.5
70	84.3	17.8	64.6	87.7	89.4	90.5
80	84.6	21.8	67.6	88.3	90.1	90.5
90	84.9	26.3	70.0	88.8	90.3	90.5
1000	85.3	30.2	73.3	89.2	90.5	90.5
10	85.6	33.8	75.8	89.7	90.5	90.5
20	85.9	38.0	78.5	90.1	90.5	90.5
30	86.2	42.1	80.2	90.5	90.5	90.5
40	86.6	46.7	82.2	90.8	90.5	90.5
50	87.0	50.6	84.1	91.3	90.5	90.5
60	87.5	54.3	87.2	91.7	90.5	90.5
70	87.8	58.3	88.1	92.2	90.5	90.5
80	88.1	60.9	88.2	92.5	90.5	90.5
90	88.3	63.1	89.1	92.7	90.5	90.5
1100	88.5	64.5	89.1	93.0	90.5	90.5

NOTE: Colorimetric values for the filters listed in the above tabulation are not given, since the predominant transmission is outside the visible region.

Table 50 Optical Density vs. Percent Transmission

D[a]	T (%)[b]	D[a]	T (%)[b]	D[a]	T (%)[b]	D[a]	T (%)[b]
0.00	100.00	0.42	38.02	0.84	14.45	1.26	5.495
0.01	97.72	0.43	37.15	0.85	14.13	1.27	5.370
0.02	95.50	0.44	36.31	0.86	13.80	1.28	5.248
0.03	93.38	0.45	35.48	0.87	13.49	1.29	5.129
0.04	91.20	0.46	34.67	0.88	13.18	1.30	5.012
0.05	89.13	0.47	33.88	0.89	12.88	1.31	4.898
0.06	87.10	0.48	33.11	0.90	12.59	1.32	4.786
0.07	85.11	0.49	32.36	0.91	12.30	1.33	4.677
0.08	83.18	0.50	31.62	0.92	12.02	1.34	4.571
0.09	81.28	0.51	30.90	0.93	11.75	1.35	4.467
0.10	79.43	0.52	30.20	0.94	11.48	1.36	4.365
0.11	77.62	0.53	29.51	0.95	11.22	1.37	4.266
0.12	75.86	0.54	28.84	0.96	10.96	1.38	4.169
0.13	74.13	0.55	28.18	0.97	10.12	1.39	4.074
0.14	72.44	0.56	27.54	0.98	10.47	1.40	3.981
0.15	70.79	0.57	26.92	0.99	10.23	1.41	3.890
0.16	69.18	0.58	26.30	1.00	10.000	1.42	3.802
0.17	67.61	0.59	25.70	1.01	9.772	1.43	3.715
0.18	66.07	0.60	25.12	1.02	9.550	1.44	3.631
0.19	64.57	0.61	24.55	1.03	9.333	1.45	3.548
0.20	63.10	0.62	23.99	1.04	9.120	1.46	3.467
0.21	61.66	0.63	23.44	1.05	8.913	1.47	3.388
0.22	60.26	0.64	22.91	1.06	8.710	1.48	3.311
0.23	58.88	0.65	22.39	1.07	8.511	1.49	3.236
0.24	57.54	0.66	21.88	1.08	8.318	1.50	3.162
0.25	56.23	0.67	21.38	1.09	8.128	1.55	2.818
0.26	54.95	0.68	20.89	1.10	7.943	1.60	2.512
0.27	53.10	0.69	20.42	1.11	7.762	1.65	2.239
0.28	52.48	0.70	19.95	1.12	7.586	1.70	1.995
0.29	51.29	0.71	19.50	1.13	7.413	1.75	1.778
0.30	50.12	0.72	19.05	1.14	7.244	1.80	1.585
0.31	48.98	0.73	18.62	1.15	7.079	1.85	1.413
0.32	47.86	0.74	18.20	1.16	6.918	1.90	1.259
0.33	46.77	0.75	17.78	1.17	6.761	1.95	1.122
0.34	45.71	0.76	17.38	1.18	6.607	2.00	1.000
0.35	44.67	0.77	16.98	1.19	6.457	2.10	0.7943
0.36	43.63	0.78	16.60	1.20	6.310	2.20	0.6310
0.37	42.66	0.79	16.22	1.21	6.166	2.30	0.5012
0.38	41.69	0.80	15.85	1.22	6.026	2.40	0.3981
0.39	40.74	0.81	15.49	1.23	5.888	2.50	0.3162
0.40	39.81	0.82	15.14	1.24	5.754	2.60	0.2512
0.41	38.90	0.83	14.79	1.25	5.623	2.70	0.1995

Table 50–Continued

2.80	0.1585	3.00	0.1000	3.40	0.0398	3.80	0.0158
2.90	0.1259	3.20	0.0631	3.60	0.0251	4.00	0.0100

[a]Optical density = D = log (1/T).
[b]% Transmittance = T.

Table 51 Mired Values of Color Temperature

T_c (K)	Mired value	T_c (K)	Mired value
1100	909	4500	222
1200	833	4600	217
1300	769	4700	213
1400	714	4800	208
1500	667	4900	204
1600	625	5000	200
1700	588	5100	196
1800	556	5200	192
1900	526	5300	189
2000	500	5400	185
2100	476	5500	182
2200	455	5600	179
2300	435	5700	175
2400	417	5800	172
2500	400	5900	169
2600	385	6000	167
2700	370	6200	161
2800	357	6400	156
2900	345	6600	152
3000	333	6800	147
3100	323	7000	143
3200	312	7200	139
3300	303	7400	135
3400	294	7600	132
3500	286	7800	128
3600	278	8000	125
3700	270	8200	122
3800	263	8400	119
3900	256	8600	116
4000	250	8800	114
4100	244	9000	111
4200	238	9200	109
4300	233	9400	106
4400	227	9600	104

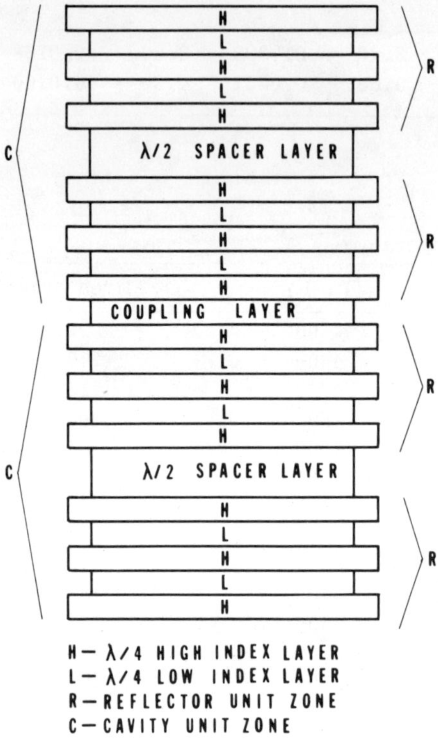

Fig. 69 Two-cavity dielectric filter material layer configurations.

central wavelength. The layer material purity and spectral transmittance influence the peak transmission achievable at any given wavelength for a given particular filter type.

Figure 69 shows the material layer configuration of a typical two-cavity dielectric filter. Each cavity unit zone is connected via a coupling layer. The number of cavities each containing two reflector unit zones can be one or many successive cavities on the same substrate. As the number of cavities increases, the relative bandwidth of the central peak decreases as shown in Fig. 70 for one, two, and three cavities. However, with each successive cavity the maximum relative transmission at the peak wavelength decreases. Figure 71 shows the relative wavelength versus the transmission percent for a one- and two-cavity dielectric filter, where the peak or central wavelength λ_p can be shifted to any desired value by the appropriate choice of the material layers and their respective thicknesses. This shifting usually occurs in the wavelength region from 600 to 1200 nm.

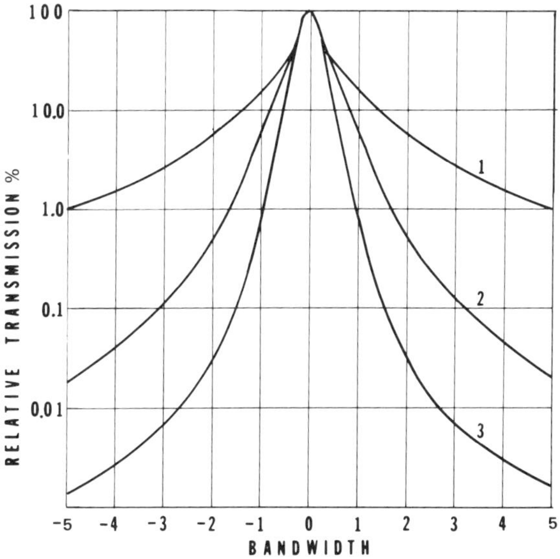

Fig. 70 Bandwidth vs. relative transmission % for multiple dielectric filter cavities.

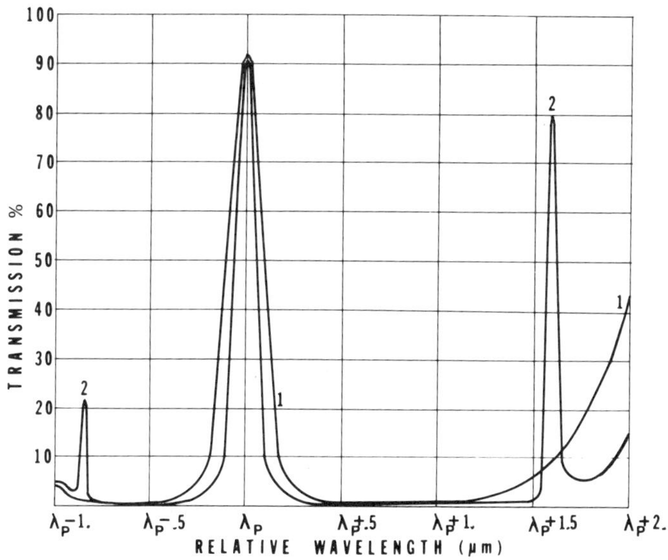

Fig. 71 Wavelength vs. transmission % for one- and two-cavity dielectric filter.

The determination of the central wavelength usually assumes a 0° incidence angle of a collimated light beam on the filter. For collimated nonorthoganal incidence beams, the relationship between the central wavelength at normal (0°) incidence λ_0 and the central wavelength at the off-normal (<30°) incidence λ_{θ_i} is given by

$$\lambda_{\theta_i} = \frac{\lambda_0(n^2 - \sin^2\theta_i)^{1/2}}{n}$$

where n is the effective refractive index of the completed filter.

The blocking of wavelengths on either side of a given filter spectrum (transmission vs. wavelength) is usually accomplished using long-pass filters (to remove short wavelengths), induced transmission filters (to remove longer wavelengths), or short-pass filters. Long-pass filters are often used as the substrate for the multilayer coatings (often colored glass).

Induced transmission filters are of a multilayer construction such as shown in Fig. 72 and require the use of a suitable substrate material. When an induced transmission filter is coupled to a dichroic filter or other types of multilayer filters, the choice of materials, layer thicknesses, incidence angles, number of layers, and layer sequence often becomes a complex problem of optimization of the desired optical properties and is usually solved by computer design and optimization. For these filters "optimization" becomes a relative term, since the design of the desired transmission properties usually involves trade-offs or compromises in the spectral characteristics versus maximum transmission percent.

Spatial Filters

A spatial filter is usually a pinhole, which can be accurately translated in the x, y, z, θ, ϕ directions using a suitable micromanipulator and which has a variable aperture and is usually coupled with a focusing lens. The radius of the hole r_h, usually used for laser beams, is given by

$$r_h = \frac{\lambda f}{r_b}$$

where

r_b = radius of the beam into the lens
λ = laser emission wavelength or the average beam wavelength
f = focal length of the lens

The fraction P(r) of the total beam power (P_t) transmitted by the spatial filter is given by

$$\frac{P(r)}{P_t} = 1 - \exp - \left(\frac{\sqrt{2}\pi r_b r_h}{\lambda f}\right)^2$$

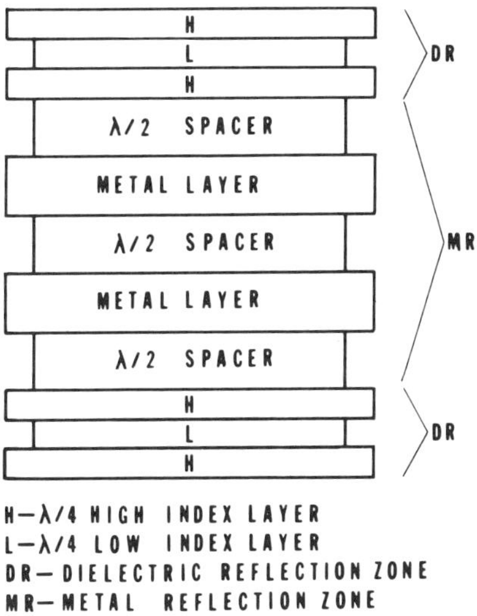

Fig. 72 Induced transmission filter material layer configurations.

The minimum frequency blocked by the spatial filter is $r_h/f\lambda$. The intensity profile of the beam is given by

$$I(r) = I_0 \exp\left[-2\left(\frac{r}{r_0}\right)^2\right]$$

where $I_0 = 2P_t/\pi r_0^2$ and r_o is the beam radius at $I_0 \exp(-2)$.

Fabry-Perot Filters

A specialized filter sometimes encountered in electro-optics is the Fabry-Perot filter. It is made with alternate layers of silver and a dielectric. The transmission (T) of the filter is given by

$$T = t^2\left[(1 - r)^2 + 4r \sin^2\frac{\delta}{2}\right]^{-1}$$

where

- r = silver reflectivity
- t = film transmission
- $\delta = 4\pi d/\lambda\ n^2 - \sin^2\theta_i + 2\phi$
- d = dielectric layer thickness
- n = dielectric index of refraction

ϕ = light phase shift at the silver-dielectric interface
ϕ_i = angle of incidence.

The points of maximum and minimum transmission are given by

$$T_{max} = \frac{t^2}{(1 - r)^2} \qquad T_{min} = \frac{t^2}{(1 + r)^2}$$

respectively. The 50% filter bandwidth (B_{50}) is given by

$$B_{50} = \frac{2\lambda_{max}}{m\pi - \phi_0} \sin^{-1} \frac{1 - r}{2r^{1/2}}$$

where

λ_{max} = peak transmission wavelength
m = order of the peak
ϕ_0 = initial phase shift

B_{50} is given between the points 50% below peak transmission. The wavelengths of peak transmission are $\lambda_m = 2nd/m \cos\theta'$, where θ' is the angle inside the etalon spacer.

Refractive Index and Physical Properties

One of the most important properties of optical materials is the index of refraction. Various standard wavelengths are used to define the refractive index vs. wavelength characteristics of a given optical material. Table 52 lists these wavelengths and the associated source and color and refractive index symbol. Another important property is the percent transmission at a specified wavelength. At transmissions above 90% some of the measured loss is due to reflections which can be reduced by using the appropriate (AR) coatings. The major concern is for transmissions below 90%. Figures 73 to 78 show the wavelength versus relative transmission percent for several important optical materials. The relative transmission is a function of the material thickness and purity. From these curves it is clear that optical materials must be carefully chosen for specific applications to minimize transmission losses by absorption. Conversely, in certain wavelength regions, an optical material may be purposely chosen to act as a filter.

In addition to refractive index and percent transmission, certain other physical properties are of importance in the selection of optical materials. Some of these properties include maximum operating temperature, density,

Table 52 Refractive Index Definitions

Symbol	Spectral line	Wavelength (nm)	Color
n_{uv}	Hg	365.0	UV
n_h	Hg	404.7	Blue
n_g	Hg	435.8	Blue
$n_{f'}$	Cd	480.0	Blue
n_f	H	486.1	Blue
n_e	Hg	546.1	Green
n_d	He	587.6	Yellow
n_D	Na	589.3	Yellow
$n_{c'}$	Cd	643.8	Red
n_c	H	656.3	Red
$n_{R'}$	He	768.2	Red
n_{ir}	Hg	1014.0	IR

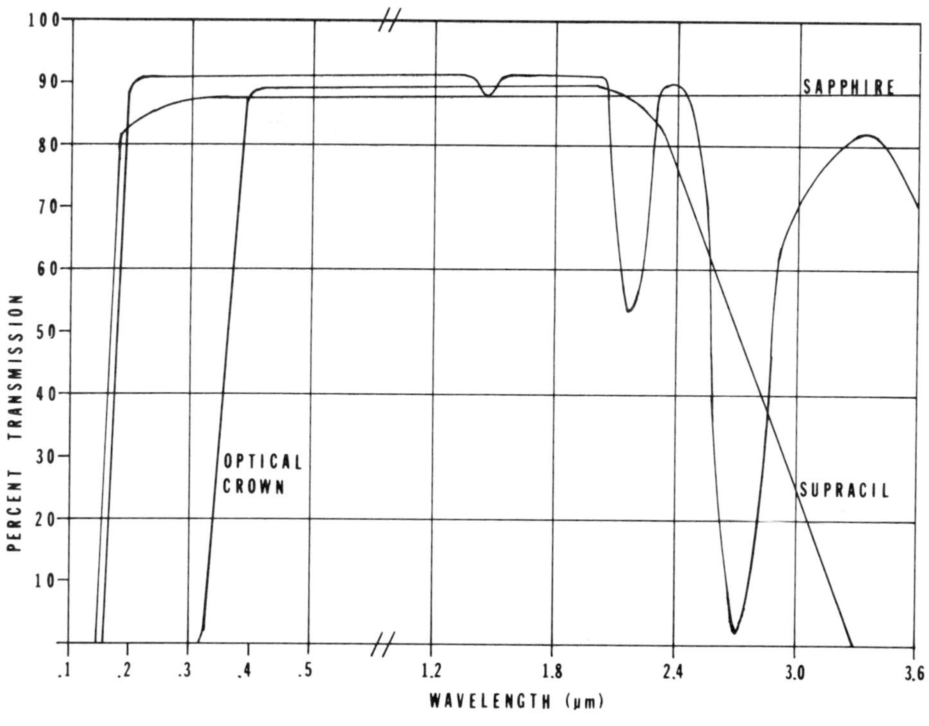

Fig. 73 Transmission vs. wavelength for various optical materials.

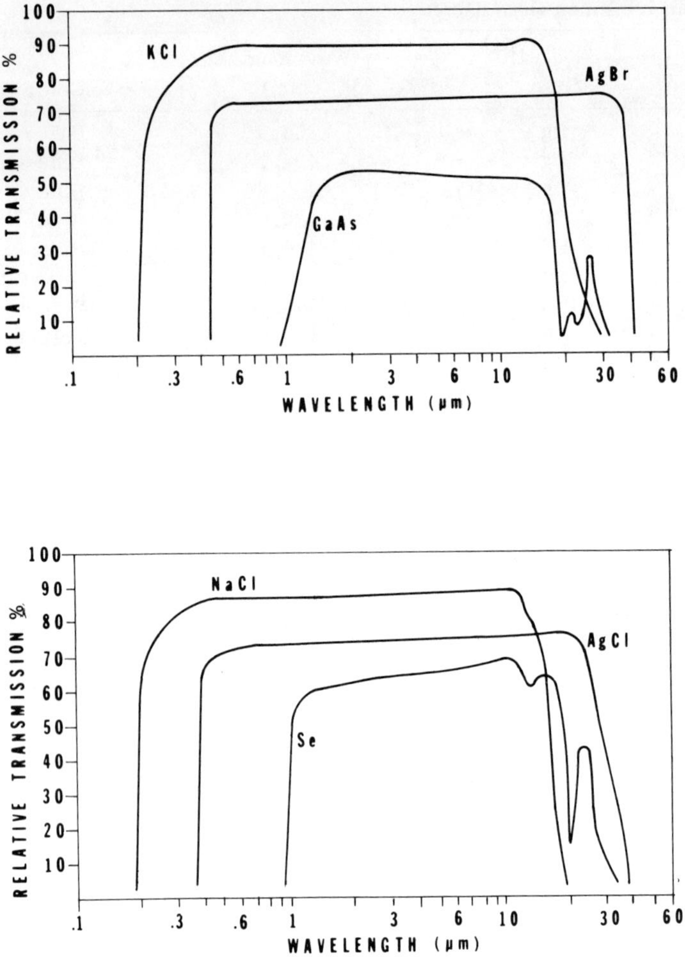

Fig. 74 Relative transmission vs. wavelength for various optical materials.

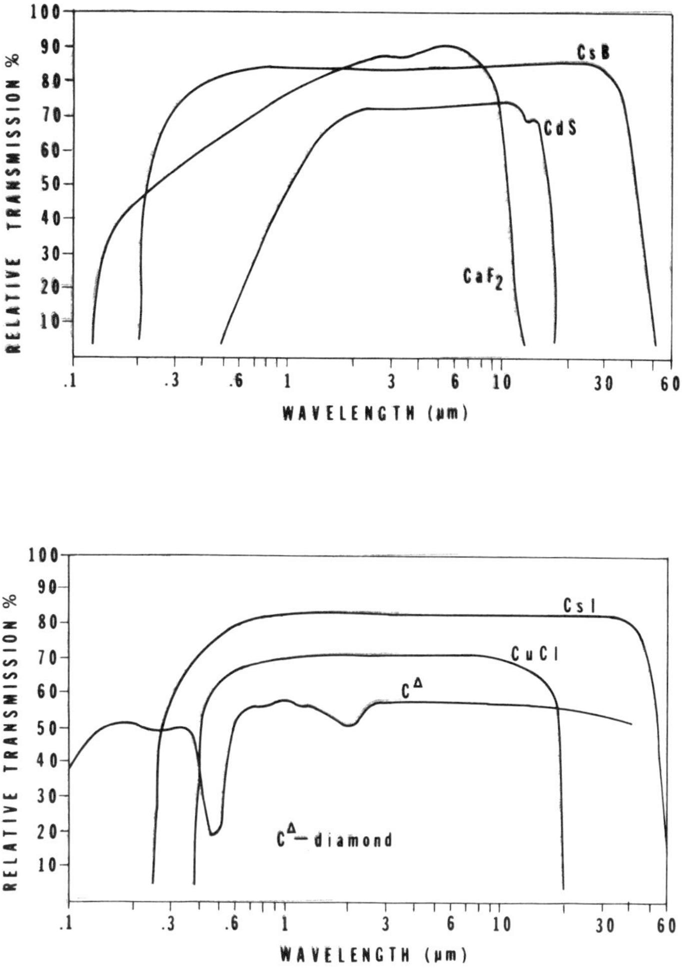

Fig. 75 Relative transmission vs. wavelength for various optical materials.

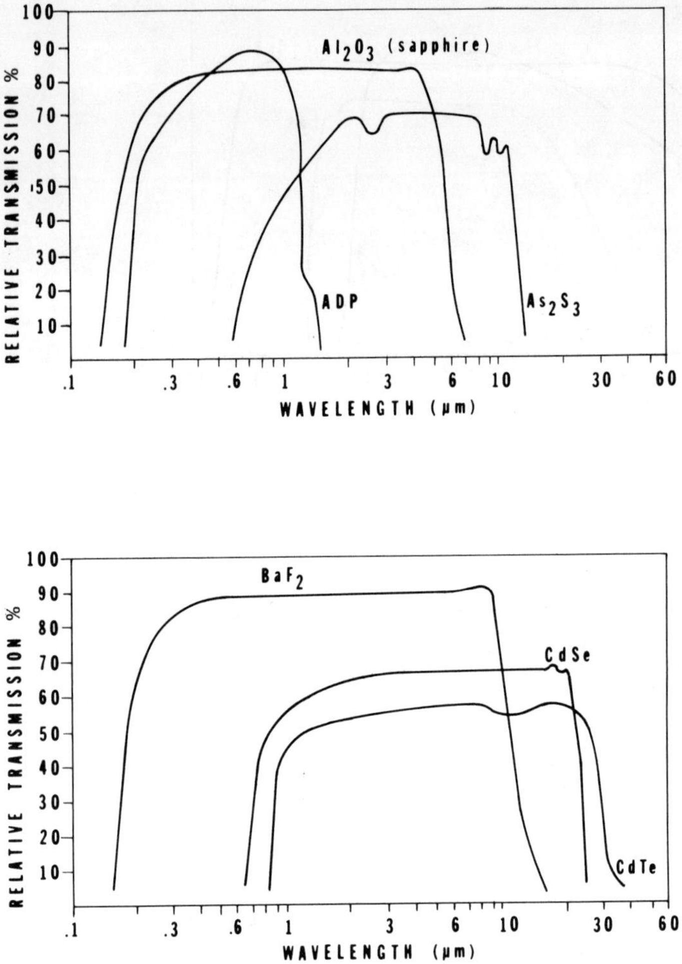

Fig. 76 Relative transmission vs. wavelength for various optical materials.

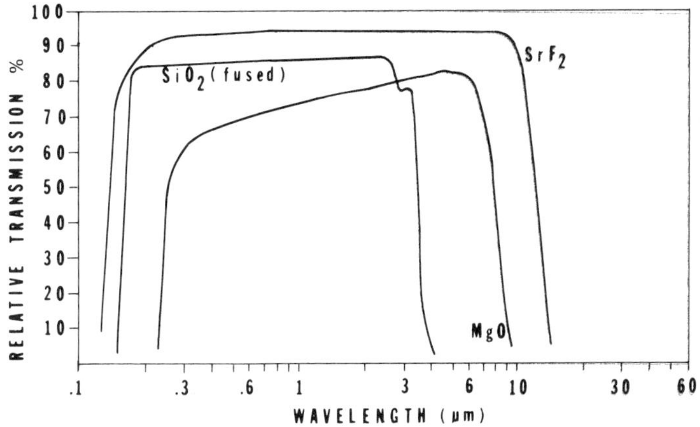

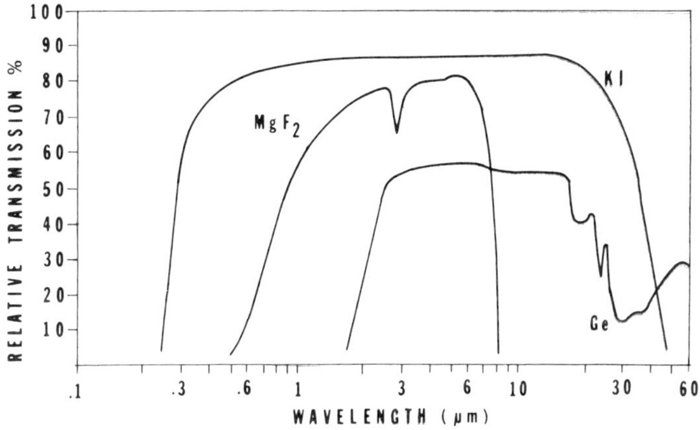

Fig. 77 Relative transmission vs. wavelength for various optical materials.

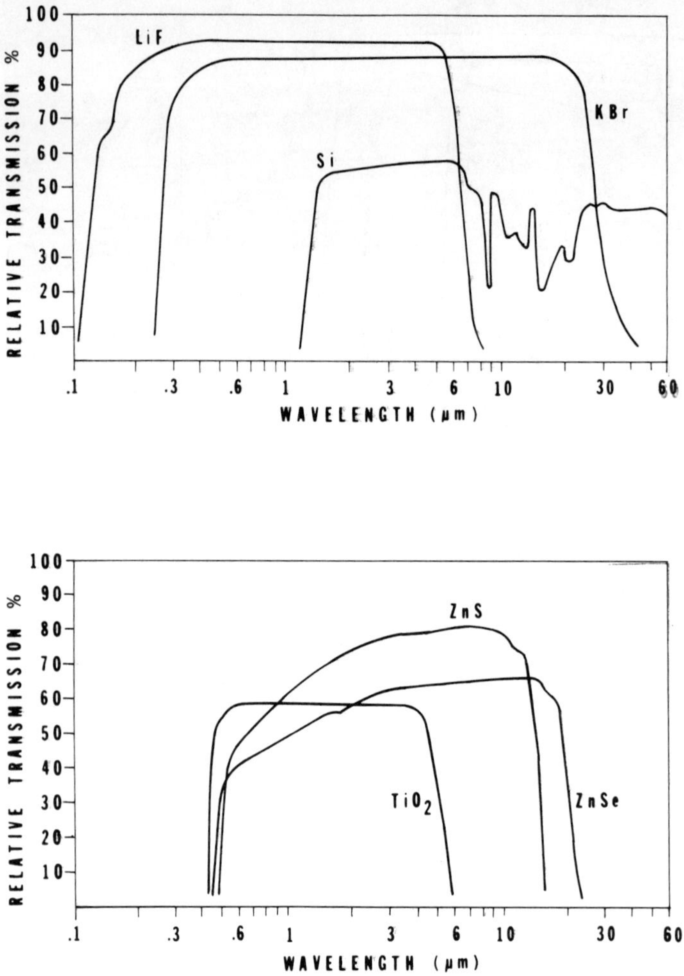

Fig. 78 Relative transmission vs. wavelength for various optical materials.

Young's modulus, Poisson's constant, the coefficient of expansion, specific heat, and for crystalline materials, the solubility in water or other media. Table 53 lists some of the physical properties of a few common optical materials. Table 56 lists some of the optical and physical properties of Schott glasses. Table 54 lists the properties of the common optical crystals including specific gravity, melting point, and solubility in water. Figure 79 shows the refractive index versus wavelength of some of the most commonly encountered optical crystals. The refractive index of certain optical crystals and materials depends on the crystal orientation; thus, an ordinary and

Table 53 Physical Properties of Common Optical Materials

Material property	Pyrex 7740	Acrylic plastic	Op. Crown B270	BK7 no. 517645	Zerodur glass-ceramic
Maximum operating T (°C)	480	80	250	350	500
Density (g/cm^3 at 20°C)	2.23	1.19	2.56	2.51	2.52
Young's modulus (kg/cm^2 X 10^5)	6.8		7.0	7.5	9.0
Poisson's constant (at 20°C)	0.20	0.35		0.21	
Thermal expansion (X 10^{-7}/°C)	32	650	93	71	.2
Refractive index (n_d)	1.474	1.491	1.523	1.517	1.542
Heat capacity (cal/g°C)	0.180	0.297	0.181	0.205	0.196
Abbe number[a] (V_d)	60.2	57.2	58.8	64.2	

[a]$V_d = (n_d - 1)/(n_f - n_e)$

extraordinary refractive index profile is often given for a crystalline material. Figure 80 shows the dual refractive index profile for crystalline quartz over the visible region.

Optical glasses are usually listed by a code number given by a six-digit number. The first three digits are given by $n_D - 1$, the refractive index at the sodium line minus 1. The second three digits are given by 10V, where V is the Abbe number defined as $(n_{D-1})/(n_f - n_c)$. Table 55 lists some selected optical glasses by code number, giving a few of the major refractive indexes at the major lines and the value of the critical angle Φ_c. Note that Φ_c is a function of wavelength and refractive index and must be specified by the wavelength used for the measurement for a particular material: $\Phi_c(\lambda, n)$. Table 56 lists the optical glass properties, including the thermal expansion coefficient and the glass transition temperature T_g, for Schott glass types. Table 57 lists the refractive index formulas of various electro-optic materials.

Table 54 Properties of Common Optical Crystals

Crystal Symbol	Index n_D	Specific gravity	Lattice symbol	Melting point (°C)	Solubility in water (g/100g at 20°C)
AgBr	2.253	6.47	Cubic	432	1.2×10^{-5}
AgCl	2.071	5.56	Cubic	455	1.5×10^{-4}
BaF_2	1.474	4.89	Cubic	1355	0.1
$CaCO_3$	1.658	2.94	Rhombic	1339	1.3×10^{-3}
$CaCO_3$	1.486	2.71	Hexagonal	1339	1.2×10^{-3}
CaF_2	1.434	3.18	Cubic	1423	1.3×10^{-3}
CdF_2	1.560	6.64	Cubic	1100	4.0
CdS	2.506	4.82	Hexagonal	1750	1.4×10^{-4}
CsBr	1.698	4.44	Cubic	636	108.3
CsCl	1.642	3.99	Cubic	645	186.1
CsF	1.478	4.12	Cubic	682	371.1
CsI	1.788	4.51	Rhombic	626	71.4
KBr	1.559	2.75	Cubic	734	65.2
KCl	1.490	1.99	Cubic	770	34.7
KF	1.363	2.48	Cubic	858	94.8
KI	1.677	3.13	Cubic	681	144.0
KH_2PO_4	(1.486; 1.510)	2.34	Tetragonal	252	30.1
LiF	1.392	2.63	Cubic	845	0.28
LiI	1.955	3.49	Cubic	449	164.2
MgF_2	(1.378; 1.390)	3.15	Tetragonal	1261	7.7×10^{-3}
MgO	1.736	3.58	Cubic	2852	6.2×10^{-4}
NaBr	1.641	3.20	Cubic	1390	91.1
NaCl	1.544	2.16	Cubic	801	36.2
NaF	1.336	2.79	Cubic	993	4.2
NaI	1.775	3.67	Cubic	661	185.3
$NaNO_3$	(1.336; 1.587)	2.26	Trig or rh.	307	88.0
$NH_4H_2PO_4$	(1.479; 1.525)	1.80	Tetragonal	190	36.8
$PbCl_2$	(2.199; 2.217)	5.85	Rhombic	501	1.0
PbF_2	2.260	8.24	Rhombic	855	6.4×10^{-2}
SrF_2	1.442	4.24	Cubic	1473	1.2×10^{-2}
TlBr	2.601	7.56	Cubic	480	4.8×10^{-2}
TlCl	2.247	7.00	Cubic	430	0.32
α-ZnS	(2.356; 2.378)	3.98	Hexagonal	1700	7.0×10^{-4}
β-ZnS	2.368	4.10	Cubic	1020	6.9×10^{-4}

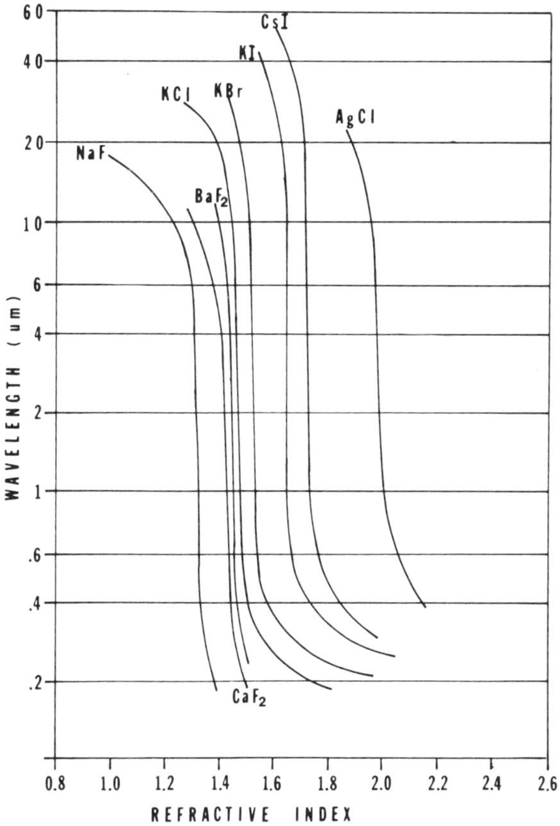

Fig. 79 Refractive index vs. wavelength for various optical materials.

Optical Plastics

Using carefully controlled molding techniques plastic optical components can be made with very close tolerances: ± 0.01 mm for linear dimensions and radii of curvature, ±0.15 mm for thickness and homogeneity, and ±1 arc minute for angular dimensions. The main problem in optical plastics is obtaining consistent and homogeneous indexes of refraction.

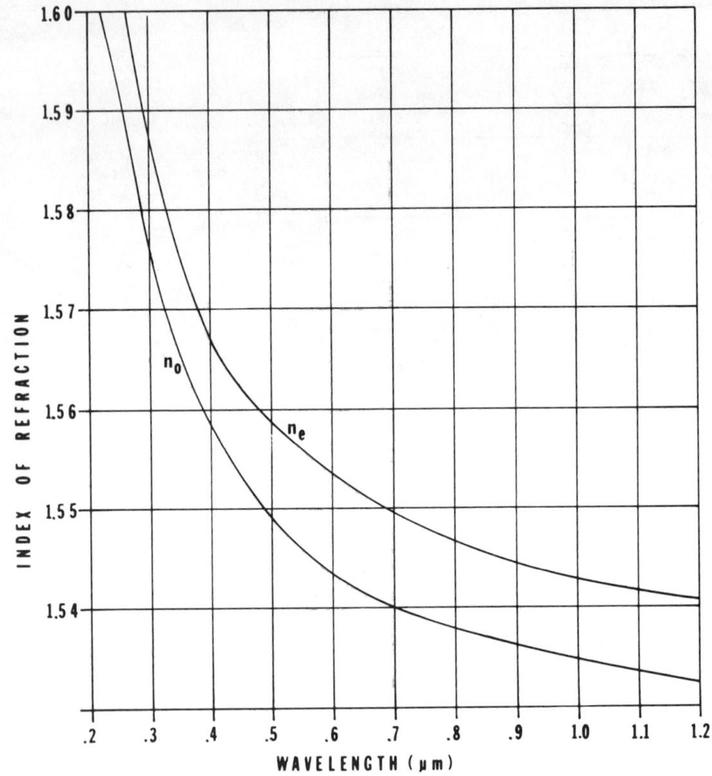

Fig. 80 Index of refraction vs. wavelength for crystalline quartz (300 K).

Depending upon the extent of impurities, molding temperature, and the concentration of various copolymers and solvents, optical plastics can have reasonably large variations in the index of refraction for a given material composition. Table 58 gives the approximate index of refraction for various optical plastics and the expected range for some materials which have known compositions made in varying degrees of polymerization or mass density.

Optical Fibers

The major characteristics to be considered in optical fibers for optical communications links are

1. Attenuation, and its variance with transmission input wavelength, modal distribution, and cable temperature

Table 55 Properties of Selected Optical Glasses

Glass types	Code number[a]	Critical angle[b]	653.3 nm n_c	546.1 nm n_e	486.1 nm n_F	434.1 nm n_G	404.7 nm n_h
Crown flint	501/567	41° 45'	1.49940	1.50417	1.50825	1.51327	1.51714
Borosilicate crown	505/665	41° 38'	1.50272	1.50688	1.51032	1.51455	1.51771
Crown flint	508/610	41° 32'	1.50551	1.51005	1.51383	1.51849	1.52201
Borosilicate crown	511/635	41° 26'	1.50860	1.51300	1.51664	1.52112	1.52450
Crown flint	513/605	41° 22'	1.51050	1.51509	1.51897	1.52375	1.52737
Borosilicate crown	517/645	41° 14'	1.51462	1.51901	1.52264	1.52712	1.53047
Crown flint	523/585	41° 03'	1.52035	1.52521	1.52930	1.53437	1.53822
Crown flint	530/518	40° 49'	1.52702	1.53251	1.53724	1.54316	1.54770
Light barium crown	541/598	40° 28'	1.53832	1.54323	1.54737	1.55250	1.55638
Light flint	549/457	40° 13'	1.54556	1.55199	1.55757	1.56468	1.57020
Light flint	573/425	39° 28'	1.56912	1.57631	1.58257	1.59059	1.59686
Light barium crown	588/533	39° 02'	1.58477	1.59071	1.59579	1.60214	1.60698
Dense barium crown	611/588	38° 22'	1.60796	1.61359	1,61835	1.62425	1.62868
Dense barium crown	613/595	38° 19'	1.60999	1.61557	1.62029	1.62614	1.63053
Barium flint	617/385	38° 12'	1.61240	1.62095	1.62845	1.63815	1.64578
Dense flint	654/340	37° 12'	1.64857	1.65872	1.66782	1.67967	1.68908
Dense barium flint	657/512	37° 07'	1.65325	1.66016	1.66612	1.67360	1.67934
Dense barium flint	700/410	36° 21'	1.69509	1.70421	1.71218	1.72246	1.73054
Extra-dense flint	728/283	35° 22'	1.72080	1.73430	1.74652	1.76276	1.77592
Dense flint	741/277	35° 01'	1.73306	1.74077	1.75980	1.79065	1.80383
Dense flint	805/255	33° 12'	1.79608	1.80518	1.82771	1.86428	1.87742
Lanthanum dense flint	808/408	33° 10'	1.80214	1.80802	1.82194	1.84284	1.86011

[a]Code number: $V = (n_D - 1)/(n_F - n_C)$

[b]Critical angle: function of wavelength and refractive index. Table 54 values given at 589.3 nm.

Table 56 Schott Optical Glass Properties

Type	n_d	ν_d	n_F-n_C	n_e	ν_e	$n_{F'}-n_{C'}$	n_C-n_r	n_d-n_C	$n_{F'}-n_e$	n_g-n_F	n_h-n_g	Density g/cm³	Knoop hardness kp/mm²	$\alpha \times 10^7 \cdot °C$ -30° to +70 °C	Tg [°C]
LaSF 7 921 357	1.921200	35.72	0.025790	1.927317	35.49	0.026129	0.004212	0.007606	0.013600	0.014899	0.012849	5.79	670	57	684
LaSF 8 * 807 316	1.807410	31.61	0.025542	1.813454	31.38	0.025920	0.004090	0.007450	0.013591	0.015117	0.013274	4.87	420	59	476
LaSF 9 850 322	1.850256	32.23	0.026378	1.856498	31.98	0.026779	0.004194	0.007670	0.014064	0.015664	0.013782	4.53	520	76	704
LaSF 11 * 802 443	1.801658	44.26	0.018111	1.805965	44.02	0.018311	0.003039	0.005418	0.009442	0.010173	0.008615	4.62	600	58	644
LaSF 12 * 836 423	1.835653	42.29	0.019759	1.840349	42.04	0.019987	0.003289	0.005887	0.010333	0.011176	0.009500	5.28	630	60	651
LaSF 13 * 855 366	1.855443	36.59	0.023381	1.860990	36.34	0.023694	0.003800	0.006882	0.012344	0.013526	0.011664	5.10	540	62	619
LaSF 15 878 382	1.878000	38.17	0.023001	1.883460	37.94	0.023287	0.003791	0.006815	0.012084	0.013165	0.011283	4.93	640	63	666
LaSF 18 913 325	1.913479	32.49	0.028113	1.920135	32.27	0.028513	0.004538	0.008232	0.014914	0.016515	0.014435	4.64	620	57	657
SF 1 717 295	1.717360	29.51	0.024307	1.723107	29.29	0.024688	0.003845	0.007045	0.012994	0.014536	0.012834	4.46	340	81	417
SF 2 648 339	1.647689	33.85	0.019135	1.652219	33.60	0.019412	0.003073	0.005592	0.010165	0.011261	0.009836	3.86	350	84	441
SF 3 740 282	1.740002	28.20	0.026244	1.746202	27.98	0.026666	0.004129	0.007506	0.014061	0.015784	0.013999	4.64	330	84	415
SF 4 755 276	1.755201	27.58	0.027383	1.761668	27.37	0.027829	0.004299	0.007905	0.014687	0.016526	0.014687	4.78	330	80	420
SF 5 673 322	1.672698	32.21	0.020884	1.677641	31.97	0.021194	0.003337	0.006086	0.011116	0.012353	0.010828	4.07	340	82	425
SF 6 805 254	1.805182	25.43	0.031660	1.812652	25.24	0.032200	0.004922	0.009092	0.017049	0.019303	0.017302	5.18	310	81	423
SF 7 640 346	1.639801	34.61	0.018487	1.644180	34.36	0.018749	0.002979	0.005412	0.009806	0.010841	0.009448	3.80	350	79	448
SF 8 689 312	1.688932	31.18	0.022098	1.694159	30.95	0.022432	0.003517	0.006428	0.011781	0.013122	0.011533	4.22	340	82	423
SF 9 654 337	1.654461	33.65	0.019447	1.659065	33.41	0.019728	0.003126	0.005685	0.010328	0.011441	0.009990	3.91	360	81	434
SF 10 728 284	1.728250	28.41	0.025634	1.734302	28.19	0.026050	0.004030	0.007402	0.013749	0.015490	0.013809	4.28	370	75	454
SF 11 785 258	1.784720	25.76	0.030468	1.791901	25.55	0.030998	0.004731	0.008734	0.016439	0.018728	0.016931	4.74	380	61	503
SF 12 648 338	1.648310	33.84	0.019158	1.652846	33.59	0.019434	0.003079	0.005601	0.010174	0.011278	0.009866	3.74	360	78	452
SF 13 741 276	1.740771	27.60	0.026841	1.747104	27.38	0.027290	0.004201	0.007730	0.014429	0.016324	0.014633	4.36	380	71	472
SF 14 762 265	1.761820	26.53	0.028718	1.768592	26.31	0.029211	0.004472	0.008245	0.015475	0.017588	0.015861	4.54	370	66	478
SF 15 699 301	1.698951	30.07	0.023246	1.704445	29.84	0.023611	0.003679	0.006739	0.012429	0.013931	0.012355	4.06	370	79	455
SF 16 646 341	1.646111	34.05	0.018975	1.650605	33.80	0.019248	0.003052	0.005548	0.010074	0.011153	0.009735	3.85	340	84	443
SF 17 650 337 *	1.650170	33.67	0.019311	1.654742	33.42	0.019591	0.003100	0.005643	0.010259	0.011369	0.009935	3.87	350	85	434
SF 18 722 293	1.721510	29.25	0.024671	1.727342	29.03	0.025059	0.003899	0.007149	0.013192	0.014768	0.013053	4.49	330	81	422
SF 19 667 330	1.666800	33.01	0.020202	1.671582	32.76	0.020498	0.003239	0.005898	0.010740	0.011917	0.010430	4.02	360	77	434
SF 50 655 329 *	1.654727	32.88	0.019914	1.659440	32.63	0.020210	0.003179	0.005803	0.010601	0.011776	0.010319	3.79	360	101	401
SF 51 660 329	1.660248	32.94	0.020044	1.664993	32.69	0.020342	0.003202	0.005841	0.010668	0.011849	0.010403	3.81	380	78	455
SF 52 689 306 *	1.688521	30.62	0.022484	1.693838	30.39	0.022832	0.003565	0.006526	0.012007	0.013407	0.011815	4.10	350	95	406
SF 53 728 287	1.728301	28.69	0.025388	1.734299	28.47	0.025793	0.004001	0.007344	0.013595	0.015261	0.013545	4.45	340	82	420
SF 54 741 281	1.740800	28.09	0.026370	1.747028	27.88	0.026797	0.004147	0.007617	0.014138	0.015905	0.014131	4.56	360	77	432
SF 55 762 270	1.761800	26.95	0.028267	1.768472	26.74	0.028739	0.004421	0.008140	0.015192	0.017151	0.015302	4.72	330	82	421
SF 56 785 261	1.784701	26.08	0.030092	1.791799	25.87	0.030602	0.004693	0.008651	0.016197	0.018342	0.016421	4.92	340	79	429
SF 57 847 238	1.846663	23.83	0.035534	1.855041	23.64	0.036165	0.005469	0.010152	0.019208	0.021861	0.019725	5.51	300	83	422
SF 58 918 215	1.917613	21.51	0.042658	1.927647	21.34	0.043461	0.006504	0.012110	0.023189	0.026699	0.024371	5.95	270	90	396
SF 59 953 204	1.952497	20.36	0.046774	1.963492	20.21	0.047684	0.007069	0.013217	0.025511	0.029534	0.027206	6.26	250	94	362
SF 61 751 275	1.750842	27.52	0.027287	1.757284	27.30	0.027735	0.004281	0.007871	0.014647	0.016510	0.014698	4.64	340	79	424
SF 62 681 319	1.681339	31.94	0.021331	1.686386	31.70	0.021650	0.003406	0.006214	0.011359	0.012635	0.011090	4.15	360	82	439

Table 56–Continued

Type	n_d	ν_d	$n_F - n_C$	n_e	ν_e	$n_{F'} - n_{C'}$	$n_C - n_r$	$n_d - n_C$	$n_{F'} - n_e$	$n_g - n_F$	$n_h - n_g$	Density g/cm³	Knoop hardness kp/mm²	$\alpha \times 10^7/°C$ −30° to +70 °C	Tg [°C]
F 14 601 382	1.601400	38.23	0.015731	1.605130	37.97	0.015938	0.002568	0.004638	0.008299	0.009103	0.007866	3.44	380	79	438
F 15 606 378	1.605651	37.83	0.016011	1.609447	37.56	0.016224	0.002608	0.004714	0.008454	0.009283	0.008027	3.48	390	81	433
BaSF 1 626 390	1.626061	38.96	0.016068	1.629871	38.68	0.016283	0.002617	0.004731	0.008484	0.009313	0.008052	3.66	400	85	493
BaSF 2 664 358	1.664460	35.83	0.018545	1.668853	35.56	0.018808	0.002989	0.005431	0.009834	0.010867	0.009468	3.90	410	82	493
BaSF 5 603 425 *	1.603230	42.48	0.014201	1.606602	42.19	0.014378	0.002339	0.004206	0.007462	0.008130	0.006972	3.49	400	79	502
BaSF 6 668 419	1.667551	41.93	0.015921	1.671330	41.64	0.016123	0.002615	0.004710	0.008374	0.009142	0.007864	3.79	450	74	570
BaSF 10 650 392	1.650160	39.15	0.016608	1.654100	38.87	0.016830	0.002702	0.004889	0.008768	0.009618	0.008306	3.91	400	86	484
BaSF 12 670 392 *	1.669979	39.20	0.017090	1.674032	38.92	0.017319	0.002784	0.005032	0.009023	0.009907	0.008579	3.90	430	79	532
BaSF 13 698 386	1.697610	38.57	0.018086	1.701897	38.29	0.018330	0.002942	0.005321	0.009557	0.010521	0.009147	3.97	450	71	584
BaSF 14 * 700 350	1.699681	34.96	0.020012	1.704418	34.70	0.020299	0.003221	0.005853	0.010626	0.011788	0.010336	4.00	420	70	538
BaSF 50 * 710 366	1.710200	36.62	0.019395	1.714797	36.37	0.019654	0.003162	0.005711	0.010243	0.011266	0.009769	4.07	450	55	500
BaSF 51 724 381	1.723728	38.11	0.018991	1.728233	37.85	0.019242	0.003097	0.005596	0.010019	0.010981	0.009478	4.31	450	54	522
BaSF 52 702 410	1.701810	41.01	0.017112	1.705872	40.75	0.017320	0.002836	0.005082	0.008977	0.009777	0.008389	3.96	480	52	533
BaSF 53 * 736 320	1.736273	31.99	0.023015	1.741717	31.74	0.023365	0.003665	0.006694	0.012272	0.013696	0.012091	4.42	400	71	507
BaSF 54 * 736 322	1.736270	32.15	0.022902	1.741687	31.90	0.023251	0.003643	0.006658	0.012215	0.013630	0.012028	4.41	440	73	515
BaSF 55 * 700 347	1.699810	34.68	0.020178	1.704587	34.42	0.020469	0.003251	0.005902	0.010714	0.011886	0.010434	3.95	450	51	509
BaSF 56 * 657 367	1.657149	36.73	0.017889	1.661388	36.47	0.018137	0.002892	0.005247	0.009473	0.010444	0.009078	3.85	430	81	491
BaSF 57 651 419	1.651475	41.90	0.015547	1.655165	41.62	0.015741	0.002562	0.004605	0.008171	0.008919	0.007671	3.73	450	71	549
LaF N2 744 448	1.744001	44.77	0.016618	1.747952	44.49	0.016812	0.002758	0.004948	0.008692	0.009395	0.007981	4.54	450	82	616
LaF N3 717 480	1.716997	47.98	0.014943	1.720553	47.69	0.015108	0.002501	0.004469	0.007788	0.008366	0.007056	4.34	460	78	620
LaF N7 750 350	1.749502	34.95	0.021445	1.754584	34.72	0.021736	0.003493	0.006309	0.011335	0.012489	0.010852	4.38	430	53	500
LaF N8 735 416	1.735197	41.59	0.017678	1.739395	41.32	0.017893	0.002925	0.005248	0.009274	0.010095	0.008653	4.02	500	66	545
LaF 9 795 284	1.795040	28.39	0.028001	1.801656	28.18	0.028451	0.004407	0.008092	0.015003	0.016847	0.014951	4.96	350	72	450
LaF 10 784 439	1.784431	43.88	0.017875	1.788681	43.64	0.018074	0.002994	0.005343	0.009325	0.010059	0.008532	4.61	600	65	614
LaF N11 757 318	1.756925	31.80	0.023799	1.762558	31.58	0.024149	0.003816	0.006946	0.012656	0.014060	0.012324	4.62	400	56	472
LaF 13 * 776 378	1.775507	37.84	0.020496	1.780369	37.58	0.020764	0.003352	0.006048	0.010804	0.011837	0.010215	4.55	490	57	582
LaF 20 682 482	1.682476	48.20	0.014159	1.685845	47.92	0.014313	0.002375	0.004238	0.007376	0.007934	0.006710	3.87	520	74	609
LaF 21 788 474	1.788309	47.39	0.016633	1.792269	47.15	0.016802	0.002821	0.005005	0.008630	0.009235	0.007764	4.44	610	59	667
LaF 22 782 371	1.781791	37.09	0.021077	1.786786	36.83	0.021365	0.003425	0.006197	0.011145	0.012283	0.010691	4.21	520	70	683
LaF 23 689 495	1.689002	49.48	0.013926	1.692316	49.18	0.014077	0.002337	0.004170	0.007252	0.007787	0.006565	4.21	420	81	601
LaF 24 757 478	1.757192	47.81	0.015836	1.760961	47.57	0.015997	0.002687	0.004765	0.008218	0.008799	0.007402	4.23	610	58	619
LaF 25 * 784 413	1.784272	41.30	0.018989	1.788781	41.04	0.019219	0.003145	0.005640	0.009959	0.010838	0.009292	4.45	580	58	588
LaF 26 746 400	1.745973	40.02	0.018642	1.750397	39.76	0.018873	0.003082	0.005528	0.009791	0.010686	0.009193	4.20	520	56	553
LaSF 1 803 468	1.802789	46.76	0.017168	1.806875	46.52	0.017345	0.002902	0.005158	0.008918	0.009552	0.008040	4.95	600	59	626
LaSF N3 808 408	1.808010	40.75	0.019827	1.812718	40.50	0.020068	0.003280	0.005887	0.010400	0.011307	0.009675	4.68	560	59	599
LaSF 5 881 410	1.880692	41.01	0.021476	1.885797	40.77	0.021727	0.003567	0.006394	0.011236	0.012151	0.010320	5.77	650	59	696
LaSF 6 961 349	1.960520	34.92	0.027503	1.967043	34.70	0.027872	0.004471	0.008093	0.014525	0.015933	0.013747	6.13	670	63	696

Table 56–Continued

Type	n_d	ν_d	n_F-n_C	n_e	ν_e	$n_{F'}-n_{C'}$	n_C-n_r	n_d-n_C	$n_{F'}-n_e$	n_g-n_F	n_h-n_g	Density g/cm³	Knoop hardness kp/mm²	$\alpha \times 10^7 \cdot °C$ - 30° to + 70 °C	Tg [°C]
LLF 3 * 560 472	1.560130	47.16	0.011876	1.562953	46.88	0.012008	0.001995	0.003552	0.006195	0.006693	0.005695	2.99	450	70	518
LLF 4 * 561 452	1.561381	45.24	0.012410	1.564329	44.95	0.012554	0.002070	0.003699	0.006491	0.007034	0.006003	3.02	400	82	465
LLF 6 532 488	1.531719	48.76	0.010905	1.534312	48.48	0.011022	0.001841	0.003270	0.005676	0.006110	0.005177	2.81	420	75	422
LLF 7 * 549 454	1.548831	45.41	0.012087	1.551703	45.13	0.012226	0.002021	0.003606	0.006317	0.006837	0.005827	2.98	390	73	422
BaF 3 583 465	1.582670	46.47	0.012538	1.585650	46.17	0.012684	0.002088	0.003736	0.006557	0.007093	0.006033	3.28	420	78	533
BaF 4 606 439	1.605620	43.93	0.013787	1.608894	43.63	0.013956	0.002276	0.004091	0.007235	0.007862	0.006721	3.50	400	79	540
BaF 5 607 494	1.607290	49.40	0.012293	1.610214	49.10	0.012427	0.002061	0.003679	0.006405	0.006890	0.005825	3.54	450	70	604
BaF 6 589 487	1.588999	48.66	0.012104	1.591878	48.36	0.012239	0.002028	0.003619	0.006312	0.006799	0.005757	3.36	420	73	578
BaF 8 624 470	1.623740	47.00	0.013270	1.626895	46.71	0.013422	0.002212	0.003957	0.006934	0.007491	0.006363	3.67	460	70	589
BaF 9 643 480	1.643277	47.96	0.013414	1.646468	47.66	0.013565	0.002239	0.004004	0.007002	0.007547	0.006393	3.85	460	65	610
BaF N10 670 471	1.670031	47.11	0.014222	1.673412	46.82	0.014384	0.002377	0.004246	0.007427	0.008024	0.006824	3.76	480	68	630
BaF N11 667 484	1.666721	48.42	0.013769	1.669995	48.13	0.013920	0.002309	0.004119	0.007178	0.007734	0.006555	3.76	470	68	631
BaF 12 * 639 452	1.639300	45.18	0.014151	1.642662	44.88	0.014318	0.002349	0.004210	0.007410	0.008041	0.006872	3.60	460	68	596
BaF 13 669 450	1.668920	44.96	0.014877	1.672454	44.67	0.015055	0.002467	0.004424	0.007794	0.008461	0.007235	3.80	470	74	606
BaF 50 683 445	1.682726	44.50	0.015341	1.686372	44.23	0.015519	0.002549	0.004569	0.008025	0.008691	0.007405	3.80	490	83	546
BaF 51 652 449	1.652242	44.93	0.014517	1.655691	44.64	0.014688	0.002411	0.004321	0.007599	0.008232	0.007017	3.42	470	84	556
BaF 52 609 464	1.608589	46.44	0.013104	1.611703	46.14	0.013257	0.002177	0.003901	0.006857	0.007422	0.006323	3.32	440	84	546
BaF 53 670 471	1.670030	47.10	0.014226	1.673412	46.81	0.014385	0.002381	0.004252	0.007423	0.008010	0.006803	3.75	510	65	630
BaF 54 667 482	1.666721	48.22	0.013826	1.670009	47.94	0.013975	0.002323	0.004141	0.007201	0.007751	0.006561	3.78	530	62	630
LF 1 * 573 426	1.573090	42.58	0.013458	1.576285	42.31	0.013622	0.002224	0.003993	0.007063	0.007690	0.006591	3.16	390	85	435
LF 2 * 589 409	1.589212	40.94	0.014392	1.592627	40.66	0.014575	0.002366	0.004258	0.007571	0.008270	0.007115	3.30	390	89	440
LF 3 582 421	1.582151	42.08	0.013836	1.585435	41.80	0.014007	0.002287	0.004104	0.007265	0.007920	0.006800	3.21	420	81	463
LF 4 * 578 416	1.578451	41.59	0.013908	1.581752	41.31	0.014081	0.002296	0.004123	0.007306	0.007968	0.006845	3.21	400	81	442
LF 5 581 409	1.581440	40.85	0.014233	1.584818	40.58	0.014413	0.002340	0.004212	0.007486	0.008177	0.007032	3.22	410	91	419
LF 6 * 567 428	1.567320	42.84	0.013243	1.570465	42.56	0.013403	0.002193	0.003933	0.006945	0.007559	0.006479	3.11	390	85	431
LF 7 575 415	1.575011	41.49	0.013860	1.578300	41.22	0.014031	0.002286	0.004108	0.007281	0.007943	0.006825	3.20	390	79	442
LF 8 564 438	1.564435	43.75	0.012900	1.567499	43.47	0.013055	0.002141	0.003835	0.006760	0.007346	0.006286	3.08	380	85	439
F 1 626 357	1.625882	35.70	0.017530	1.630035	35.45	0.017773	0.002837	0.005144	0.009282	0.010240	0.008905	3.65	350	87	432
F 2 620 364	1.620040	36.37	0.017050	1.624081	36.11	0.017284	0.002765	0.005007	0.009021	0.009933	0.008617	3.61	370	82	432
F 3 613 370	1.612931	37.04	0.016549	1.616854	36.78	0.016772	0.002691	0.004868	0.008745	0.009618	0.008333	3.55	360	80	438
F 4 617 366	1.616591	36.63	0.016834	1.620581	36.37	0.017063	0.002734	0.004948	0.008901	0.009799	0.008501	3.58	360	83	436
F 5 603 380	1.603420	38.03	0.015868	1.607183	37.76	0.016078	0.002590	0.004676	0.008373	0.009188	0.007943	3.47	380	80	438
F 6 636 353	1.636359	35.34	0.018005	1.640624	35.09	0.018259	0.002907	0.005276	0.009544	0.010537	0.009169	3.76	360	85	439
F 7 625 356	1.625361	35.56	0.017586	1.629527	35.30	0.017833	0.002838	0.005153	0.009321	0.010293	0.008957	3.62	360	98	429
F 8 596 392	1.595510	39.18	0.015200	1.599115	38.91	0.015398	0.002490	0.004488	0.008010	0.008774	0.007571	3.38	370	82	446
F 9 620 381	1.620451	38.08	0.016293	1.624313	37.81	0.016510	0.002658	0.004800	0.008602	0.009454	0.008193	3.56	400	77	468
F 11 621 359	1.620962	35.90	0.017298	1.625055	35.63	0.017541	0.002808	0.005074	0.009170	0.010190	0.008981	2.66	520	75	572
F 13 622 360	1.622372	36.04	0.017267	1.626463	35.79	0.017506	0.002797	0.005069	0.009140	0.010078	0.008758	3.62	370	87	434

Table 56–Continued

Type	n_d	v_d	n_F-n_C	n_e	v_e	$n_{F'}-n_{C'}$	n_C-n_r	n_d-n_C	$n_{F'}-n_e$	n_g-n_F	n_h-n_g	Density g/cm³	Knoop hardness kp/mm²	$\alpha \times 10^7$/°C −30° to +70 °C	Tg [°C]
BaLF 6 589 530 *	1.589041	53.01	0.011111	1.591686	52.73	0.011222	0.001886	0.003347	0.005761	0.006154	0.005164	3.32	470	67	577
BaLF 8 554 512 *	1.553609	51.18	0.010816	1.556183	50.89	0.010930	0.001827	0.003247	0.005623	0.006032	0.005087	2.99	420	83	513
BaLF 50 589 514	1.588931	51.37	0.011464	1.591659	51.08	0.011582	0.001941	0.003446	0.005953	0.006376	0.005368	3.11	480	83	555
BaLF 51 574 521 *	1.573932	52.10	0.011017	1.576555	51.81	0.011129	0.001868	0.003314	0.005717	0.006120	0.005149	3.03	460	81	527
SSK 1 617 539	1.617201	53.91	0.011448	1.619928	53.62	0.011561	0.001949	0.003453	0.005928	0.006322	0.005295	3.63	450	63	621
SSK 2 622 532	1.622301	53.16	0.011707	1.625089	52.87	0.011824	0.001988	0.003526	0.006069	0.006482	0.005438	3.67	460	62	636
SSK 3 615 512	1.614836	51.16	0.012018	1.617697	50.86	0.012144	0.002026	0.003606	0.006248	0.006697	0.005641	3.61	460	66	615
SSK 4 618 551	1.617650	55.14	0.011201	1.620319	54.86	0.011308	0.001913	0.003384	0.005792	0.006161	0.005145	3.63	460	61	639
SSK N5 658 509	1.658440	50.88	0.012940	1.661520	50.59	0.013076	0.002183	0.003883	0.006727	0.007213	0.006085	3.71	470	68	641
SSK N8 618 498	1.617720	49.77	0.012412	1.620672	49.48	0.012545	0.002092	0.003721	0.006461	0.006951	0.005888	3.33	460	71	597
SSK 50 618 526	1.617954	52.61	0.011745	1.620751	52.32	0.011865	0.001990	0.003533	0.006094	0.006516	0.005475	3.42	480	74	612
SSK 51 604 536	1.603608	53.63	0.011254	1.606288	53.35	0.011365	0.001914	0.003392	0.005831	0.006225	0.005223	3.28	470	76	616
SSK 52 658 509	1.658439	50.90	0.012936	1.661519	50.61	0.013070	0.002189	0.003888	0.006717	0.007190	0.006047	3.76	500	67	639
LaK 3 694 535 *	1.693500	53.45	0.012975	1.696592	53.17	0.013100	0.002210	0.003916	0.006713	0.007144	0.005965	4.35	450	82	610
LaK N6 643 580	1.642499	57.96	0.011085	1.645142	57.70	0.011181	0.001915	0.003368	0.005706	0.006037	0.005015	3.80	460	70	634
LaK N7 652 585	1.651601	58.52	0.011134	1.654255	58.27	0.011228	0.001931	0.003390	0.005724	0.006050	0.005023	3.84	460	71	618
LaK 8 713 538	1.713003	53.83	0.013245	1.716160	53.61	0.013359	0.002297	0.004028	0.006815	0.007218	0.006008	3.78	590	56	640
LaK N9 691 547	1.691002	54.71	0.012631	1.694013	54.48	0.012740	0.002186	0.003839	0.006501	0.006880	0.005722	3.55	580	63	630
LaK 10 720 504	1.719999	50.41	0.014282	1.723400	50.17	0.014418	0.002449	0.004318	0.007385	0.007875	0.006603	3.81	580	57	620
LaK 11 658 573	1.658302	57.26	0.011497	1.661043	57.01	0.011595	0.001993	0.003498	0.005913	0.006256	0.005196	3.79	520	72	616
LaK N12 678 552	1.677900	55.20	0.012280	1.680826	54.93	0.012394	0.002101	0.003715	0.006343	0.006735	0.005610	4.10	470	76	621
LaK N13 694 533	1.693501	53.33	0.013004	1.696599	53.05	0.013130	0.002211	0.003922	0.006732	0.007166	0.005985	4.24	470	84	614
LaK N14 697 554	1.696800	55.41	0.012575	1.699798	55.19	0.012679	0.002190	0.003832	0.006460	0.006828	0.005671	3.63	600	55	661
LaK N16 734 517	1.733500	51.65	0.014202	1.736884	51.42	0.014332	0.002443	0.004303	0.007329	0.007790	0.006504	4.02	600	53	611
LaK 17 788 505	1.788470	50.48	0.015620	1.792190	50.24	0.015767	0.002674	0.004723	0.008074	0.008592	0.007181	4.88	650	59	674
LaK N18 729 542 *	1.728750	54.16	0.013455	1.731958	53.94	0.013570	0.002337	0.004095	0.006919	0.007321	0.006087	4.19	650	53	623
LaK N19 755 531 *	1.754960	53.05	0.014231	1.758351	52.82	0.014358	0.002455	0.004318	0.007335	0.007780	0.006479	4.46	650	59	658
LaK 20 693 516	1.693487	51.56	0.013450	1.696690	51.28	0.013587	0.002275	0.004047	0.006977	0.007444	0.006232	4.13	440	82	592
LaK 21 641 601	1.640496	60.10	0.010658	1.643038	59.85	0.010744	0.001861	0.003255	0.005466	0.005765	0.004778	3.74	460	68	627
LaK 22 651 559	1.651129	55.89	0.011650	1.653905	55.62	0.011757	0.001998	0.003527	0.006014	0.006388	0.005330	3.86	450	72	621
LaK 23 669 574	1.668815	57.38	0.011655	1.671594	57.12	0.011757	0.002015	0.003541	0.006000	0.006348	0.005274	4.00	460	79	608
LaK 24 697 562	1.696800	56.18	0.012404	1.699758	55.96	0.012504	0.002166	0.003786	0.006364	0.006717	0.005572	3.78	630	59	654
LaK 25 691 548	1.691001	54.78	0.012613	1.694007	54.51	0.012731	0.002159	0.003815	0.006515	0.006917	0.005762	4.32	470	79	635
LaK 26 692 586	1.692491	58.56	0.011825	1.695311	58.32	0.011922	0.002061	0.003606	0.006071	0.006409	0.005316	4.15	550	69	608
LaK 28 744 508	1.744287	50.77	0.014659	1.747779	50.54	0.014795	0.002513	0.004435	0.007573	0.008059	0.006738	4.09	620	57	625
LLF 1 548 458	1.548140	45.75	0.011980	1.550987	45.47	0.012118	0.002003	0.003574	0.006261	0.006778	0.005776	2.94	390	81	448
LLF 2 541 472	1.540719	47.17	0.011464	1.543444	46.88	0.011591	0.001925	0.003428	0.005980	0.006457	0.005488	2.87	410	79	442

Table 56–Continued

Type	n_d	ν_d	$n_F - n_C$	n_e	ν_e	$n_{F'} - n_{C'}$	$n_C - n_r$	$n_d - n_C$	$n_{F'} - n_e$	$n_g - n_F$	$n_h - n_g$	Density g/cm³	Knoop hardness kp/mm²	$\alpha \times 10^7 \cdot °C$ -30° to +70 °C	Tg [°C]
ZK N7 508 612	1.508469	61.19	0.008310	1.510452	60.98	0.008371	0.001470	0.002552	0.004244	0.004462	0.003690	2.49	450	45	528
BaK 1 573 575	1.572500	57.55	0.009948	1.574871	57.27	0.010038	0.001707	0.003014	0.005132	0.005442	0.004528	3.19	460	76	602
BaK 2 540 597	1.539960	59.71	0.009043	1.542116	59.44	0.009120	0.001566	0.002750	0.004652	0.004918	0.004082	2.86	450	80	562
BaK 4 569 561	1.568830	56.13	0.010135	1.571245	55.85	0.010229	0.001740	0.003069	0.005232	0.005558	0.004639	3.10	470	70	552
BaK 5 557 587	1.556710	58.65	0.009492	1.558973	58.37	0.009576	0.001637	0.002882	0.004889	0.005177	0.004303	3.02	460	78	580
BaK 6 574 564	1.574440	56.40	0.010186	1.576867	56.12	0.010280	0.001750	0.003085	0.005258	0.005587	0.004667	3.10	480	73	578
BaK 50 568 580	1.567737	57.99	0.009790	1.570071	57.75	0.009872	0.001698	0.002980	0.005033	0.005322	0.004421	2.93	520	37	629
SK 1 * 610 567	1.610250	56.71	0.010761	1.612815	56.43	0.010860	0.001844	0.003257	0.005555	0.005896	0.004911	3.56	490	61	650
SK 2 607 567	1.607381	56.65	0.010721	1.609936	56.38	0.010819	0.001837	0.003246	0.005534	0.005871	0.004890	3.55	460	60	654
SK 3 609 589	1.608810	58.92	0.010332	1.611273	58.66	0.010421	0.001787	0.003141	0.005317	0.005624	0.004672	3.53	480	63	644
SK 4 613 586	1.612720	58.63	0.010451	1.615212	58.35	0.010543	0.001806	0.003176	0.005380	0.005694	0.004732	3.57	470	64	643
SK 5 589 613	1.589130	61.27	0.009615	1.591424	61.03	0.009691	0.001683	0.002939	0.004927	0.005191	0.004296	3.30	480	55	658
SK 6 614 564	1.613750	56.40	0.010882	1.616344	56.11	0.010984	0.001862	0.003292	0.005620	0.005965	0.004968	3.60	450	62	648
SK 7 607 595	1.607290	59.46	0.010214	1.609726	59.19	0.010301	0.001770	0.003108	0.005252	0.005551	0.004606	3.51	510	64	636
SK 8 611 559	1.611171	55.92	0.010929	1.613776	55.64	0.011031	0.001869	0.003305	0.005646	0.005997	0.005000	3.57	470	60	638
SK 9 614 552	1.614050	55.17	0.011130	1.616702	54.88	0.011237	0.001901	0.003362	0.005756	0.006127	0.005118	3.60	480	60	641
SK 10 623 569	1.622801	56.90	0.010945	1.625410	56.62	0.011045	0.001879	0.003316	0.005647	0.005991	0.004990	3.66	480	70	624
SK 11 564 608	1.563839	60.80	0.009273	1.566051	60.55	0.009348	0.001614	0.002828	0.004759	0.005018	0.004158	3.08	510	65	610
SK 12 583 595	1.583131	59.45	0.009808	1.585470	59.19	0.009892	0.001699	0.002984	0.005044	0.005331	0.004424	3.27	490	64	633
SK 13 592 583	1.591809	58.30	0.010151	1.594229	58.02	0.010241	0.001747	0.003079	0.005232	0.005539	0.004604	3.37	470	68	620
SK 14 603 606	1.603110	60.60	0.009952	1.605484	60.35	0.010033	0.001734	0.003036	0.005107	0.005387	0.004465	3.44	490	60	649
SK 15 623 581	1.622990	58.06	0.010731	1.625548	57.79	0.010825	0.001852	0.003259	0.005527	0.005852	0.004866	3.64	450	69	634
SK 16 620 603	1.620411	60.33	0.010284	1.622864	60.08	0.010368	0.001794	0.003138	0.005277	0.005566	0.004613	3.58	490	63	638
SK N18 639 554	1.638541	55.42	0.011521	1.641286	55.15	0.011628	0.001975	0.003487	0.005950	0.006328	0.005289	3.64	470	64	643
SK 19 613 574	1.613421	57.37	0.010693	1.615970	57.08	0.010791	0.001838	0.003240	0.005516	0.005849	0.004870	3.56	500	64	649
SK 20 * 560 612	1.559630	61.21	0.009143	1.561811	60.96	0.009216	0.001595	0.002790	0.004690	0.004942	0.004092	3.03	530	64	605
SK 51 621 603	1.620900	60.31	0.010295	1.623355	60.02	0.010386	0.001775	0.003125	0.005303	0.005614	0.004667	3.52	380	89	597
SK 52 639 555	1.638540	55.53	0.011499	1.641281	55.27	0.011603	0.001981	0.003487	0.005929	0.006288	0.005239	3.30	520	60	624
KF 1 * 540 511	1.540411	51.10	0.010575	1.542928	50.81	0.010685	0.001793	0.003178	0.005493	0.005897	0.004981	2.78	430	88	478
KF 3 515 547	1.514540	54.70	0.009406	1.516781	54.43	0.009494	0.001613	0.002846	0.004859	0.005172	0.004328	2.56	410	81	470
KF 6 517 522	1.517421	52.20	0.009913	1.519780	51.93	0.010010	0.001693	0.002991	0.005135	0.005493	0.004624	2.67	420	69	446
KF 9 523 515	1.523412	51.49	0.010166	1.525831	51.22	0.010267	0.001731	0.003062	0.005272	0.005648	0.004762	2.71	440	68	445
KF 50 531 511	1.530881	51.12	0.010386	1.533353	50.84	0.010491	0.001765	0.003126	0.005389	0.005771	0.004859	2.70	430	73	475
BaLF 2 * 571 509	1.570989	50.88	0.011223	1.573660	50.58	0.011342	0.001891	0.003366	0.005837	0.006266	0.005284	3.18	420	82	542
BaLF 3 * 571 529	1.571353	52.93	0.010794	1.573923	52.64	0.010902	0.001829	0.003249	0.005598	0.005981	0.005018	3.18	450	81	543
BaLF 4 580 537	1.579569	53.71	0.010790	1.582139	53.43	0.010896	0.001837	0.003253	0.005589	0.005962	0.004995	3.17	460	64	569
BaLF 5 547 536	1.547391	53.63	0.010207	1.549821	53.33	0.010309	0.001735	0.003075	0.005291	0.005652	0.004744	2.95	440	81	529

Table 56–Continued

Type	n_d	ν_d	n_F-n_C	n_e	ν_e	$n_{F'}-n_{C'}$	n_C-n_r	n_d-n_C	$n_{F'}-n_e$	n_g-n_F	n_h-n_g	Density g/cm³	Knoop hardness kp/mm²	$\alpha \times 10^7$·°C -30° to +70 °C	Tg [°C]
FK 1 471 673 *	1.470690	67.34	0.006990	1.472358	67.14	0.007035	0.001249	0.002157	0.003556	0.003717	0.003057	2.31	370	84	390
FK 3 465 658	1.464500	65.77	0.007063	1.466186	65.57	0.007110	0.001258	0.002176	0.003597	0.003767	0.003102	2.27	340	82	362
FK 5 487 704	1.487490	70.41	0.006924	1.489143	70.22	0.006966	0.001246	0.002144	0.003513	0.003662	0.003006	2.45	450	92	464
FK 51 487 845	1.486561	84.47	0.005760	1.487936	84.07	0.005804	0.001007	0.001762	0.002948	0.003090	0.002542	3.73	360	136	405
FK 52 486 818	1.486052	81.81	0.005941	1.487471	81.41	0.005988	0.001034	0.001814	0.003044	0.003196	0.002633	3.64	360	144	434
PK 1 504 669	1.503781	66.92	0.007528	1.505578	66.73	0.007576	0.001349	0.002326	0.003826	0.003998	0.003288	2.44	570	62	572
PK 2 518 651	1.518211	65.05	0.007966	1.520112	64.85	0.008020	0.001416	0.002453	0.004059	0.004252	0.003502	2.51	520	69	568
PK 3 525 647	1.525419	64.66	0.008126	1.527358	64.46	0.008181	0.001443	0.002501	0.004142	0.004340	0.003578	2.59	570	71	567
PK 50 521 697	1.520539	69.71	0.007467	1.522321	69.50	0.007515	0.001331	0.002303	0.003800	0.003973	0.003268	2.59	350	88	496
PSK 2 569 631	1.568730	63.08	0.009016	1.570882	62.85	0.009083	0.001590	0.002765	0.004608	0.004841	0.003999	3.06	490	64	608
PSK 3 552 635	1.552322	63.46	0.008704	1.554399	63.23	0.008768	0.001537	0.002671	0.004447	0.004671	0.003859	2.91	510	62	602
PSK 50 558 673 *	1.557529	67.29	0.008286	1.559507	67.03	0.008347	0.001462	0.002542	0.004233	0.004441	0.003664	2.93	390	86	543
PSK 52 603 654	1.603100	65.41	0.009220	1.605300	65.14	0.009292	0.001611	0.002817	0.004724	0.004967	0.004105	3.38	400	85	597
PSK 53 620 635	1.620140	63.52	0.009763	1.622469	63.23	0.009845	0.001692	0.002972	0.005017	0.005291	0.004382	3.60	370	94	614
BK 1 510 635	1.510090	63.46	0.008038	1.512007	63.23	0.008098	0.001416	0.002465	0.004110	0.004319	0.003570	2.46	480	77	547
BK 3 498 651 *	1.498311	65.06	0.007659	1.500139	64.89	0.007707	0.001374	0.002367	0.003892	0.004065	0.003342	2.37	510	53	553
BK 6 531 621	1.531130	62.15	0.008546	1.533168	61.92	0.008611	0.001503	0.002618	0.004373	0.004601	0.003805	2.69	490	78	537
BK 7 517 642	1.516800	64.17	0.008054	1.518722	63.96	0.008110	0.001429	0.002477	0.004108	0.004309	0.003554	2.51	520	71	559
UBK 7 517 643	1.516800	64.29	0.008039	1.518719	64.08	0.008095	0.001427	0.002472	0.004100	0.004302	0.003549	2.51	500	70	563
BK 8 520 637 *	1.520152	63.68	0.008168	1.522101	63.46	0.008227	0.001446	0.002510	0.004169	0.004378	0.003616	2.57	520	74	544
BK 10 498 670	1.497821	66.95	0.007436	1.499597	66.77	0.007482	0.001335	0.002299	0.003776	0.003944	0.003242	2.39	490	58	532
BK 13 521 628 *	1.521221	62.77	0.008304	1.523202	62.56	0.008363	0.001471	0.002553	0.004238	0.004451	0.003678	2.54	550	68	563
BaLK 1 526 600 *	1.526421	60.03	0.008770	1.528513	59.75	0.008845	0.001520	0.002668	0.004510	0.004768	0.003959	2.70	430	91	509
BaLK 3 518 603	1.518350	60.28	0.008599	1.520400	60.03	0.008669	0.001500	0.002624	0.004413	0.004657	0.003863	2.61	460	83	506
K 3 518 590	1.518230	58.98	0.008787	1.520324	58.71	0.008863	0.001521	0.002672	0.004522	0.004783	0.003977	2.54	470	83	521
K 4 519 574 *	1.518951	57.40	0.009041	1.521106	57.15	0.009118	0.001568	0.002750	0.004651	0.004929	0.004108	2.63	460	73	507
K 5 522 595	1.522489	59.48	0.008784	1.524583	59.22	0.008858	0.001524	0.002673	0.004517	0.004777	0.003969	2.59	450	82	543
K 7 511 604	1.511121	60.41	0.008461	1.513138	60.16	0.008530	0.001474	0.002580	0.004344	0.004587	0.003807	2.53	470	84	513
K 10 501 564	1.501371	56.41	0.008888	1.503488	56.15	0.008967	0.001538	0.002700	0.004580	0.004866	0.004068	2.52	440	65	459
K 11 500 614 *	1.500129	61.44	0.008140	1.502070	61.21	0.008203	0.001428	0.002488	0.004172	0.004405	0.003661	2.50	460	64	493
K 50 523 602	1.522572	60.18	0.008683	1.524642	59.94	0.008753	0.001515	0.002650	0.004456	0.004702	0.003899	2.62	450	70	553
UK 50 523 604 *	1.522571	60.38	0.008654	1.524635	60.14	0.008723	0.001512	0.002642	0.004439	0.004683	0.003884	2.62	460	70	554
K 51 505 596	1.505180	59.55	0.008483	1.507203	59.34	0.008548	0.001495	0.002599	0.004341	0.004580	0.003804	2.47	460	43	521
ZK 1 533 580	1.533149	57.98	0.009196	1.535341	57.71	0.009276	0.001589	0.002793	0.004736	0.005017	0.004176	2.71	430	75	562
ZK 5 534 553	1.533750	55.31	0.009651	1.536050	55.02	0.009743	0.001648	0.002915	0.004990	0.005308	0.004433	2.76	430	87	534

Table 56–Continued

Type	n_d	ν_d	$n_F - n_C$	n_e	ν_e	$n_{F'} - n_{C'}$	$n_C - n_r$	$n_d - n_C$	$n_{F'} - n_e$	$n_g - n_F$	$n_h - n_g$	Density g/cm³	Knoop hardness kp/mm²	$\alpha \times 10^7 \cdot$°C -30° to +70 °C	Tg [°C]
SF 63 748 277	1.748400	27.71	0.027004	1.754775	27.50	0.027445	0.004239	0.007792	0.014491	0.016321	0.014521	4.62	340	82	430
TiK 1 479 587 *	1.478690	58.70	0.008155	1.480634	58.44	0.008224	0.001420	0.002484	0.004192	0.004437	0.003693	2.39	330	103	340
TiF 1 511 510 *	1.511181	51.01	0.010022	1.513562	50.70	0.010130	0.001697	0.003006	0.005219	0.005644	0.004823	2.47	440	91	443
TiF 2 533 460 *	1.532559	45.99	0.011581	1.535306	45.67	0.011720	0.001931	0.003445	0.006072	0.006641	0.005753	2.51	450	86	453
TiF 3 548 422 *	1.547649	42.20	0.012976	1.550724	41.89	0.013146	0.002132	0.003830	0.006843	0.007536	0.006575	2.61	390	101	419
TiF 4 584 370	1.584064	37.04	0.015767	1.587794	36.75	0.015993	0.002551	0.004617	0.008369	0.009310	0.008212	2.68	410	89	463
TiF N 5 594 355	1.593555	35.51	0.016713	1.597508	35.24	0.016956	0.002694	0.004885	0.008884	0.009909	0.008763	2.71	410	90	472
TiF 6 617 310 *	1.616501	30.97	0.019904	1.621184	30.67	0.020256	0.003106	0.005704	0.010756	0.012382	0.011392	2.79	310	139	410
TiSF 1 673 289	1.673393	28.90	0.023304	1.678890	28.66	0.023689	0.003672	0.006725	0.012515	0.014196	0.012812	2.81	450	90	495
KzF 1 551 497	1.551149	49.68	0.011095	1.553789	49.41	0.011208	0.001887	0.003339	0.005758	0.006173	0.005206	2.72	440	69	486
KzF 2 529 517	1.529440	51.68	0.010244	1.531879	51.45	0.010338	0.001765	0.003104	0.005289	0.005639	0.004731	2.55	400	60	447
KzF 5 521 527 *	1.521300	52.74	0.009685	1.523654	52.50	0.009974	0.001710	0.003001	0.005097	0.005425	0.004543	2.49	390	57	440
KzF 6 527 511 *	1.526819	51.13	0.010303	1.529272	50.90	0.010399	0.001774	0.003119	0.005323	0.005682	0.004771	2.54	380	55	444
KzFS 1 613 443 *	1.613099	44.34	0.013826	1.616386	44.13	0.013966	0.002360	0.004164	0.007176	0.007716	0.006538	3.14	370	50	472
KzFS 2 558 539	1.557809	53.85	0.010359	1.560277	53.66	0.010442	0.001819	0.003166	0.005314	0.005626	0.004693	2.68	380	50	479
KzFS N4 613 443	1.613401	44.30	0.013848	1.616693	44.07	0.013994	0.002349	0.004158	0.007203	0.007763	0.006591	3.20	380	45	492
KzFS 5 653 397	1.653319	39.71	0.016454	1.657225	39.49	0.016643	0.002759	0.004911	0.008603	0.009354	0.008027	3.44	370	52	469
KzFS 6 592 485 *	1.591965	48.51	0.012203	1.594869	48.29	0.012319	0.002101	0.003692	0.006310	0.006752	0.005694	3.03	370	51	482
KzFS 7 681 374 *	1.680649	37.39	0.018205	1.684967	37.16	0.018432	0.003013	0.005397	0.009567	0.010467	0.009033	3.77	380	53	473
KzFS 8 720 346	1.720470	34.61	0.020814	1.725402	34.39	0.021094	0.003394	0.006127	0.010997	0.012108	0.010510	4.13	390	53	470
LgSK 2 586 610	1.585990	61.04	0.009600	1.588277	60.66	0.009698	0.001615	0.002884	0.004983	0.005308	0.004432	4.15	340	121	515
KzFS N2 558 542	1.558361	54.16	0.010310	1.560818	53.97	0.010391	0.001814	0.003153	0.005285	0.005593	0.004665	2.56	430	45	499
KzFS N5 654 396	1.654117	39.63	0.016507	1.658036	39.40	0.016702	0.002756	0.004915	0.008644	0.009408	0.008073	3.46	420	45	501
KzFS N7 681 372	1.680639	37.18	0.018305	1.684979	36.95	0.018540	0.003015	0.005413	0.009639	0.010585	0.009179	3.51	460	48	512
KzFS N9 599 469	1.598557	46.90	0.012762	1.601592	46.68	0.012887	0.002185	0.003852	0.006612	0.007091	0.005991	3.01	430	41	512
FN 11 621 362	1.620962	36.18	0.017163	1.625023	35.91	0.017405	0.002786	0.005034	0.009099	0.010111	0.008911	2,66	520	75	572
SF 64 706 308	1.705850	30.75	0.022952	1.711271	30.51	0.023309	0.003654	0.006663	0.012266	0.013785	0.012266	3,00	500	88	546
LaK N22 651 559	1.651130	55.89	0.011650	1.653907	55.62	0.011756	0.002000	0.003530	0.006010	0.006376	0.005307	3.73	530	66	679
LaK 31 697 564	1.696732	56.42	0.012349	1.699677	56.21	0.012448	0.002159	0.003770	0.006335	0.006686	0.005549	3.68	630	57	653
LaSF 30 803 464	1.803182	46.38	0.017318	1.807304	46.14	0.017497	0.002931	0.005204	0.008995	0.009637	0.008109	4.56	630	60	684

Table 57 Refractive Index Formulas[a]

---

1. Ammonium dihydrogen phosphate (ADP):

$$\eta_0 = \left(2.302484 + \frac{0.0117089}{\lambda^2 - 0.0131486} - \frac{3.751806\lambda^2}{100.0 - 0.25\lambda^2}\right)^{1/2}$$

$$\eta_e = \left(2.163077 + \frac{0.0096703}{\lambda^2 - 0.0128447} - \frac{1.451540\lambda^2}{100.0 - 0.25\lambda^2}\right)^{1/2}$$

$$\Delta\eta_0 = (\eta_0^2 - 3.0297n_0 + 2.3004)(0.713)(10^{-2})(298 - T)$$

$$\Delta\eta_e = (\eta_e^2)(0.675)(10^{-6})(298 - T)$$

2. Barium sodium niobate ($Ba_2NaNb_5O_{15}$):

$$\eta_\alpha = \left(1.0 + \frac{3.6008\lambda^2}{\lambda^2 - 0.0321987}\right)^{1/2}$$

$$\eta_\beta = \left(1.0 + \frac{3.9495\lambda^2}{\lambda^2 - 0.0401401}\right)^{1/2}$$

$$\eta_\tau = \left(1.0 + \frac{3.9495\lambda^2}{\lambda^2 - 0.0403889}\right)^{1/2}$$

3. Barium titanate ($BaTiO_3$):

$$\eta_0 = \left(1.0 + \frac{4.2390\lambda^2}{\lambda^2 - 0.4968}\right)^{1/2}$$

$$\eta_e = \left(1.0 + \frac{4.0854\lambda^2}{\lambda^2 - 0.04356}\right)^{1/2}$$

4. Beryllium oxide (BeO):

$$\eta_0 = \left(1.0 + \frac{1.919087\lambda^2}{\lambda^2 - 0.00727575} - \frac{3.972323\lambda^2}{199.31087 - \lambda^2}\right)^{1/2}$$

Table 57–Continued

$$\eta_e = \left(1.0 + \frac{1.972142\lambda^2}{\lambda^2 - 0.00748564} - \frac{17.5887\lambda^2}{779.49122 - \lambda^2}\right)^{1/2}$$

5. Bismuth germanium oxide (Bi_4GeO_{12}):

$$\eta = \left(1.0 + \frac{3.08959\lambda^2}{\lambda^2 - 0.01337}\right)^{1/2}$$

6. BK-7 glass (SiO_2):

$$\eta = \left[2.2718927 - 1.0108077\,(10^{-2})\lambda^2 + 1.0592509\,(10^{-2})\lambda^{-2} + 2.0816965\,(10^{-4})\lambda^{-4} - 7.6472538\,(10^{-6})\lambda^{-6} + 4.9240991\,(10^{-7})\lambda^{-8}\right]^{1/2}$$

7. Cadmium telluride (CdTe):

$$\eta = \left(5.681 + \frac{1.532\lambda^2}{\lambda^2 - 0.336}\right)^{1/2}$$

8. Gadolinium molybdate ($Gd_2(M_0O_4)_3$):

$$\eta_\alpha = \left(1.0 + \frac{2.2450\lambda^2}{\lambda^2 - 0.022693}\right)^{1/2}$$

$$\eta_\beta = \left(1.0 + \frac{2.24654\lambda^2}{\lambda^2 - 0.0226803}\right)^{1/2}$$

$$\eta_\tau = \left(1.0 + \frac{2.41951\lambda^2}{\lambda^2 - 0.0245458}\right)^{1/2}$$

9. Lead niobate ($PbNb_4O_{11}$):

$$\eta_\alpha = \left(1.0 + \frac{4.1241\lambda^2}{\lambda^2 - 0.040804}\right)^{1/2}$$

Table 57-Continued

$$\eta_\beta = \left(1.0 + \frac{4.1398\lambda^2}{\lambda^2 - 0.040441}\right)^{1/2}$$

$$\eta_\tau = \left(1.0 + \frac{4.2461\lambda^2}{\lambda^2 - 0.040562}\right)^{1/2}$$

10. Lead titanate ($PbTiO_3$):

$$\eta_o = \left(1.0 + \frac{5.3589\lambda^2}{\lambda^2 - 0.050176}\right)^{1/2}$$

$$\eta_e = \left(1.0 + \frac{5.3651\lambda^2}{\lambda^2 - 0.047089}\right)^{1/2}$$

11. Lithium iodate ($LiIO_3$):

$$\eta_o = \left(1.0 + \frac{2.40109\lambda^2}{\lambda^2 - 0.021865}\right)^{1/2}$$

$$\eta_e = \left(1.0 + \frac{1.91359\lambda^2}{\lambda^2 - 0.019401}\right)^{1/2}$$

12. Lithium niobate ($LiNbO_3$):

$$\eta_o = \left[4.9130 + \frac{1.173(10^5) + 1.65(10^{-2})T^2}{\lambda^2 - \left[2.12(10^{+2}) + 2.71(10^{-5})T^2\right]^2} - 2.78(10^{-8})\lambda^2\right]^{1/2}$$

$$\eta_e = \left[4.5567 + 2.605(10^{-7})T^2 + \frac{0.97(10^5) + 2.7(10^{-2})T^2}{\lambda^2 - \left[2.01(10^2) + 5.4(10^{-5})T^2\right]^2} - 2.24(10^{-8})\lambda^2\right]^{1/2}$$

Table 57-Continued

13. Potassium dihydrogen phosphate (KDP):

$$\eta_o = \left(2.259276 + \frac{0.01008956}{\lambda^2 - 0.012943} - \frac{3.251305\lambda^2}{100 - 0.25\lambda^2}\right)^{1/2}$$

$$\eta_e = \left(2.132668 + \frac{0.008637494}{\lambda^2 - 0.7011098} - \frac{0.8069981\lambda^2}{100 - 0.25\lambda^2}\right)^{1/2}$$

$$\eta_o\Delta = (n_o^2 - 1.432)\,(0.402)\,(10^{-4})\,(298 - T)$$

$$\eta_e\Delta = (n_e^2 - 1.105)\,(0.221)\,(10^{-4})\,(298 - T)$$

14. Potassium lithium niobate ($K_3Li_2Nb_5O_{15}$):

$$\eta = \left(1.0 + \frac{3.708\lambda^2}{\lambda^2 - 0.04601}\right)^{1/2}$$

$$\eta_e = \left(1.0 + \frac{3.349\lambda^2}{\lambda^2 - 0.03564}\right)^{1/2}$$

15. Rubidium dihydrogen phosphate (RDP):

$$\eta_o = \left(1.0 + \frac{1.2068\lambda^2}{\lambda^2 - 0.01539}\right)^{1/2}$$

$$\eta_e = \left(1.0 + \frac{1.15123\lambda^2}{\lambda^2 - 0.01005}\right)^{1/2}$$

16. Silicon carbide (SiC):

$$\eta_o = \left(1.0 + \frac{5.5515\lambda^2}{\lambda^2 - 0.0264062}\right)^{1/2}$$

$$\eta_e = \left(1.0 + \frac{5.7382\lambda^2}{\lambda^2 - 0.0285508}\right)^{1/2}$$

Table 57–Continued

17. Sodium bromate ($NaBrO_3$):

$$\eta = \left(1.0 + \frac{1.3194\lambda^2}{\lambda^2 - 0.0081} + \frac{0.2357\lambda^2}{\lambda^2 - 0.0398} - 0.0174\lambda^2\right)^{1/2}$$

18. Sodium chlorate ($NaClO_3$):

$$\eta = \left(1.0 + \frac{1.1825\lambda^2}{\lambda^2 - 0.0081} + \frac{0.07992\lambda^2}{\lambda^2 - 0.034225} - 0.00864\lambda^2\right)^{1/2}$$

19. Terbium molybdate ($Tb_2(M_0O_4)_3$):

$$\eta_\alpha = \left(1.0 + \frac{2.27241\lambda^2}{\lambda^2 - 0.023359}\right)^{1/2}$$

$$\eta_\beta = \left(1.0 + \frac{2.273955\lambda^2}{\lambda^2 - 0.02330}\right)^{1/2}$$

$$\eta_\tau = \left(1.0 + \frac{2.443016\lambda^2}{\lambda^2 - 0.025133}\right)^{1/2}$$

20. Tourmaline:

$$\eta_o = \left(1.0 + \frac{1.6346\lambda^2}{\lambda^2 - 0.010734}\right)^{1/2}$$

$$\eta_e = \left(1.0 + \frac{1.57256\lambda^2}{\lambda^2 - 0.011346}\right)^{1/2}$$

21. Zinc selenide (ZnSe):

$$\eta = \left(3.855 + \frac{2.045\lambda^2}{\lambda^2 - 0.109}\right)^{1/2}$$

Table 57-Continued

22. Zinc telluride (ZnTe):

$$\eta = \left(4.271 + \frac{3.012\lambda^2}{\lambda^2 - 0.142}\right)^{1/2}$$

[a]λ is in units of micrometers; T is in kelvin.

Table 58 Electro-Optic Plastics: Index of Refraction

Plastic name	n_D	Plastic name	n_D
Acetals	1.48	Chlorofluoroethylene	1.42
Acrylics	1.46-1.52	Chlorophenyl ethyl methacrylate	1.56
Acrylonitrile butadiene styrene	1.54	Chlorostyrene	1.61
Allyl cinnamate	1.57	Cinnamyl methacrylate	1.60
Allyl methacrylamide	1.55	Cresyl methacrylate	1.57
Allyl methacrylate	1.52	Cyclohexanedid dimethacrylate	1.51
Allyl resins	1.50	Cyclohexyl ethoxyacrylate	1.50
Amino ethyl methacrylate	1.54	Cyclohexyl methacrylate	1.51
Benz hydryl methacrylate	1.59	Cyclohexyl phenyl methacrylate	1.56
Benzyl methacrylamide	1.60	Decamethylene glycol dimethacrylate	1.50
Benzyl methacrylate	1.57	Diacetin methacrylate	1.49
Bornyl methacrylate	1.51	Dibromopropyl methacrylate	1.57
Borno ethyl methacrylate	1.54	Dichloropropyl methacrylate	1.53
Borno phenyl methacrylate	1.60	Dichlorostyrene	1.62
Butodiene styrene	1.56	Diethyl-amino ethyl methacrylate	1.52
Butyl mercaptyl methacrylate	1.54	Dioxolymethyl methacrylate	1.51
Butyl methacrylate	1.48	Diphenyl-ethyl methacrylate	1.58
Cellulosics	1.46-1.51	Divinyl benzene	1.62
Chlorbenzhydryl methacrylate	1.60	Epoxy compounds	1.55-1.61
Chlorobenzyl methacrylate	1.58		
Chlorocyclohexy methacrylate	1.52		

Table 58-Continued

Ethoxy-ethyl methacrylate	1.48
Ethyl acrylate	1.47
Ethyl cellulose	1.47
Ethylene chlorodrin methacrylate	1.52
Ethylene dimethacrylate	1.51
Ethylene glycol mono-methacrylate	1.51
Ethyl glycolate methacrylate	1.49
Ethylidene di-methacrylate	1.48
Ethyl sulfide di-methacrylate	1.55
Eugenol methacrylate	1.57
Fluoroplastics	1.34-1.42
Fluorenyl methacrylate	1.63
Furfuryl methacrylate	1.54
Hexy methacrylate	1.48
Ionomers	1.51
Isopropyl methacrylate	1.47
Isopropyl styrene	1.55
Lead methacrylate	1.64
Menthyl methacrylate	1.49
Methacrylic anhydride	1.52
Methacryl-phenyl salicylate	1.60
Methallyl methacrylate	1.49-1.51
Methoxy ethyl methacrylamide	1.52
Methoxy phenyl methacrylate	1.57
Methoxy styrene	1.59-1.60
Methyl acrylate	1.48
Methyl bromoacrylate	1.51
Methyl chloroacrylate	1.52
Methyl chclohexyl methacrylate	1.50-1.51
Methylene butyrolactone	1.54
Methylene valeroactone	1.54
Methylisopropenyl ketone	1.52
Methyl methacrylamide	1.54
Methyl methacrylate	1.49
Methyl methoxy styrene	1.59
Methyl methylene butyrolactone	1.51
Methyl pentene polymers	1.46
Methyl styrene	1.59
Naphthyl carbinyl methacrylate	1.63
Naphthyl methacrylate	1.63-1.64
Nitrile barrier resins	1.51-1.53
Nitrobenzyl methacrylate	1.58
Nitromethyl propyl methacrylate	1.49
Pentachlorophenyl methacrylate	1.61
Phenolic casting resins	1.58-1.66
Phenolic molding compound	1.50-1.69
Phenyl amyl methacrylate	1.54
Phenyl cellosolve methacrylate	1.56
Phenyl cyclohexyl methacrylate	1.56
Phenyl ethyl methacrylate	1.59
Phenyl ethyl methacrylate	1.55-1.56
Phenyl methacrylate	1.57
Phenyl sulfone ethyl methacrylate	1.57
Phenyl vinyl ketone	1.59
Polyacrylonitrile	1.50-1.52
Polyallomers	1.49
Polyaryl sulfone	1.67
Polybutadiene-co-styrene	1.53-1.54
Polybutylene	1.50
Polycarbonates	1.58-1.59
Polychloroprene	1.55-1.56
Polyesters	1.52-1.57
Polyether sulfone	1.65
Polyethylenes	1.49-1.58
Polyethylene terephthalate	1.64
Polyhexamethylene adipamide	1.52-1.58

Table 58-Continued

Polyisobutene-co-isoprene	1.51	Triethoxy-silicon methacrylate	1.44
Polyisoprene	1.52-1.53	Triethyl carbinyl methacrylate	1.49
Polymethylpentane	1.47		
Polypropylene	1.49	Trifluoropropyl methacrylate	1.48
Polystyrene	1.56-1.60		
Polysolfone	1.63-1.64	Trimethyl cyclohexyl methacrylate	1.49
Polytetrafluorethylene	1.35		
Polyurethane	1.50-1.60	Urea formaldehyde	1.54-1.56
Polyvinyl chloride	1.52-1.55	Vinyl benzoate	1.58
Silicones	1.43	Vinyl butyral	1.47-1.49
Styrene	1.59	Vinyl carbazole	1.68
Styrene actylonitrile	1.56-1.57	Vinyl cyclohexane dioxide	1.53
Styrene butadiene	1.52-1.55		
Terpineyl methacrylate	1.51	Vinyl formate	1.48
Tert-butyl methacrylate	1.46	Vinylidene chloride	1.60-1.63
Tetrahydrofurfuryl methacrylate	1.51	Vinyl methacrylate	1.51
		Vinyl naphthalene	1.68
Thermoplastic elastomers	1.47-1.55	Vinyl phenyl sulfide	1.66
		Vinyl phthalimide	1.62

2. Distortion, and its variance with bandwidth, modal distribution, the amplitudes and wavelength of the input light, length of the fiber, and environmental temperature
3. Radiation, and its variance with bend radius and fiber temperature
4. Physical parameters, including size, weight, total volume, ease of installation, splicing, and coupling
5. Environmental parameters, including resistance to water, stress and chemical corrosion, mechanical stresses, and temperature

Numerical Aperture. The numerical aperture of an optical fiber is defined as

$$NA = \left(N_{core}^2 - N_{clad}^2\right)^{1/2}$$

where N_{core} is the refractive index of the cord and N_{clad} is the refractive index of the cladding.

The numerical aperture determines the coupling efficiency between the LED or laser light source and the optical fiber. Small values of NA tend to give low pulse dispersions; but, on the other hand, narrow confinement of the light to the fiber core enhances losses due to micro-bending. For long-distance telecommunications fibers, an NA of about 0.2 is usually acceptable. For short-distance applications, higher NA values are sometimes desirable.

The numerical aperture is determined by the type and concentration of dopants in the fiber. The oxides of germanium, boron, phosphorous, titanium, and aluminum are commonly used as dopants. For graded index fibers, NA values of 0.15 to 0.25 are common with plastic-clad silica fibers, having values of about 0.20. For silica fibers doped with several percent fluorine, numerical aperture values can be as high as 0.25 with attenuations less than 4 dB/km. In general, fibers with large NA values have high losses but are easier and cheaper to manufacture, since the fiber drawing process is simpler and the demand on chemical purity and imperfections is not as great as in lower loss fibers.

Index of Refraction. Choosing the absolute values of the index of refraction for core and cladding materials is not a simple matter, and depends in part upon the specific communications application. A trade-off exists in choosing these values in that increasing the difference in the refractive index of the core and cladding permits better source coupling but, at the same time, increases the degree of intermodal dispersion (pulse spreading).

Using an approximate parabolic profile for the index of refraction minimizes modal propagation time delay differences (often called modal delay distortion), thus increasing bandwidth capabilities. A typical radial profile can be defined by

$$N_r = N_0 \left[1 - \frac{(NA)^2}{n_0} \left(\frac{r}{r_c} \right)^A \right]^{1/2}$$

where

n_0 = refractive index on the central axis
r_c = radius of the fiber core
A = profile parameter whose typical value is about 2

Losses due to fiber bending decrease as the refractive index differential between the core and cladding increases and as the core radius decreases. However, as the index differential increases, the wave-guide bandwidth decreases, and the core radius decreases, the coupling efficiency decreases as well.

Attenuation. Attenuation losses in fibers are caused by material absorption and scattering, wave-guide scattering, and radiation losses. Material absorption is mostly caused by OH^- ions and transition metal ions in the glass. Wave-guide scattering is caused mainly by geometric irregularities at the core-cladding interface. Carefully controlled fabrication procedures can keep wave-guide scattering losses down to less than 1 dB/km. Radiation losses are caused by the bending of fibers, particularly at small radii of curvature. Radiation losses can be particularly large in cable without plastic cushioning material surrounding the fibers. These losses can be minimized in part by using high numerical apertures for the fibers, which is difficult, especially for graded index fibers, since doping techniques to increase the numerical aperture to 0.20 or beyond must be employed.

To minimize fiber attenuations, certain material properties are desirable in the wave guide. First, obtain a low concentration of impurities, particularly those which absorb in the visible and near ir such as certain transition metals and especially OH^- groups. Second, match the dielectric properties of all components, especially in multicomponent glasses, to minimize scattering from compositional fluctuations. Third, achieve a relatively low glass transition temperature to minimize density fluctuations caused by loss of volatile components and to reduce the broadening of the uv absorption band.

The effects of contamination by certain transition elements in the fiber can be somewhat controlled by optimizing the redox state of the glass to achieve a balance of the effects of various metal ions in various oxidation states. For the design of specific transmission systems, the fiber composition can be optimized for a particular wavelength to achieve the greatest transmission of the light source along the fiber. When designing fibers for certain laser frequencies, it is possible to add buffering agents such as As_2O_3 to the glass to control and stabilize the redox state of the transition element contaminants. For Fe^{2+} and Co^{3+} ions, the total wave-guide concentration cannot exceed 1 ppb if attenuation levels lower than 1-dB/km are to be obtained. The major impurity of concern in silica-based optical wave guides is the OH^- radical which has weak absorption bands at 725, 825, and 875 nm and a strong absorption band at 950 nm. Figure 81 shows the fiber attenuation vs. wavelength for typical low-loss fibers. This larger band contributes to transmission losses at a rate of about 1 dB/km/ppm. To minimize the OH^- absorption band in fibers, starting materials are often heated and dried and used in very low humidity controlled production areas. The limitation of using totally dry materials and manufacturing environments is that defect absorption effects can occur from oxygen defects initiated by the fiber drawing process.

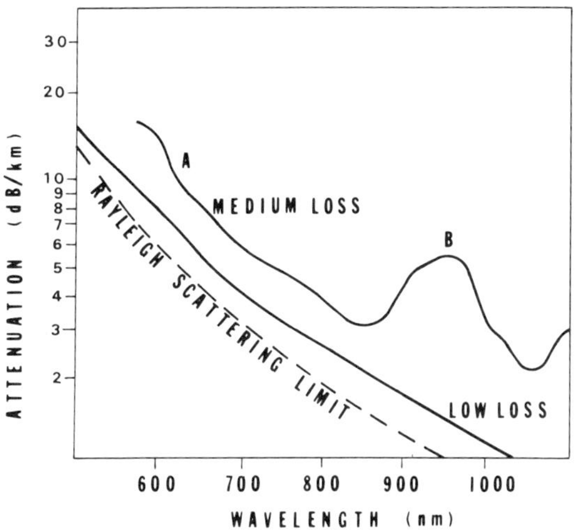

Fig. 81 Typical fiber attenuation vs. wavelength for production optical fibers: A, drawing-induced absorption; B, OH^- absorption region.

Other losses are caused by density fluctuations within the fiber. This intrinsic type of scattering is the fundamental limit to attenuation in wave guides. The magnitude of this scattering loss is given by

$$\alpha_S = \frac{8\pi^3}{3\lambda^2}(n^2 - 1)KT\beta$$

where

T = transition temperature at which the fluctuations are frozen into the glass
β = isothermal compressibility
λ = wavelength of the transmitted light.

It is clear from the $1/\lambda^4$ factor that this type of scattering loss decreases rapidly with increasing wavelength. For fused silica with T ~ 1700 K, the value of α_S is about 1-6 dB/km at 830 nm. In multicomponent glasses, the transition temperature can be lowered, thus reducing the associated density fluctuation losses.

Pulse Dispersion. The two types of dispersion in optical wave guides are multimode group delay spreading and material dispersion, both of which limit the data rate capacity. Light that enters a fiber at an angle other than along the central axis takes longer to transverse the length of the fiber, since it must undergo numerous internal reflections per unit length of fiber with internal cores ranging from 10- to 100-nm diameter. The difference in the total distrance traveled by central axis and off-axis rays causes a time lag to occur which broadens the input pulse. This pulse-dispersion or pulse-broadening effect limits the effective bandpass and information capabilities of any fiber. By carefully grading the change in the refractive index between the cladding and core, pulse broadening can be minimized. This decreases the effective path length differential between low- and high-order modes, since larger angle rays travel faster through the lower refractive index material portion of the wave guide, reducing the time lag.

To minimize pulse dispersion, the graded index profile in fibers should be carefully tailored to the particular fiber material being used. For plastic-clad high-silica fibers with step-index profiles, pulse dispersions of about 20 nsec/km can easily be achieved for multimode fibers. Carefully prepared, parabolically graded index profiles can achieve dispersions as low as 0.2 ns/km. Graded index fibers made by the double crucible method using controlled metal alkali in exchange at the draw nozzle have produced fibers with pulse dispersions of about 1 ns/km. Although the parabolic index profile gives the smallest group delay distortion compared to any other refractive index profile, it is very sensitive to physical deviations from the parabolic shape, making manufacturing procedures very stringent.

Group delay distortion in multimode fibers can be effectively reduced by making the differential in the index of refraction of the core and cladding less than 0.007. This, however, makes the coupling of the input light source very difficult. In single-mode fibers with a refractive index differential of less than 0.01, the signal dispersion becomes dominated by material dispersion instead of wave-guide mode dispersion.

Pulse spreading for step-index (SI) multimode fibers is given by

$$\Delta_{SI} = n_{core} \frac{\Delta nL}{c}$$

where Δn is the refractive index difference between the core and cladding and L is the fiber length. For graded index (GI) fibers, the pulse spread is given by

$$\Delta_{GI} = An_{core} (\Delta n) \frac{L}{8c}$$

where c is the speed of light and A is a constant which relates the different qualities of graded index profiles, where the maximum value is 1.0. The ratio of pulse spreading for step index versus graded index fibers can theoretically (using A = 1.0) achieve a value of $8/\Delta n$. In practice, this ratio ranges from 200 in single research-grade fibers to 6 for production cables. Typical values of mode pulse spreading in graded index cables are 0.2 to 4.0 ns/km.

Material dispersion is caused by the nonlinear aspects of the refractive index with respect to transmission wavelength. The pulse spreading from material dispersion is given by

$$\Delta_{mat} = \frac{\lambda(\Delta\lambda)}{c}\left(\frac{d^2 n_{core}}{d\lambda^2}\right) L$$

where $\Delta\lambda$ is the spectral width of the light source and λ is the average wavelength of the transmitted light. Typical material dispersion pulse spreading values for glass fibers using LEDs in the 820-nm window region are 3.0–3.5 ns/km ($\Delta\lambda$ = 280 Å) and for lasers operating in the 1040-nm window are 0.2–0.3 ns/km ($\Delta\lambda$ = 20 Å).

For long fibers and cables (in excess of 300 m), the pulse spreading per kilometer can be approximately extrapolated by relating the pulse spread to the square root of the fiber length. For short fibers, this relationship is linear (directly proportional), since there is little energy exchange between higher order modes over short transmission lengths. These relationships can be used to estimate the critical length or maximum length of a fiber wave guide which is determined by a pulse spread equal to the original pulse width for different bit rates (bits per second, bps).

Fiber Bending. The macro-bending of optical wave guides causes radiation losses which become increasingly severe with decreasing bend radius. The smallest permissible curvature radius is limited to the actual fiber strength. Different fibers and cables have different strain-to-failure properties depending upon construction methods and materials and overall design. If a fiber is bent around a surface of radius R (bend radius), the outermost edge of the fiber cladding of radius r will be strained relative to the fiber axis by a certain percentage σ_S, where

$$\sigma_S = \left(\frac{R + 2r}{R + r} - 1\right) \times 100\%$$

For a bend radius of 1 cm and a cladding radius of 70 μm, σ_S = 0.7%. For fibers and cables to maintain longevity, the highest permissible differential bending strain should not exceed 0.2%, which for the above fiber would mean a maximum bend radius of about 5 cm.

Fiber Materials. Advances in optical materials technology and fiber fabrication procedures have resulted in fiber wave guides with low transmission losses resulting from controlled fabrication techniques and transition metal impurity levels less than several parts per billion. The three major material systems used in the manufacture of fibers are silica, glass, and plastic.

The silica system is basically made of silicon dioxide with other metal oxidex to establish a difference in the refractive index between the core and cladding. Various dopants have been used to increase the refractive index of silica including TiO_2, Al_2O_3, GeO_2, and P_2O_5. Compound glasses used for cores and claddings in optical fibers should have similar coefficients of thermal expansion, similar viscosities at the drawing temperature, low material dispersion (high purity), low Rayleigh scattering (low compositional and density fluctuations), low melting temperatures, and long-term chemical stability. The cladding is generally of lower refractive index and has a higher coefficient of thermal expansion than the core. Many dopants and materials used in wave-guide production facilities are highly toxic and special precautions must be taken to handle and apply materials safely.

Chemical purity in silica systems is more difficult to maintain at high levels than glass systems, since silica is insoluble in water and thus cannot employ the common purification processes such as those used in exchange and recrystallization techniques. Sodium aluminum silicate materials have a lower intrinsic scattering and can be manufactured more easily than pure silica. Table 59 lists a few material systems and the associated fiber properties.

Fiber Fabrication. A variety of fabrication techniques are available for optical fibers depending upon the core and cladding materials to be used and the desired refractive index profiles and attenuation limits. The major processes used are double crucible (DC), stratified melt (SM), chemical vapor disposition (CVD), rod in tube, and direct fiber drawing.

All processes require ultrapurified starting materials. Some processes are batch-type operations which first make preforms and subsequently make fibers by various drawing methods. A few processes can be designed to be continuous given sufficient supplies of raw materials, manpower, and furnaces capable of maintaining constant temperatures for long periods of time. Such continuous processes are advantageous economically in terms of producing a large volume of long-length fibers.

High-silica fibers are usually prepared by a gas-phase reaction either by flame hydrolysis or by plasma-induced chemical vapor deposition where doped layers of SiO_2 are deposited on the outside of a mandrel or on the

Table 59 Materials System and Fiber Properties

Material system	Attenuation (dB/km)	Fabrication method	NA	Fiber core diameter (μm)
SiO_2-P_2O_5	3	CVD	0.18	50
SiO_2-GeO_2	4	CVD	0.14	20
Soda-borosilicate	15	Double crucible	0.15	20
Alkali-lead-silicate	20	Cladded rod	0.45	50
Soda-lime-silicate	25	Cladded rod	0.20	30

inside of a fused silica tube. The prepared preform is subsequently reduced to a rod and drawn out into the final fiber form and coated with an appropriate material. Various drawing methods can be used. Multicomponent fibers are prepared by mixing the core and cladding glasses separately and then drawing the rods directly from the melt. The rods are subsequently drawn to fibers.

The chemical vapor-deposition fabrication method can be done by applying dopants externally to a silica mandrel by flame reactions of gaseous compounds or internally to a silica tube. The amorphous evenly coated mandrels or tubes are converted into a clear glass preform in a carefully controlled high-temperature furnace, and the preforms are then drawn into fibers. The preparation of preforms using gas-phase reactions can be carried out with very high puriey levels, using the best available quality materials. Using carefully controlled preparation with controlled atmospheres, fibers can be made with losses approaching the Rayleigh scattering limit in the neighborhood of 1 dB/km at the 840-nm window.

The DC method uses purified glasses in separate crucibles in a controlled atmosphere furnace, as shown in Fig. 82. In some facilities about 1 kg of glass is kept at the controlled melt temperature in each crucible, maintained by continuous feeding of additional glass rod material. The attenuations of some fibers have been field optimized by controlling the oxygen in the furnace atmosphere to balance the relative oxidation states of iron, copper, cobalt, and other transition metal ions. Some technical problems that occur with the DC method are bubbles forming in the melts and variations in the draw nozzle conditions causing minor fiber diameter fluctuations. When properly executed, the DC method can achieve fibers with graded refractive index profiles of 100 to 125-μm diameter with ±1% geometric variations over the length of the fiber at production

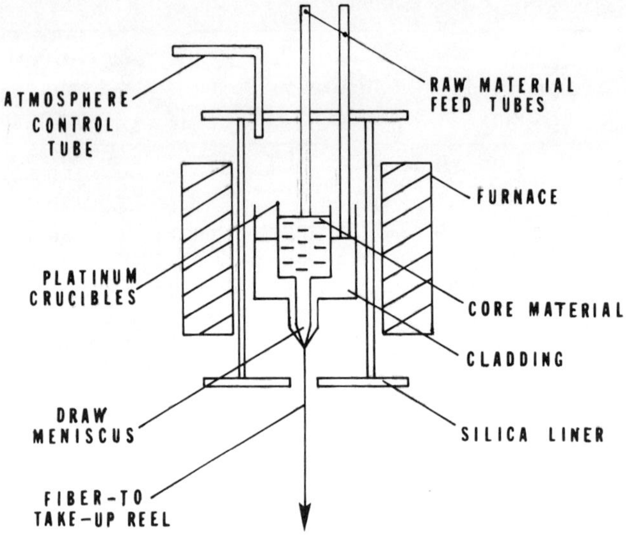

TYPICAL STARTING MATERIALS		
As_2O_3	$CaCO_3$	Na_2CO_3
B_2O_3	K_2CO_3	$PbSiO_3$
$BaCO_3$	Li_2CO_3	SiO_2

Fig. 82 Double-crucible fiber drawing process.

drawing rates of 1-10 km/h. The DC method has two advantages over the CVD method in that fiber production is continuous and large values of the refractive index differential between core and cladding can be easily achieved. The DC method is not the panacea of fiber fabrication, however, in that the refractive index profile cannot be controlled as precisely as in other methods such as the CVD method and minor geometric variations at the cladding-core interface can initiate scattering losses. Polymer-clad high-silica fibers have been prepared with losses of about 4 dB/km, and multicomponent glasses have been made with attenuations less than 5 dB/km. Utilization of the DC method in making graded index borosilicate fibers has produced fibers with attenuations of about 5 dB/km and step-index fibers of about 10 dB/km.

The SM fabrication method works by floating the cladding material on the core material in a polished crucible and then drawing a preform from the doubled-layer melt. Since one glass must essentially float on the other, the stratified melt method is used mostly for step-index fibers. By continually feeding the crucible with core and cladding raw materials, it is possible to make this a continuous process by drawing the fiber directly from the melt instead of using the preform intermediate step. This method can produce fibers of a relatively large numerical aperture.

Coatings. In the coating of polymer materials onto optical fibers, the coating must not induce any micro-bends into the fiber along the entire length of the fiber, and the particular plastics chosen for the coating must protect the fiber physically and chemically from the outside environment as well as mechanically from applied stresses.

Most optical fibers which contain silica have a small relative elongation compared to polymeric coatings. Thus, the larger the Young's modulus of the plastic coating, the thinner the applied coating should be on the wave guide. The maximum coating thickness is mostly determined by the coefficient of thermal expansion of the applied polymer and the technical capabilities to apply uniform thickness throughout the length of the fiber.

The total mechanical strength of the fiber is mainly determined by the condition of the surface. Initial manufacturing-induced cracks propagate in the presence of applied mechanical stress. Corrosion by surface moisture and certain acids accelerates this process. The purpose of coating fibers is to protect the surface from chemical attacks and mechanical damage or to at least minimize these effects. Fibers of the best mechanical integrity are obtained by coating the fiber during the drawing process with a thin plastic film of carefully controlled diameter to avoid introducing fiber micro-bending. When properly applied, coatings cause no additional attenuation and increase mechanical strength and stability compared to the naked fiber.

Numerous polymeric materials can be employed as coatings for optical wave guides applied in a variety of methods including dip-coating in solutions, extrusion, spray coating, and electrostatic coating. Typical plastics include Teflon, perfluorinated ethylene propylene (FEP), PFA, and polyurethane. Plastic coated silica fibers, often made with a FEP coating on silica, are relatively inexpensive to manufacture, attain reasonably high NA values (0.20-0.25), and have attenuations ranging from 10 to 50 dB/km. All plastic fibers (PCP) have very high attenuations no matter what coating material is used, but have high NA values (0.4-0.6). One advantage of PCP fibers and cables is total immunity to any induced water effects such as the stress-corrosion problems associated with fibers containing silica. Special plastic coatings can be used to protect the fiber from unique environmental and corrosion problems such when fibers are run in locations containing acids or alkalis.

The strength of a given coated optical fiber depends upon the combination choice of plastic coating and the type of glass used. Given the same applied stress to a set of optical fibers, the effectiveness of several coatings ranked in increasing strength protection are Teflon, acrylics, silicones, and epoxies. The glasses ranked in increasing time before failure to a given applied stress are soda-lime, borosilicate, lithium aluminosilicate, and fused silica. The difference in the type of coating used can make a difference of about a factor of 10 in the failure time rating of the internal fiber-coating combination. The differences in glasses, however, are very large compared to the coating type. For the four glasses listed, there is an approximate order of magnitude increase between each one in the time before failure ratings to high applied stresses.

Geometric Variation. The important geometric parameters to be considered during the drawing process and fiber formation are to maintain constant core and cladding diameters and exact coincidence of the circle centers along the entire length of the fiber. Fluctuations of the core diameter causes mode-coupling. Changes in the outer diameter of the fiber cause mechanical stresses and subsequent micro-bending during the cabling process. The geometric variations using the DC method are mostly determined by the physical drawing process, while other methods that use intermediate preforms have added potential irregularities.

Variations in fiber diameter and cross-sectional area can be caused by minute changes in drawing tension, preform diameter, draw speed, material composition, coating and curing, and take-up spooling. Fiber drawing using induction furnaces can achieve rates that exceed 10 m/s while achieving geometric variations of less than 0.2% using carefully controlled drawing with controlled atmospheres in the heat furnace.

Plastic Fibers. Optical plastic fibers have higher attenuations than glass and silica material systems. They are commonly used for short-distance computer applications with information capabilities of about 6 Mbps over distances of 50 to 200 m. One distinct mechanical advantage of plastic fibers is simpler and more reliable connections and coupling, since the plastic cladding can be gripped directly and crimped with various connector ferrules and rings without cracking or splitting. Plastic fiber couplings have attenuation losses which are minute compared to the total fiber attenuation. Polymer wave guides can be easily manufactured to have high numerical apertures.

Most plastic fibers and cables have operational temperature ranges less than those of wire and coax cables. But, for some low-temperature applications, plastic fibers have a slight mechanical strength advantage when subjected to dynamically applied stresses, especially compared to some low-loss glass and silica fibers. For high-temperatures or long-distance applications, plastic fibers and cables cannot be used.

Fiber Strength. The integrity of glass and silica optical fibers is highly susceptible to flaws in their structure, since there is minimal yield under

under tension. Any defect whose chemical or physical structure leads to large local stresses is a weak spot or is a potential point of failure. In general, flaws in optical fiber can be characterized by

$$\sigma_{max} = 2\sigma_{as}\left(\frac{c}{r_c}\right)^{1/2}$$

where

σ_{max} = maximum local stress
σ_{as} = applied stress
c = equivalent flaw depth
r_c = effective crack or flaw radius

When the local stress reaches a critical value determined by the weakest flaw in the fiber length, fracture occurs. The longer the fiber length, the weaker the flaws that determine any given failure rate, since flaws are not evenly distributed throughout the entire length of the fiber. Since optical fibers can be up to 10 km in length, it is very important to minimize the generation of flaws during manufacture, especially random weak points caused by imprecise controlling of all of the fiber drawing and coating parameters.

There are numerous means of determining some of the parameters used for lifetime and failure probability predictions, some of which give a diversity of predicted values. Differences arise in testing methods due to nonuniformities in sample lengths, testing environments, and the means for gripping the fibers. Some significant differences also occur in using static and dynamic stress testing. The type and environment of the fiber strength testing should come as close as possible to simulating the actual intended field environment and application, instead of attempting to invent static testing methods which may be inappropriate for real cable situations.

Available fibers and cables have a very wide range of failure stress ratings, ranging from a few thousand to more than 100,000 psi. Failure is usually caused by the presence of surface defects and flaws ranging in size from 0.01 to 10 μm. To relate failure rates to applied stresses with various lengths of fiber over a period of time, Weibul statistics are commonly used. These statistics define a probability of failure (F) from 0 to 1, for fibers of length (L), under a stress (σ), over a period of time (t).

$$F = 1 - \exp\left[-\left(\frac{\sigma}{\sigma_0}\right)^m\left(\frac{t}{t_0}\right)^b\left(\frac{L}{L_0}\right)\right]$$

where m is a distribution shape parameter and b is a parameter related to susceptibility to stress corrosion. Once the values of m and b have been established empirically over a variety of stresses, times, and fiber lengths, extrapolation can be made for any combination of these factors. This is particularly valuable in extrapolating to long time periods, which could not be done in short-term laboratory experiments.

The applied stress to optical fibers also effects the distortion losses as a function of fiber numerical aperture. For most fibers the higher the numerical aperture, the lesser the effects of stress-induced distortion losses. The magnitude of stress-induced distortion losses also depends upon the fiber quality and cabling manufacturing methods used to protect the inner fiber core.

BIBLIOGRAPHY

Bennett, J. M., Appl. Opt., vol. 15, 1976, p. 2705.

Elim, G. R., and H. A. Elim, <u>Fiber Optics in Communications Systems</u>, Marcel Dekker, New York, 1978.

<u>Handbook of Plastics and Elastomers</u>, McGraw Hill, New York, 1975.

<u>Kodak Filters for Scientific and Technical Uses</u>, B-3, Eastman Kodak Company, Rochester, N. Y., 1976.

<u>Modern Plastics and Elastomers</u>, McGraw Hill, New York, 1976.

Moses, A. J., <u>Optical Materials Properties</u>, Plenum Press, New York, 1971.

Newport Research Corp., 1978 Catalogue, Fountain Valley, Calif., 1978.

Optical glass, Schott Optical Glass Inc., Catalog No. 3050e and Mini-Catalogue 3060/72/e, Duryea, Pa., 1978.

Optics for Industry Catalogue, Rolyn Optics Co., Arcadia, Calif., 1977.

<u>Polymer Handbook</u>, Interscience, Wiley, New York, 1966.

Raine, H. C., Optical glasses, Proceedings of the London Conference on Optical Instruments, 1950, p. 243.

Thin Film Optical Filter Guide, Corion Corporation, Holliston, Mass. 1978.

<u>U-72 Kodak Irtran Intrared Optical Materials</u>, Eastman Kodak Company, Rochester, N. Y., 1977.

Chapter 9

HUMAN-ELECTRO-OPTICS INTERACTION

Color

Color can be considered to be three-dimensional and may be represented by a set of spatial coordinates called a color solid. The coordinate system usually chosen is cylindrical coordinates, where the lightness (brightness) is plotted along the vertical axis, the saturation (purity) is plotted in the radial direction from the central axis, and the hue (color) is plotted along the azimuth angle (θ). For this system red is usually chosen to be started at $\theta = 0$, blue at $\theta = \pi/2$, green at $\theta = \pi$, and yellow at $\theta = 3\pi/2$. The location of a specified color in a plane perpendicular to the lightness axis is called the chromaticness.

Color System

In the Munsell color system the hue (H) is divided into 10 basic units which are further split into 10 subdivisions. A given hue is specified by the number of the subdivision, followed by the letters describing the basic unit. Lightness is defined by the value (V) which ranges from 0 for black to 10 for white. Saturation is defined by the chroma (C). The complete color description is then given by HV/C. In the ISCC-NBS color system the common color names are used for hues with the appropriate adjectives for saturation (grayish, moderate, strong, vivid) and lightness (dark, moderate, light). The abbreviations used by both systems are given in Table 60.

Tristimulus Colorimetry

The method for generating a wide variety of colors using three color sources (primaries) is called tristimulus colorimetry. The amount of primary color required to generate a specific color is given by the

Table 60 Color System Hue Abbreviations

Munsell system	ISCC-NBS system
Red, R	Red, R
	Reddish orange, rO
Yellow-red, YR	Orange, O
	Orange-yellow, OY
Yellow, Y	Yellow, Y
Green-yellow, GY	Greenish yellow, gY
	Yellow-green, YG
	Yellowish green, yG
Green, G	Green, G
Blue-green, BG	Bluish green, bG
	Greenish blue, gB
Blue, B	Blue, B
Purple-blue, PB	Purplish blue, pB
	Violet, V
Purple, P	Purple, P
Red-Purple, RP	Reddish Purple, rP
	Purplish red, pR
	Purplish pink, pPk
	Pink, Pk
	Yellowish pink, yPk
	Brownish pink, brPk
	Brownish orange, brO
	Reddish brown, rBr
	Brown, B
	Yellowish brown, yBr
	Olive-brown, OlBr
	Olive, Ol
	Olive-green, OlG

tristimulus values. The tristimulus system uses a three-dimensional coordinate system where the length of a vector in the coordinate space corresponds to luminance and the direction corresponds to the chromaticity. In the CIE tristimulus system the luminances of the red (x) and blue (z) primaries are zero, so that the green (y) tristimulus curve of the equal-energy spectrum is similar to the spectral luminous efficiency curve. The units are chosen such that the plots of the tristimulus values of the equal-energy spectrum colors contain equal areas. The wavelength-dependent tristimulus values are given by $\bar{x}_\lambda$, $\bar{y}_\lambda$, and $\bar{z}_\lambda$. Their chromaticity coordinates (x, y, z) are given by

$$x = \frac{\bar{x}}{\bar{x}_\lambda + \bar{y}_\lambda + \bar{z}_\lambda}$$

$$y = \frac{\bar{y}_\lambda}{\bar{x}_\lambda + \bar{y}_\lambda + \bar{z}_\lambda}$$

$$z = \frac{\bar{z}_\lambda}{\bar{x}_\lambda + \bar{y}_\lambda + \bar{z}_\lambda}$$

Table 61 lists the CIE chromaticity coordinates for the standard observer with the corresponding tristimulus values.

The total tristimulus values are given by

$$X = \int \bar{x}_\lambda \Phi_{e\lambda} \, d_\lambda$$

$$Y = \int \bar{y}_\lambda \Phi_{e\lambda} \, d_\lambda$$

$$Z = \int \bar{z}_\lambda \Phi_{e\lambda} \, d_\lambda$$

where $\Phi_{e\lambda}$ is the spectral radiant flux. The chromaticity coordinates of the total tristimulus values are given by

$$x = \frac{X}{X + Y + Z}$$

$$y = \frac{Y}{X + Y + Z}$$

$$z = \frac{Z}{X + Y + Z}$$

When the spectral surface reflectance R is given, the total tristimulus values are given by

$$X = \int \bar{x}_\lambda R_\lambda \, d_\lambda$$

$$Y = \int \bar{y}_\lambda R_\lambda \, d_\lambda$$

$$Z = \int \bar{z}_\lambda R_\lambda \, d_\lambda$$

Table 61 CIE Chromaticity Coordinates for the Standard Observer

Wavelength (nm)	Chromaticity coordinates x	y	z	Tristimulus values $\bar{x}_\lambda$	$\bar{y}_\lambda$	$\bar{z}_\lambda$
380	0.1741	0.0050	0.8209	0.0014	0.0000	0.0065
385	0.1740	0.0050	0.8210	0.0022	0.0001	0.0105
390	0.1738	0.0049	0.8213	0.0042	0.0001	0.0201
395	0.1736	0.0049	0.8215	0.0076	0.0002	0.0362
400	0.1733	0.0048	0.8219	0.0143	0.0004	0.0679
405	0.1730	0.0048	0.8222	0.0232	0.0006	0.1102
410	0.1726	0.0048	0.8226	0.0435	0.0012	0.2074
415	0.1721	0.0048	0.8321	0.0776	0.0022	0.3713
420	0.1714	0.0051	0.8235	0.1344	0.0040	0.6456
425	0.1703	0.0058	0.8239	0.2148	0.0073	1.0391
430	0.1689	0.0069	0.8242	0.2839	0.0116	1,3856
435	0.1669	0.0086	0.8245	0.3285	0.0168	1.6230
440	0.1644	0.0109	0.8247	0.3483	0.0230	1.7471
445	0.1611	0.0138	0.8251	0.3481	0.0298	1.7826
450	0.1566	0.0177	0.8257	0.3362	0.0380	1.7721
455	0.1510	0.0227	0.8263	0.3187	0.0480	1.7441
460	0.1440	0.0297	0.8263	0.2908	0.0600	1.6692
465	0.1355	0.0399	0.8246	0.2511	0.0739	1.5281
470	0.1241	0.0578	0.8181	0.1954	0.0910	1.2876
475	0.1096	0.8686	0.8036	0.1421	0.1126	1.0419
480	0.0913	0.1327	0.7760	0.0956	0.1390	0.8130
485	0.0687	0.2007	0.7306	0.0580	0.1693	0.6162
490	0.0454	0.2950	0.6596	0.0320	0.2080	0.4752
495	0.0235	0.4127	0.5638	0.0147	0.2586	0.3533
500	0.0082	0.5384	0.4534	0.0049	0.3230	0.2720
505	0.0039	0.6548	0.3413	0.0024	0.4073	0.2123
510	0.0139	0.7502	0.2359	0.0093	0.5030	0.1582
515	0.0389	0.8120	0.1491	0.0291	0.6082	0.1117
520	0.0743	0.8338	0.0919	0.0633	0.7100	0.0782
525	0.1142	0.8262	0.0596	0.1096	0.7932	0.0573
530	0.1547	0.8059	0.0394	0.1655	0.8620	0.0422
535	0.1929	0.7816	0.0255	0.2257	0.9149	0.0298
540	0.2296	0.5743	0.0161	0.2904	0.9540	0.0203
545	0.2658	0.7243	0.0099	0.3597	0.9803	0.0134
550	0.3016	0.6923	0.0061	0.4334	0.9950	0.0087
555	0.3373	0.6589	0.0038	0.5121	1.0002	0.0057
560	0.3731	0.6245	0.0024	0.5945	0.9950	0.0039
565	0.4087	0.589	0.0017	0.6784	0.9786	0.0027
570	0.4441	0.5547	0.0012	0.7621	0.9520	0.0021
575	0.4788	0.5202	0.0010	0.8425	0.9154	0.0018
580	0.5125	0.4866	0.0009	0.9163	0.8700	0.0017

Table 61-Continued

585	0.5448	0.4544	0.0008	0.9786	0.8163	0.0014
590	0.5752	0.4242	0.0006	1.0263	0.7570	0.0011
595	0.6029	0.3965	0.0006	1.0567	0.6949	0.0010
600	0.6270	0.3725	0.0005	1.0622	0.6310	0.0008
605	0.6482	0.3514	0.0004	1.0456	0.5668	0.0006
610	0.6658	0.3340	0.0002	1.0026	0.5030	0.0003
615	0.6801	0.3197	0.0002	0.9384	0.4412	0.0002
620	0.6915	0.3083	0.0002	0.8544	0.3810	0.0002
625	0.7006	0.2993	0.0001	0.7514	0.3210	0.0001
630	0.7079	0.2920	0.0001	0.6424	0.2650	0.0000
635	0.7140	0.2859	0.0001	0.5419	0.2170	0.0000
640	0.7190	0.2809	0.0001	0.4479	0.1750	0.0000
645	0.7230	0.2770	0.0000	0.3608	0.1382	0.0000
650	0.7260	0.2740	0.0000	0.2835	0.1070	0.0000
655	0.7283	0.2717	0.0000	0.2187	0.0816	0.0000
660	0.7300	0.2700	0.0000	0.1649	0.0 10	0.0000
665	0.7311	0.2689	0.0000	0.1212	0.0446	0.0000
670	0.7320	0.2680	0.0000	0.0874	0.0320	0.0000
675	0.7327	0.2673	0.0000	0.0636	0.0232	0.0000
680	0.7334	0.2666	0.0000	0.0468	0.0170	0.0000
685	0.7340	0.2660	0.0000	0.0329	0.0119	0.0000
690	0.7344	0.2656	0.0000	0.0227	0.0082	0.0000
695	0.7346	0.2654	0.0000	0.0158	0.0057	0.0000
700	0.7347	0.2653	0.0000	0.0114	0.0041	0.0000
705	0.7347	0.2653	0.0000	0.0081	0.0029	0.0000
710	0.7347	0.2653	0.0000	0.0058	0.0021	0.0000
715	0.7347	0.2653	0.0000	0.0041	0.0015	0.0000
720	0.7347	0.2653	0.0000	0.0029	0.0010	0.0000
725	0.7347	0.2653	0.0000	0.0020	0.0007	0.0000
730	0.7347	0.2653	0.0000	0.0014	0.0005	0.0000
735	0.7347	0.2653	0.0000	0.0010	0.0004	0.0000
740	0.7347	0.2653	0.0000	0.0007	0.0003	0.0000
745	0.7347	0.2653	0.0000	0.0005	0.0002	0.0000
750	0.7347	0.2653	0.0000	0.0003	0.0001	0.0000
755	0.7347	0.2653	0.0000	0.0002	0.0001	0.0000
760	0.7347	0.2653	0.0000	0.0002	0.0001	0.0000
765	0.7347	0.2653	0.0000	0.0001	0.0000	0.0000
770	0.7347	0.2653	0.0000	0.0001	0.0000	0.0000
775	0.7347	0.2653	0.0000	0.0000	0.0000	0.0000
780	0.7347	0.2653	0.0000	0.0000	0.0000	0.0000

Sources: 1931 Commission Internationale de l'Eclairage (CIE) Tables, and D. B. Judd, J. Opt. Soc. Amer., vol. 23, 1933, p. 359.

When both the spectral irradiation of the illumination ($E_{e\lambda}$) and the spectral surface reflectance (R_λ) are known, the total tristimulus values are given by

$$X = \int \bar{x}_\lambda E_{e\lambda} R_\lambda \, d\lambda$$

$$X = \int \bar{y}_\lambda E_{e\lambda} R_\lambda \, d\lambda$$

$$Z = \int \bar{z}_\lambda E_{e\lambda} R_\lambda \, d_\lambda$$

Standard Sources

The various common standard sources of illumination are A (with a spectral distribution of a tungsten lamp at 2856 K), B (source A with a filter to simulate noon sunlight), and C (source A with a filter that simulates total daylight conditions), and D_{55}, D_{69}, and D_{75} which correspond to natural daylight with correspondent color temperatures of 5500, 6500, and 7500 K, respectively. Table 62 lists the spectral energy distributions and chromaticity coordinates of the CIE standard light sources. Figure 83 shows the chromaticity diagram for emission standards, color regions, and wavelengths. Each point on the diagram represents a specific chromaticity (hue and saturation) and does not depend on the luminance. The locus of blackbodies at various temperatures is plotted from Table 63 which gives the chromaticity coordinates of blackbodies as a function of temperature.

Table 62[a] Spectral Energy Distributions and Chromaticity Coordinates of CIE Standard Light Sources

Wavelength (nm)	A	B	C	D_{55}	D_{69}	D_{75}
330	27	5	4	207	371	550
340	36	23	26	240	400	573
350	47	54	66	279	450	627
360	62	93	123	307	467	630
370	78	148	203	344	522	703
380	98	218	313	326	500	668
390	121	304	450	382	547	700
400	147	402	601	610	828	1019
410	177	507	765	686	916	1119
420	210	615	932	716	935	1128
430	247	711	1067	679	868	1033
440	287	786	1154	856	1049	1211

Table 62[a]-Continued

450	331	831	1178	981	1171	1330
460	378	859	1169	1004	1178	1323
470	429	895	1176	999	1149	1272
480	483	926	1177	1026	1159	1269
490	539	939	1146	980	1088	1177
500	599	916	1065	1007	1094	1165
510	661	882	972	1008	1078	1137
520	725	871	920	1000	1049	1086
530	791	897	931	1042	1077	1105
540	860	943	970	1021	1044	1063
550	929	982	999	1030	1040	1049
560	1000	1000	1000	1000	1000	1000
570	1072	998	972	973	964	956
580	1144	982	929	977	957	942
590	1217	965	885	914	886	870
600	1290	953	852	944	900	873
610	1363	958	840	951	896	862
620	1436	970	837	942	876	836
630	1508	982	836	904	833	787
640	1480	994	834	923	837	785
650	1650	1011	838	889	800	748
660	1720	1021	835	903	802	745
670	1788	1020	820	940	822	755
680	1854	1011	798	900	783	717
690	1919	988	762	797	697	640
700	1983	964	725	829	716	652
710	2048	936	688	849	743	681
720	2104	904	649	702	616	565
730	2161	870	612	793	699	643
740	2217	845	584	850	751	692
750	2270	829	562	719	636	587
760	2321	824	552	528	464	427
770	2370	831	553	759	668	614
780	2417	847	562	716	634	584
			Chromaticity coordinates			
x	0.4465	0.3485	0.3101	0.3324	0.3127	0.2991
y	0.4075	0.3518	0.3163	0.3475	0.3291	0.3150

[a]Tables 62 and 63 reproduced with permission by the Optical Society of America. [Optical Society of America, The Science of Color, Thomas Y. Crowell, New York, 1953, pp. 256 and 280.]

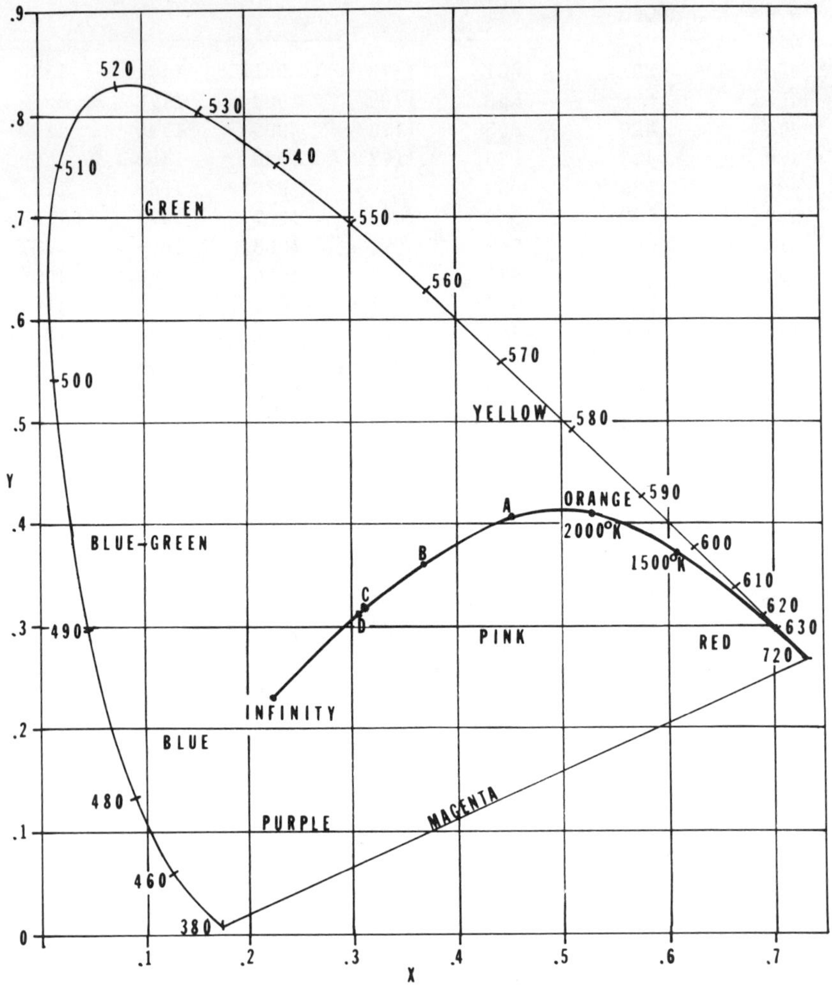

Fig. 83 Chromaticity diagram for emission standards, color regions, and wavelengths.

Photopic and Scotopic Vision

Photopic and scotopic vision differ in their spectral sensitivity, directional sensitivity, and rate of dark adaptation. The measured dark-adaptation response of the eye depends upon the intensity and the duration of the preadaptation radiation and the intensity and spectral distribution of the light

Table 63[a] Chromaticity Coordinates of Blackbodies

T_B (K)	x	y
1000	0.6526	0.3446
1500	0.5856	0.3932
2000	0.5266	0.4133
2500	0.4769	0.4137
3000	0.4368	0.4041
3500	0.4053	0.3906
4000	0.3804	0.3767
4500	0.3607	0.3635
5000	0.3450	0.3516
5500	0.3324	0.3410
6000	0.3220	0.3317
6500	0.3135	0.3236
7000	0.3063	0.3165
7500	0.3003	0.3103
8000	0.2952	0.3048
8500	0.2907	0.2999
9000	0.2869	0.2956
9500	0.2836	0.2918
10000	0.2806	0.2883
20000	0.2565	0.2577
∞	0.2399	0.2342

[a]Tables 62 and 63 reproduced with permission by the Optical Society of America. [Optical Society of America, The Science of Color, Thomas Y. Crowell, New York, 1953, pp. 256 and 280.]

used to measure the adaptation thresholds. With scotopic conditions the human eye has a sensitivity maximum at 500 nm. With photopic conditions the maximum sensitivity is at 555 nm. The minimum radiance for the detection of any light ($R_{L,min}$) is called the absolute threshold. The minimum radiance for the detection of color ($R_{C,min}$) is called the chromatic threshold. The photochromatic interval is defined by $R_{C,min} - R_{L,min}$.

When the eye is photopic (cone vision), it is light-adapted to vision above 0.1 mL. For vision below 0.01 mL, the eye vision is scotopic (rod vision). The spectral responsivities of the eye in the light- and dark-adapted states are shown in Table 64 with luminosity as a function of wavelength. Figure 84 shows the threshold illumination as a function of background luminance necessary for the human eye to distinguish point sources with high accuracy.

Table 64 Photopic and Scotopic Luminosity Data

Wavelength (nm)	Photopic $\bar{y}$	Scotopic V'	Wavelength (nm)	Photopic $\bar{y}$	Scotopic V'
380	0.0000	0.00059	580	0.8700	0.1212
385	0.0001	0.00111	585	0.8163	0.0899
390	0.0001	0.00221	590	0.7570	0.0655
395	0.0002	0.00453	595	0.6949	0.0469
400	0.0004	0.00929	600	0.6310	0.03325
405	0.0006	0.01850	605	0.5668	0.02312
410	0.0012	0.03484	610	0.5030	0.01593
415	0.0022	0.06046	615	0.4412	0.01088
420	0.0040	0.09666	620	0.3810	0.00737
425	0.0073	0.1436	625	0.3210	0.00497
430	0.0116	0.1998	630	0.2650	0.003335
435	0.0168	0.2625	635	0.2170	0.002235
440	0.0230	0.3281	640	0.1750	0.001497
445	0.0298	0.3931	645	0.1382	0.001005
450	0.0380	0.4550	650	0.1070	0.000677
455	0.0480	0.5129	655	0.0816	0.000459
460	0.0600	0.5672	660	0.0610	0.0003129
465	0.0739	0.6205	665	0.0446	0.0002146
470	0.0910	0.6756	670	0.0320	0.0001480
475	0.1126	0.7337	675	0.0232	0.0001026
480	0.1390	0.7930	680	0.0170	0.0000716
485	0.1693	0.8509	685	0.0119	0.0000502
490	0.2080	0.9043	690	0.0082	0.00003533
495	0.2586	0.9491	695	0.0057	0.00002502
500	0.3230	0.9817	700	0.0041	0.00001780
505	0.4073	0.9984	705	0.0029	0.00001273
510	0.5030	0.9966	710	0.0021	0.00000914
515	0.6082	0.9750	715	0.0015	0.00000660
520	0.7100	0.9352	720	0.0010	0.00000478
525	0.7932	0.8796	825	0.0007	0.000003482
530	0.8620	0,8110	730	0.0005	0.000002546
535	0.9149	0.7332	735	0.0004	0.000001870
540	0.9540	0.6497	740	0.0003	0.000001379
545	0.9803	0.5644	745	0.0002	0.000001022
550	0.9950	0.4808	750	0.0001	0.000000760
555	1.0002	0.4015	755	0.0001	0.000000567
560	0.9950	0.3288	760	0.0001	0.000000425
565	0.9786	0.2639	765	0.0000	0.000000320
570	0.9520	0.2076	770	0.0000	0.000000241
575	0.9154	0.1602	775	0.0000	0.000000183

[a]Source: Optical Society of America, The Science of Color, Thomas Y. Crowell, New York, 1953, p. 309.

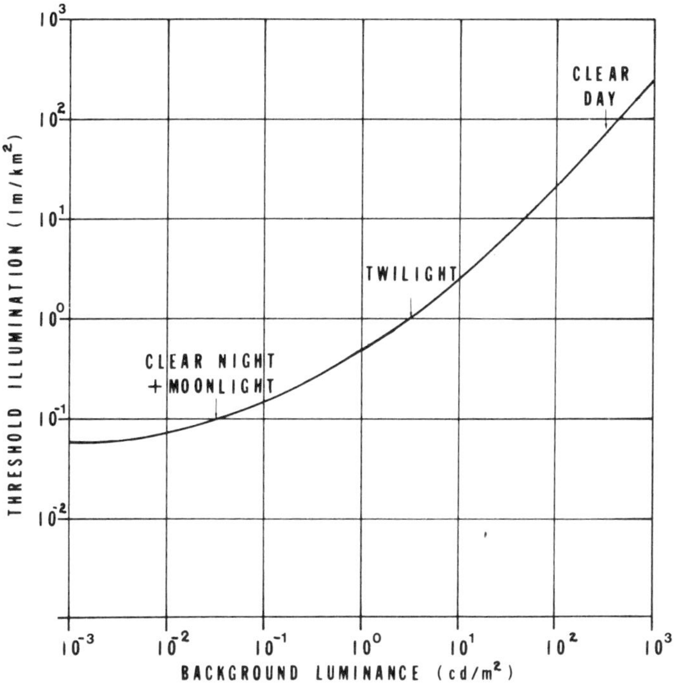

Fig. 84 Human eye point source threshold illumination vs. background luminance.

The spectral luminous efficacy of radiant flux $K(\lambda)$ for photopic vision is given by

$$K(\lambda) = \frac{\Phi_{v\lambda}}{\Phi_{e\lambda}}$$

where $\Phi_{v\lambda}$ is the luminous flux and $\Phi_{e\lambda}$ is the radiant flux at a specified wavelength. The spectral luminous efficiency of radiant flux $V(\lambda)$ is given by

$$V(\lambda) = \frac{K(\lambda)}{K(\lambda_{max})}$$

where $K_{\lambda max}$ is the maximum luminous efficacy ($K_{\lambda max} = 673$ lm/W). The spectral luminous efficacy of radiant flux $K'(\lambda)$ for scotopic vision and the spectral luminous efficiency of radiant flux $V'(\lambda)$ are given by

$$V'(\lambda) = \frac{K'(\lambda)}{K'(\lambda_{max})}$$

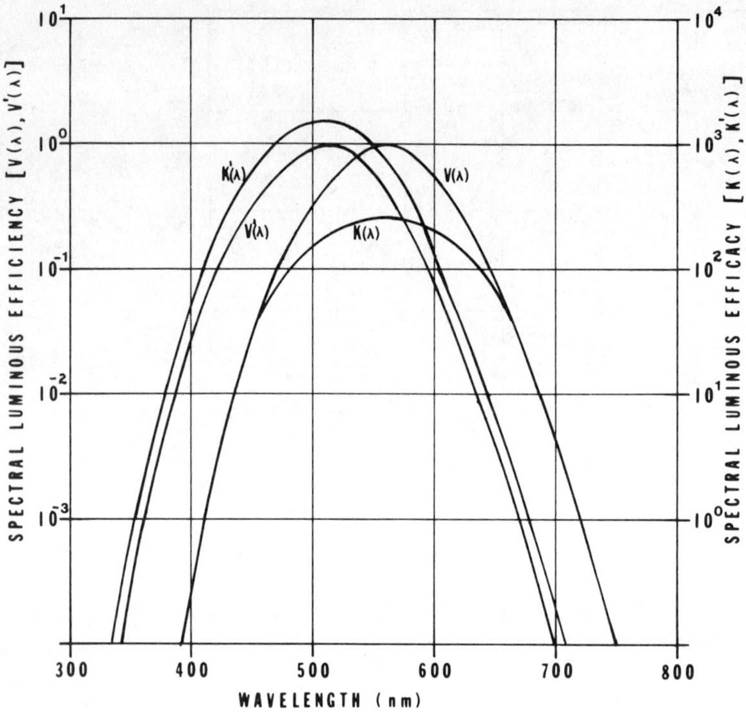

Fig. 85 The relative and absolute spectral response of the human eye.

where $K'(\lambda_{max})$ is the maximum luminous efficacy ($K'_{\lambda\,max}$ = 1725 lm/W). Figure 85 shows the spectral luminous efficiency and efficacy as a function of wavelength.

BIBLIOGRAPHY

Billmeyer, F. W., and M. Seltzman, Principles of Color Technology, Wiley-Interscience, New York, 1966.

Consweet, T. N., Visual Perception, Academic Press, New York, 1970.

Evans, R. M., The Perception of Color, Wiley, New York, 1974.

Graham, C. H. (Ed.), Vision and Visual Perception, Wiley, New York, 1965.

ISCC-NBS Method of Designation Colors, Circular No. 553, U. S. Department of Commerce, National Bureau Of Standards, Washington, D.C., 1955.

Judd, D. B., and G. Wyszecki, Color in Business, Science and Industry, Wiley, New York, 1975.

Kingslake, R., Applied Optics and Optical Engineering, Vol. 1, Academic Press, New York, 1965.

Newhall, S. M., D. Nickerson, and D. B. Judd, Final report of the OSA subcommittee on the spacing of the Newhall colors, J. Opt. Soc. Amer., vol. 33, 1943.

Nineroff, I., Spectral tristimulus values for the CIE uniform spacing system, J. Opt. Soc. Amer., vol. 54, 1964.

Science of Color, Optical Society of America, Committee on color industry, Washington, D.C., 1953.

Sheppard, J. J., Human Color Perception, American Elsevier, New York, 1968.

Wright, W. D., The Measurement of Colour, Van Nostrand, New York, 1969.

Wyszecki, G., and W. S. Stiles, Color Science, Wiley, New York, 1967.

Chapter 10

ATMOSPHERIC AND UNDERWATER OPTICAL TRANSMISSION

Atmospheric Transmission

The transmission of radiation in the atmosphere is affected by absorption, scattering, and refraction. In the earth's atmosphere absorption is dominated by water vapor, carbon dioxide, ozone, nitrous oxide, and other gases. Scattering is caused by numerous sources of solid and liquid particulate matter including dust, clouds, and precipitation. Light refraction in the atmosphere is caused predominantly by the turbulent nature of the air and sky. Attenuation and absorption of radiation in industrial areas can be very significant and is caused by excess water vapor, carbon dioxide, and other gases and particulate matter. Absorption is a major factor for wavelengths less than 0.28 μm because of the atmospheric ozone. At wavelengths beyond the visible, water vapor, carbon dioxide, and the other gases cause very high attenuations. Figure 86 shows the transmittance of a typical 2-km atmosphere path length at sea level, as a function of wavelength, containing 10 mm of water at 20° C and an average concentration of carbon dioxide and nitrogen oxides.

Aerosols contribute, to a major degree, to the scattering of light in the atmosphere. Aerosols include clouds, fog, mist, dust, and smoke. Other than natural causes such as forest fires, volcanic activity, and surface winds, most dust in the atmosphere is man-made. The particle size is described by a size parameter α given by

$$\alpha = \frac{2\pi r}{\lambda}$$

where r is the effective spherical particle radius. For small particles where $r < 0.1$ μm, the scattering is given by Raleigh scattering

$$\frac{2\pi r}{\lambda} < 1$$

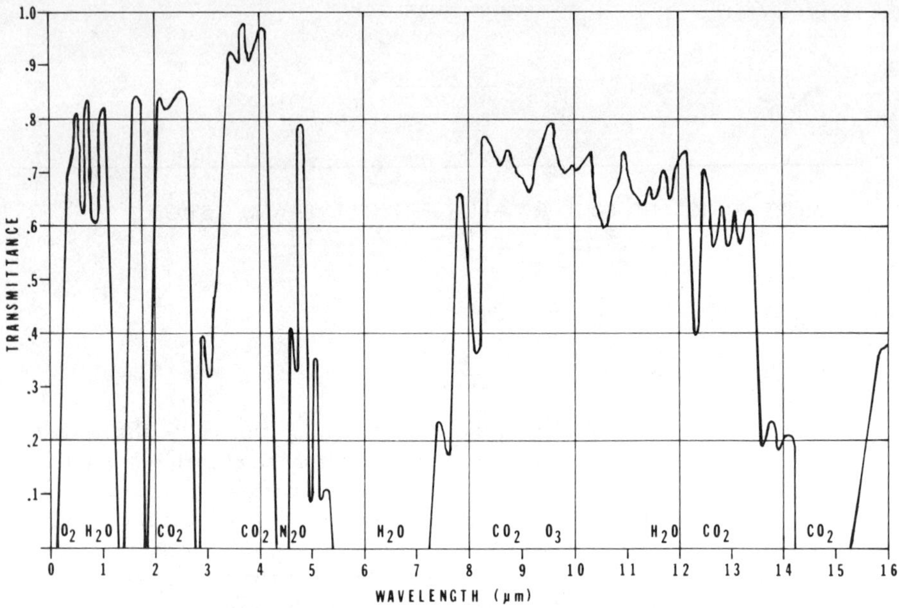

Fig. 86 Transmittance vs. wavelength for 2-km atmospheric path length at sea level containing 10 mm of water at 20° C. Source: R. D. Hudson, Infrared System Engineering, Wiley, New York, 1969.

where the total particulate (molecular) scattering decreases as r^6. The contribution of Rayleigh scattering increases with the increasing ratio of large particles to small particles in the size distribution. Scattering by aerosol particles is called Mie scattering

$$\frac{2\pi r}{\lambda} > 2$$

As the particle size increases, the amount of light scattering in the forward direction increases, and vice versa.

The polarization characteristics of scattered light depend upon the size parameter (α), the angle at which the scattering is measured (ϕ), the index of refraction (n), and the initial polarization of the light. The presence of aerosols can affect the degree of polarization of radiation transmitted in the atmosphere. When the aerosol concentration is relatively low, the degree of polarization at the point of maximum polarization increases with wavelength. As the aerosol concentration in the atmosphere increases, the point of maximum polarization is shifted to the shorter wavelengths.

The illumination (E) of a given light source in the atmosphere when the background light is small compared to the average object luminance is given by

$$E(\lambda) = I_\lambda T^x(\lambda)\, x^{-2}$$

where

I_λ = luminous intensity
$T(\lambda)$ = transmittance
x = source to detector distance

$T(\lambda)$ is a strong function of the radiation wavelength, the atmospheric composition, temperature pressure, and path length:

$$T = \frac{\int_{\lambda 1}^{\lambda 2} T(\lambda)\, d\lambda}{\lambda_2 - \lambda_1}$$

For horizontal or short path lengths where atmospheric variations are small, the transmittance can be given as a function of the extinction coefficient $\sigma(\lambda)$ given by

$$T(\lambda) = \exp\,[-\sigma(\lambda)x]$$

The extinction coefficient is the sum of the ozone absorption coefficient, the Rayleigh scattering coefficient, and the aerosol scattering coefficient.

The <u>visual range</u> $V(\lambda)$ is defined as the maximum distance at which an object of illumination E_m is perceived by a detector:

$$E_m(\lambda) = I_\lambda T^V(\lambda) V^2(\lambda)$$

When the background is comparable in magnitude to the object, then

$$E(\lambda) = \frac{[I_\lambda - (L_b(\lambda) - L_o(\lambda)A]\, T^x(\lambda)}{x^2}$$

where

$L_b(\lambda)$ = luminance of the background light
$L_o(\lambda)$ = natural body luminance of the object
A = projected object area into a plane normal to the detector line of vision

Table 65 Optical Ranges, Transmissivities, and Attenuation Coefficients for Various Atmospheric Conditions

Condition	Range V_o(km)	Trans-missivity	σV_o
Dense fog	0.04-0.07	10^{-21}	56-98
Thick fog	0.07-0.25	10^{-10}	16-56
Moderate fog	0.25-0.50	10^{-5}	8-16
Light fog	0.50-1.0	0.0025	4-8
Thin fog	1.0-2.0	0.05	2-4
Haze	2.0-4	0.25	1-2
Light haze	4-10	0.55	0.4-1
Clear	10-20	0.75	0.2-0.4
Very clear	20-50	0.90	0.08-0.2
Extremely clear	50-150	0.99	0.026-0.08
Rayleigh scattering limit	310		0.0126

The distance V_o at which an object can be seen against a given background is defined by the minimum perceptible contrast ϵ:

$$\epsilon = \frac{\{L_o(\lambda) - L_H(\lambda) - [L_b(\lambda) - L_H(\lambda)]\, T^d(\lambda)\}\, T^V(\lambda)}{[L_b(\lambda) - L_H(\lambda)]\, T^{(d+V)}(\lambda) + L_H(\lambda)}$$

where $L_H(\lambda)$ is the luminance of the horizon sky and d is the distance between the object and the background. When the object is black and viewed against the horizon sky, the distance V_o is given by

$$\epsilon = T^{V_o}(\lambda)$$

Some typical values of V_o for various atmospheric conditions are given in Table 65.

The spectral emittance of distant objects can be shifted when atmospheric attenuation is significant especially for particulate scattering. For distant objects traveling at very high speeds (stars, etc.) the shifting of light through the various atmospheric layers and doppler effects can cause dramatic apparent source color changes. The threshold illumination for dim light sources is a function of the luminance of the background light and various object factors such as geometric configuration and color.

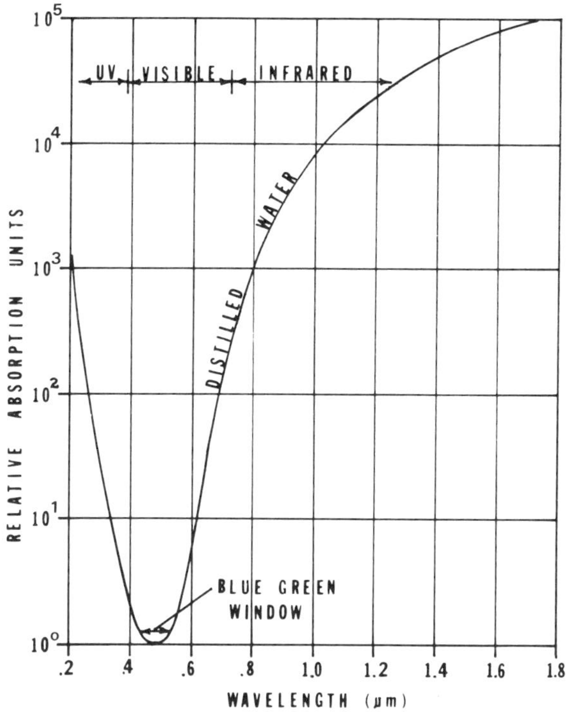

Fig. 87 Relative absorption spectrum for distilled water.

Underwater Transmission

Light attenuation in water is caused by molecular absorption, absorption by particulate matter (scattering), and absorption by colored dissolved matter. For wavelengths below the blue-green attenuation minimum, the increased attenuation is caused by small-particle scattering. For wavelengths above the blue-green minimum, increased attenuation is dominated by absorption. Figure 87 shows the relative absorption spectrum for distilled water. For ocean water types this curve is shifted upward. Light transmitted in water is partially linearly polarized. The light transmission for various water types as a function of path length is shown in Fig. 88.

The intensity (I) of sunlight as a function of depth is given by

$$I = I_0 S(\theta)\, e^{-\gamma L}$$

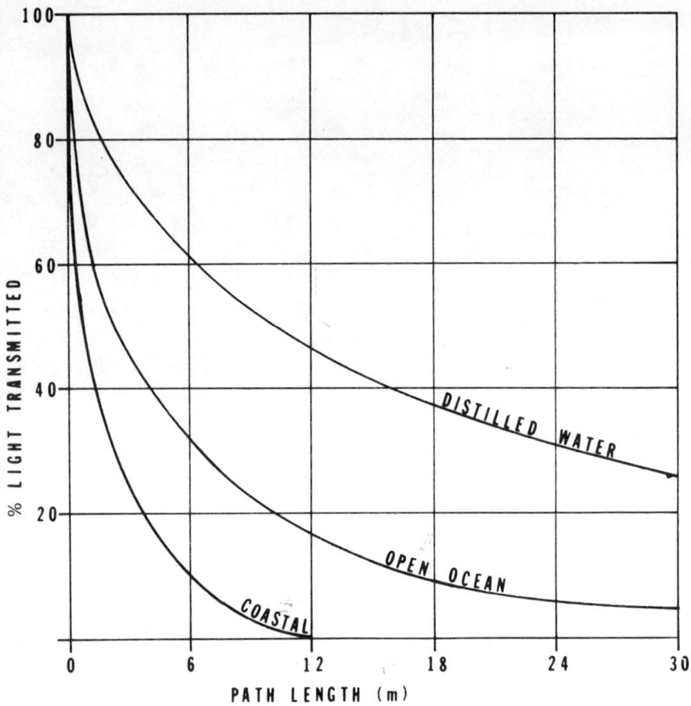

Fig. 88 Light transmission as a function of path length for average water types.

where

I_o = intensity of light at the surface (L = 0)
$S(\theta)$ = surface reflection constant
γ = extinction coefficient
L = path length (depth)

The value of $S(\theta)$ which is a function of the angle of the sun (or light source) with the surface is given in Fig. 89. Light directed toward the surface of a body of water undergoes reflection defined by Fresnel's laws.

The experimental attenuation of light transmitted in water makes the selection of the proper transmission wavelength an important factor in the design of practical underwater systems. A significant increase in depth penetration can be achieved by matching the signal wavelength to the minimum transmission window value which can be accomplished in a continuous mode using tunable laser sources in the 430- to 570-mm wavelength range.

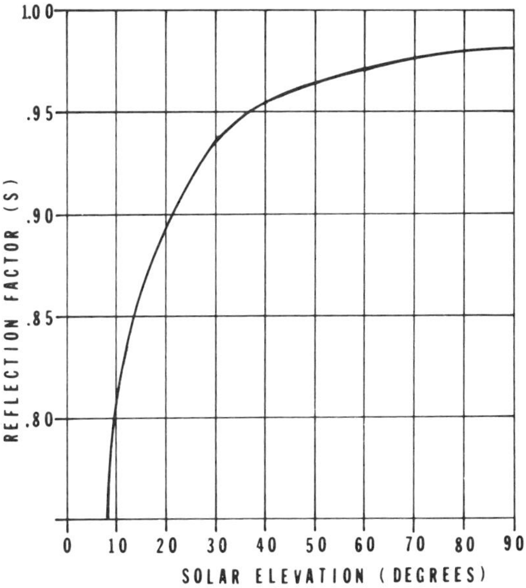

Fig. 89 Reflection factor vs. angular elevation of above-surface light sources.

For fixed systems, an average emission wavelength must be chosen for a variety of seasonal and daily fluctuations in the water composition and transmission parameters. Table 66 lists some of the transmission window values for various types of ocean water.

For lighting used for underwater imaging systems, the sensitivity curve of the image tube, the output spectrum of the light source, and the wavelength of minimum attenuation should all be matched as closely as possible to maximize range and resolution. The effective range of an underwater imaging or communications system is highly sensitive to the detector type and sensitivity. An image orthicon is commonly used in underwater imaging systems, and high-gain photomultipliers are commonly used for communications detectors.

The beam-volume attenuation coefficient $\alpha(\lambda)$ and the diffuse attenuation coefficient $K(\lambda)$ are used to characterize the optical transmission properties of water. The beam attenuation coefficient is a measure of the decay of the unscattered light and is defined in the equation

$$P_r(\lambda) = P_o(\lambda)e^{-\alpha(\lambda)r}$$

Table 66 Transmission Windows for Various Types of Ocean Waters

Ocean type	Transmission window values		
	$K(m^{-1})$	Loss (dB/m)	Range-λ (nm)
Exceptionally clear	0.02–0.03	0.09–0.13	430–470
Open ocean	0.04–0.07	0.17–0.30	470–490
Continental shelf	0.09–0.12	0.39–0.45	490–510
Coastal shallows	0.14–0.18	0.65–0.75	510–550
Shore and harbor	0.37–0.43	1.38–1.75	550–570

where

$P_o(\lambda)$ = initial beam radiant power
$P_r(\lambda)$ = measured beam radiant power
r = water path length

The beam attenuation coefficient is the sum of the absorption coefficient $a(\lambda)$ and the scattering coefficient $S(\lambda)$:

$$\alpha(\lambda) = a(\lambda) + S(\lambda)$$

The transmittance of an underwater beam is defined by

$$T(\lambda) = \frac{P_r(\lambda)}{P_o(\lambda)}$$

α is usually given in units of m^{-1} and can be defined in terms of the transmittance by

$$\alpha(\lambda) = \frac{1}{r} \ln \frac{1}{T(\lambda)}$$

The transmittance, hence the value of $\alpha(\lambda)$, usually varies with depth and water temperature due to the nonhomogeneous nature of large bodies of water, particularly the ocean.

The diffuse attenuation coefficient $K(\lambda)$ is a measure of the decay of diffuse light and is defined in the equation

$$E_z = E_o e^{-K(\lambda)z}$$

where E_z is the residual down-welling irradiance in a horizontal plane at a depth z and E_o is the irradiance at the sea surface. The diffuse attenuation coefficient is dependent upon the absorption and scattering properties of the water:

$$K(\lambda) = a(\lambda) + \frac{S(\lambda)}{n}$$

where n is an empirically determined value (usually $5 < n < 7$). The value of K, usually given in units of m^{-1}, depends on the degree of the diffuse nature of the light, the means of measurement, and the water depth. The field of view used to mean the value of K is usually large (170-180°), while the field of view used for α determination is very small (0.5-1.0°). $K(\lambda)$ and $\alpha(\lambda)$ are usually assigned depth average values given by

$$\overline{K}(\lambda) = \frac{\int_{z_1}^{z_2} K(\lambda)\, dz}{z_2 - z_1}$$

$$\overline{\alpha}(\lambda) = \frac{\int_{z_1}^{z_2} \alpha(\lambda)\, dz}{z_2 - z_1}$$

The volume scattering function $\sigma(\theta)$ is given by

$$S = 2\pi \int_0^{\pi} \sigma(\theta) \sin\theta \, d\theta$$

or by

$$\sigma(\theta) = \frac{dI(\theta)}{E\, dV}$$

where $I(\theta)$ is the light intensity in the θ direction for the elemental volume dV irradiated at E. Figure 90 shows the volume scattering function for distilled and clear ocean water as a function of the angle θ.

When a narrow beam of light is used to transmit a signal from underneath a given water surface to an object in the air space above the surface, numerous factors control the beam intensity and divergence including water values of $S(\lambda)$ and $\alpha(\lambda)$, the degree of surface wave action (hence internal reflection properties), the transmittance of the air at the given beam wavelength, and the degree of beam scattering caused by water spray just above the surface. The greater the wave action, the transmitter depth, and the values of $S(\lambda)$ and $\alpha(\lambda)$ and the lower the transmittance of the air space, the greater the beam width in air will be. A plot of the average beam width in air versus the transmitter path length times the water attenuation

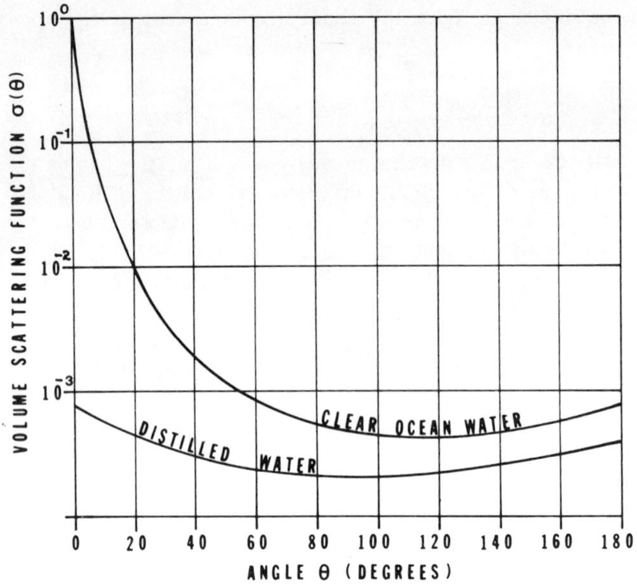

Fig. 90 Volume scattering function for clear ocean and distilled water.

coefficient are given in Fig. 91 for expected values in the ocean, starting with a beam width of 5°. For initial beam width of 1°, such as with laser sources, the anticipated beam width in air of the water-to-air emergent signal will be correspondingly smaller. The received beam intensity for most practical cases will be very low compared to the initial transmission intensity. The potential information bandwidth for subsurface-to-air optical transmission is very limited due to the surface scattering properties of the water which can cause beam interruption, thus nulling the intended transmission.

The intensity $I(\theta, \lambda)$ of a subsurface-to-air optical beam can be approximated by

$$I(\theta, \lambda) = \frac{P_o e^{-\gamma r}}{\pi (r_a \tan \theta)^2}$$

where

P_o = initial beam power
γ = extinction coefficient

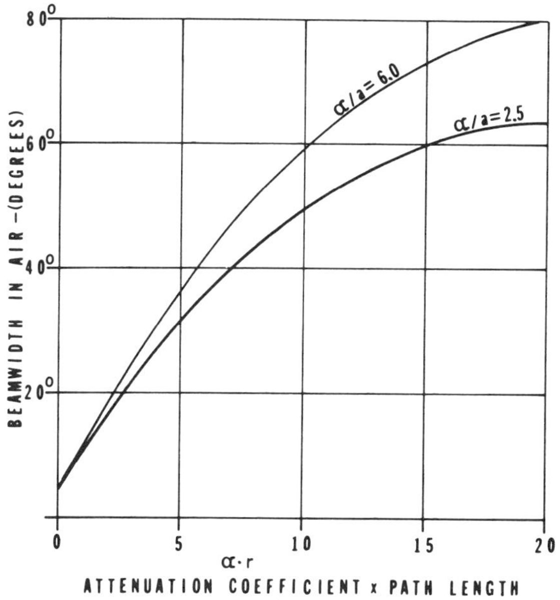

Fig. 91 Subsurface-to-air optical beamwidth vs. transmitter depth. Source: Proc. SPIE, Ocean Opt., vol. 64, 1975.

r = water path length
r_a = air path length

The beam width is given by

$$\theta = \left(\theta_o^{\,2} + \theta_w^{\,2} + \theta_s^{\,2}\right)^{1/2}$$

where

θ_o = beam width of the transmitter in air
θ_w = beam width divergence caused by surface and spray effects
θ_s = beam width component increase caused by scattering in the water

BIBLIOGRAPHY

Arnush, D., Underwater light-beam propagation in the small-angle scattering approximation, J. Opt. Soc. Amer., vol. 62, 1972.

Barnes, K. R., The Optical Transfer Function, American Elsevier, New York, 1971.

Brookner, E., Atmospheric propagation and communication channel model for laser wavelengths, IEEE Trans. Commun. Technol., COM-18, 1970.

Deirmendjian, D., Scattering and polarization properties of water clouds and hazes in the visible and infrared, Appl. Opt., vol. 3, 1964.

DeVelis, J. B., and G. B. Parrent, J. Opt. Soc. Amer., vol. 59, 1967.

Duntley, S. Q., Light in the sea, J. Opt. Soc. Amer., vol. 53, no. 2, 1963.

Duntley, S. Q., Visibility in the oceans, Opt. Spectra, 4th qtr., 1967.

Funk, G. J., Multiple scattering calculation of light propagation in ocean water, Appl. Opt., vol. 12, 1973.

Gordon, H. R., Simple calculation of the diffuse reflectance of the oceans, Appl. Opt., vol. 12, 1973.

Gordon, H. R., and O. B. Brown, Irradiance reflectivity of a flat ocean as a function of its optical properties, Appl. Opt., vol. 12, 1973.

Hill, M. N. (Ed.), The Sea, Interscience, New York, 1963.

Jerlov, N. G., Optical Oceanography, Elsevier, New York, 1968.

Jerlov, N. G., and G. Steerman (Eds.), Optical Aspects of Oceanography, Academic Press, New York, 1974.

Kerr, J. R., P. J. Tillerton, A. R. Kraemer, and C. R. Cooke, Atmospheric optical communications systems, Proc. IEEE, vol. 58, 1970.

Lawrence, R. S., and J. W. Strohbehn, A survey of clean-air propagation effects relevant to optical communications, Proc. IEEE, vol. 58, 1970.

Merten, L. E., In-Water Photography: Theory and Practise, Wiley-Interscience, New York, 1970.

Middleton, W. E. K., Vision through the Atmosphere, University of Toronto Press, Toronto, Canada, 1958.

Ocean optics, Proc. SPIE, vol. 64, 1975.

Replogle, F. S., and J. B. Steiner, Resolution measurements in natural water, J. Opt. Soc. Amer., vol. 55, 1965.

Tatarski, V. I., Wave Propagation in a Turbulant Medium, McGraw-Hill, New York, 1965.

Tricker, R. A. R., Introduction to Meteorological Optics, American Elsevier, New York, 1970.

Valley, S. L., Handbook of Geophysics and Space Environments, McGraw-Hill, New York, 1965.

Wells, W. H., Loss of resolution in water as a result of multiple small-angle scattering, J. Opt. Soc. Amer., vol. 59, 1969.

Williams, J., Optical Properties of the Sea, U. S. Naval Institute, Annapolis, Md., 1970.

Yura, H. T., Small-angle scattering of light by ocean water, Appl. Opt., vol. 10, 1971.

Chapter 11

ELECTRO-OPTIC DISPLAYS AND ARRAYS

Display Systems

Alphanumeric display devices are commonly grouped into four categories, depending on the means of forming the numerical or alphabetical characters: the seven-segment display, the 14- to 16-segment display, the 5 X 7 dot matrix, and the 7 X 9 dot matrix. The seven-segment display commonly used for digital clock displays can produce only a limited number of alphabetical characters including A, C, E, F, H, I, L, O, P, S, and U. Table 67 lists some typical seven-segment display electrical and optical parameters. Fourteen to sixteen display segment units can produce all numbers, alphabetical characters, and a few additional specialized symbols such as mathematical symbols, punctuation, and others. The dot matrix displays can produce all alphanumeric characters and are chosen for specific applications on the basis of size and projected clarity. Figure 92 shows two commonly used alphanumeric display devices with high character potential.

The most common alphanumeric display systems are the light-emitting-diode (LED) displays, the nixie tube (large display noble gas tubes), liquid crystal displays (low power consumption devices), vacuum incandescent and fluorescent displays, and gas-discharge display units. Table 68 compares the important electrical and optical parameters of the major digital display technologies. For large-screen displays such as the CRT, the major considerations in systems applications include resolution, contrast, brightness, memory capabilities, and gray-scale and color potentials. Table 69 lists some of the major large-screen display technologies with the appropriate system parameters.

Types of LED Numeric Packages

The three common types of LED numeric packages are the bar-segment, monolithic, and reflector packages. The bar-segment displays consist of

Table 67 Typical Seven-Segment Display Parameters

Parameter	Symbol	Typical value range
Response time	tr	10–90 ns
Temperature coefficient of V_F/segment	$\Delta V_F/°C$	-1.5 to -2.5 mV/° C
Operating temperature range	T_A	-40 to 85°C
DC power dissipation/segment	P_D	40–80 mW at 20° C
Forward voltage/segment	V_F	1.6–2.1 V at I_F = 20 mA
Reverse voltage/segment	V_R	5–6 V
Peak forward current/ segment	I_p	50–150 mA at 20°C
Average forward current/ segment	I_{avg}	5–25 mA at 20° C
Reverse current/segment	I_R	10–20 A at V_R = 5 V
Luminous intensity/segment	I_v	
	650 nm (red)	0.3–1.7 mcd at 20 mA
	585 nm (yellow)	0.2–1.5 mcd at 20 mA
	565 nm (green)	0.1–0.7 mcd at 20 mA
Character height	h	3–2.0 mm

seven bars of semiconductor material ($GaAs_{1-x}P_x$, for example) that are placed symmetrically on a circuit-board substrate as shown in Fig. 93. This type of numeric display uses less semiconductor material than the monolithic displays. The bar-segment displays have been found to be less cost effective in mass production than monolithic chip displays. The reflector display typically uses light, small-square chips of LED material which is typically doped with nitrogen. The reflector cavities of these devices are often filled with an epoxy containing finely powdered, suspended particles (which act to scatter light) and dyes to absorb incident ambient light. High-brightness monolithic displays can also be made using slanting black-surface facets which serve as self-reflections for the emitted junction radiation.

Display Visibility and Brightness

The visibility or viewability of displays and arrays is a very important parameter to consider in design. The total brightness of any display or

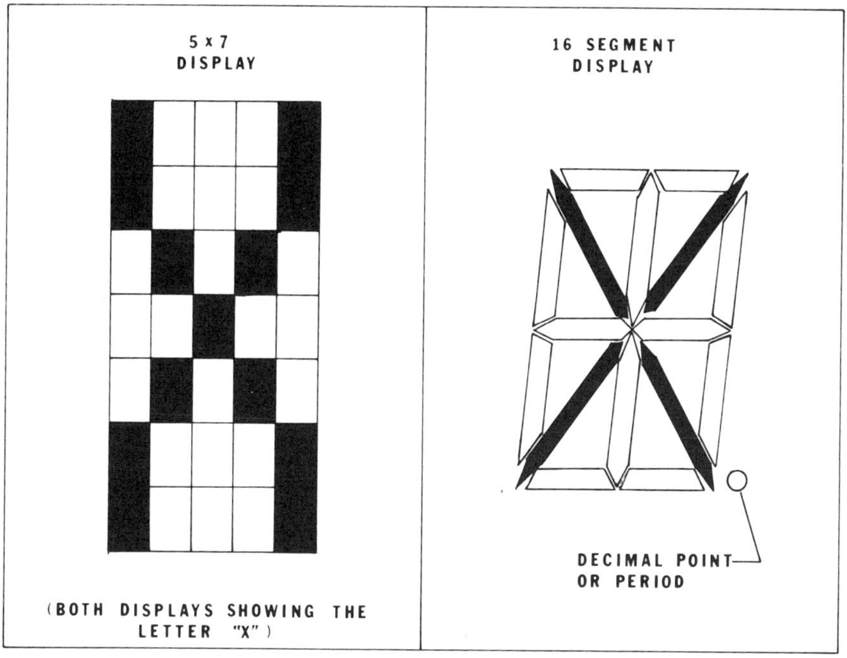

Fig. 92 Two basic alphanumeric display unit types.

information system is critical when determining the suitability of the display for information transfer (viewability) and must be simultaneously considered along with the level of ambient light and the reflectivity of the display surface. High contrast and high ambients can be achieved for observation by eliminating ambient reflections by using a light-absorbing black layer (~0.3 μm thick) on the rear electrode. For thin-film electroluminescent emitters a properly designed black layer application can eliminate ambient reflections from both the rear (bottom) electrode and from the space between electrodes. The actual reflectance from the applied black layer is a function of the index of refraction of the black material and the transparent emitter above it.

Watch displays are one important commercial application requiring attention to display visibility and brightness. Table 70 lists some of the parameters of LCD and LED watch displays to compare both display techniques. The relative light scattering intensity as a function of viewing angle for various common displays is shown in Fig. 94.

Table 68 Comparison of Digital Display Technologies

	LED	Gas discharge	Planar gas discharge	Fluorescent	Incandescent (direct-view filament)	Dynamic scattering LCD	Field effect LCD
Power/digit (0.3 to 0.6 in.)	10 to 140 mW depending on color	350 mW	30 to 100 mW	100 mW	250 mW to 1 W	100 μW	1 to 10 μW
Voltage	2 to 5 V	175 V	180 V	15 to 25 V	5 V	18 V	3 to 7 V
Temperature range (°C)	-55 to 125	0 to 70	0 to 55	-55 to 100	-55 to 100	0 to 80	0 to 70
Switching speed	1 μs	150 μs	1 ms	1 ms	150 μs	300 ms	0.1 to 0.3 s
Life in hours	$>10^5$	10^5	3×10^4	10^5	10^4 to 10^5	2×10^4	2×10^4
Colors	Red, orange, yellow, green	Neon orange	Orange; others with filters	Blue-green; others with filters	All with filters	Depends on illumination	Depends on illumination
Brightness	Excellent	Excellent	Good	Good	Excellent		
Contrast ratio	10:1	10:1	20:1	10:1	20:1	20:1	20:1

Viewing angle	150°	100°	120°	150°	150°	90° to 150°	90° to 120°
Font	7 and 16 segments, 5 X 7 dot matrix	Individual characters	7 segment	7 segment	7 and 16 segment shaped characters	7 and 16 segment	7 and 16 segment
Vertical size	0.1 to 1 in.	0.3 to 12 in.	0 to 2 in.	0.5 to 0.75 in.	Up to 1 in.	0.2 to 8 in.	0.2 to 2 in.
MOS compatibility	No	No	Yes	Yes	No	Yes	Yes

Adapted from Arthur D. Little, World Electro-Optics Industry 1975-1980, E-O Components Update, May, 1976.

Table 69 Comparison of Major Large-Screen Display Technologies

Type	Advantages	Disadvantages
CRT	Bright, gray scale, adequate reliability, image storage capability available at expense of gray scale, colors	Bulk, high voltage, nondigital addressing, separate memory required for refresh, flicker
AC plasma discharge	Bright, colors, inherent memory, switching threshold, selective erase, potential for long life	Gray scale doubtful, probably incompatible with ICs, high drive power, expensive overall system today
DC plasma discharge	Bright at expense of resolution, colors gray scale, functional scanning, potential for long life	High drive power, line memory required, limited number (≤250) of scan lines
Liquid crystal	Good contrast in bright ambient, low power, IC-compatible, some intrinsic storage	Limited temperature range, reliability uncertain, usually requires separate memory, slow switching speeds
Electroluminescent	Gray scale, colors, initial high brightness	No internal memory, address thresholds absent, not compatible with ICs, lack of uniformity
Scanned laser	High resolution, functional scattering, good contrast, colors	Low brightness, bulk, low power efficiency
Light valves	High resolution, uniform gray scale, inherent memory, functional scanning	Bulk, requires special optics, slow response speed

Adapted from Arthur D. Little, World Electro-Optics Industry 1975-1980, E-O Components Update, May 1976.

Gas-Discharge Displays

Gas-discharge displays can be operated in either an ac or dc mode, using various types of gases and electrode material configurations. Both the ac and dc discharge matrices can be operated in a memory or nonmemory mode. In dc gas-discharge devices the electrodes are immersed in the

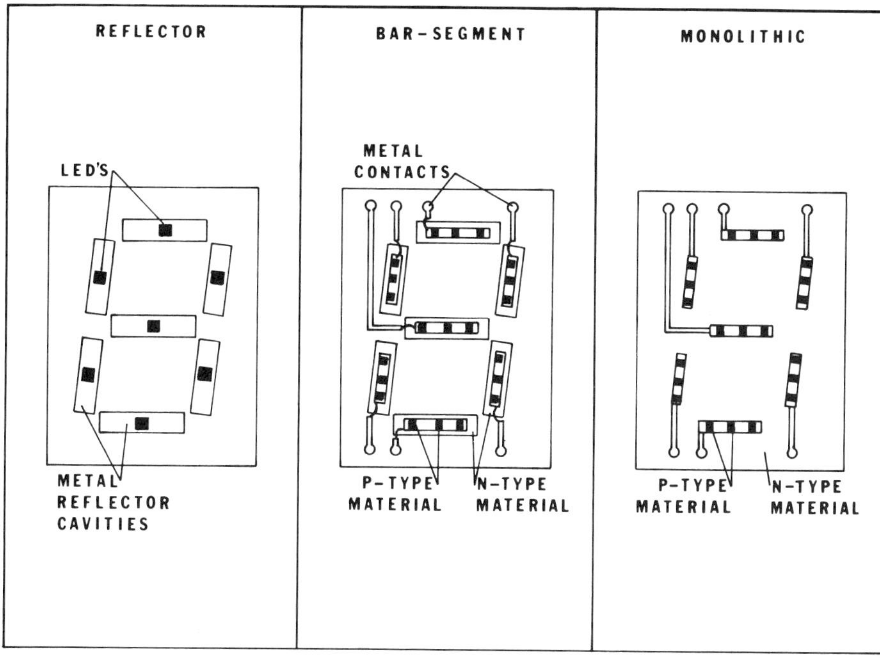

Fig. 93 Three basic types of LED numeric units. Source: M. G. Craford, Recent developments in light-emitting-diode technology, Proc. Soc. Information Display, vol. 18, no. 2, 1977.

display gas, while in ac devices the electrodes are usually insulated from the gas by their dielectric films. In ac displays the electrode configuration is usually symmetrical, while dc displays often have asymmetrical cathode-anode configurations. To minimize response times and maximize sweep rates, a background or sustaining voltage is usually applied to the display matrix so that only a small additional pulse is needed at a particular intersection to initiate the gas discharge. The actual required firing voltage is basically a function of the gas composition, the gas pressure and temperature, and the device spatial configuration. To achieve consistent firing voltages at all elements in large arrays, the electrode size and spacing must be kept to within very close tolerances.

To reduce the required firing voltage and the time lag to achieve the discharge, electrons or ions can be introduced into the electrode gap spaces from nearby discharges by a process called priming. Priming can be used in ac displays to spatially translate information from one display element to another. One of the major advantages of the gas-discharge

Table 70 Comparison of LCD and LED Watch Displays

Features or factors	LCD (with backlight, etc.)	LED (with color filter)
Appearance	Good	Better
Contrast	Medium	High (except in sun)
Jewel-like	No	Yes
Color, normal	No (silver/dark)	Red
Color options	Pastel, gold, silver	Orange, yellow, green
Backlight required	Yes, at night	No
Visibility, sunlight	Excellent	Good except in brightest
Visibility, night	Fair with backlight	Excellent
Need hand for switch	No, except at night	Yes
Other options	Inertia, proximity	Inertia, proximity
Display	Continuous (exc. light)	Momentary
Power drain	Low (except the light)	High if switched often
Number of battery cells	One	Two
Ease of driving display	Good	Less than LCD
Speed of response	0.25 (poor for display of seconds and colon)	<1 ms

Adapted from Arthur D. Little, World Electro-Optics Industry 1975-1980, E-O Components Update, May, 1976.

matrix displays is the ability to effectively alter the position of the neighboring discharge sites.

Gas-discharge display units can be used for small and large area numerical displays and for ac, dc, and interactive plasma panel computer graphic systems. The use of gas-discharge panels for television applications is limited by the high required luminance levels of commercial screens.

Electrophoretic Image Displays

The electrophoretic image display (EPID) is a passive display technique based upon the transport of charged pigment particles in a highly stable colloidal suspension in the presence of an applied electric field and the

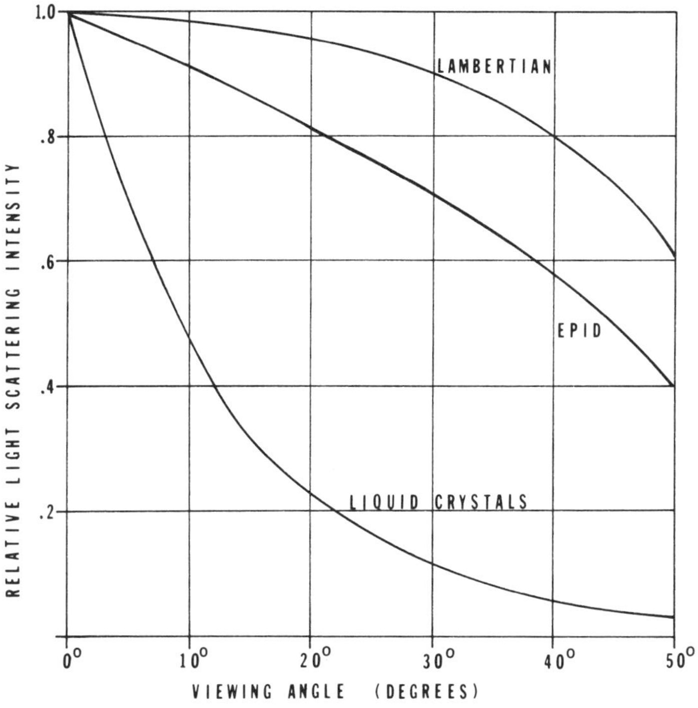

Fig. 94 Relative light-scattering intensity vs. viewing angle for common displays. Source: A. L. Dalisa and R. J. Seymour, Proc. IEEE, vol. 61, 1973.

subsequent absorption and scattering of ambient light. The colloidal suspension which must be chemically and electrostatically stable, consists of submicrometer-sized pigment particles in a nonaqueous colored suspension liquid (dye). The mobility, scattering power, color, and opacity are the most important considerations in pigment selection. By selecting various combinations of the pigment and dye, a wide variety of colors can be generated including black and white. The basic EPID cell consists of a glass plate coated with a transparent electrode (indium oxide) which contacts the colloidal suspension. The top coating of the back plate of the cell is a conducting electrode. The cell plate spacing is in the range of 10 to 70 μm and is usually attained by using thin plastic film spacers.

The average current is in the range of 0.08 to 0.10 $\mu A/cm^2$, with a power consumption of about 3 to 5 $\mu W/Am^2$. The speed of the pigment transport to the electrodes is a function of the applied voltage, the cell spacing, and

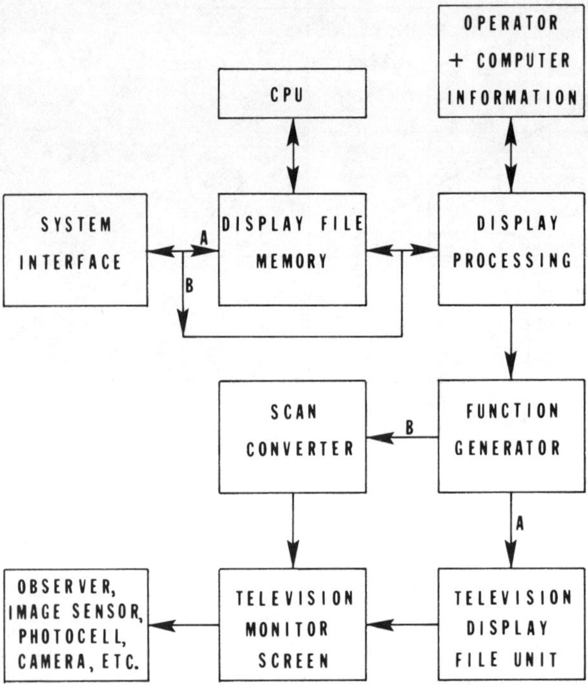

Fig. 95 Basic block diagram of computer-type television display systems.

the material components of the dye-pigment suspension. Typical transportation times are in the range of 5 to 30 ms for applied potentials of 30 to 50 V. When the cell polarity is reversed, the color tone is altered, since the pigment position changes. If the cell voltage potential is disrupted, the pigment will remain in contact with the electrode for a long period of time, giving the device a memory period of about 100 h. The operational temperature range is in part a function of the colloid materials and the duty cycle of the display cell. Typically, the operational temperature rating is from -10 to 80°C. The lifetime of those devices is usually given in terms of total number of switches before failure, typically in the range of 10^8 to 10^9 switches.

Computer Displays

The two basic types of computer display units are the graphic plotters and the CRT display devices. The graphic plotters are convenient when transforming computer data and generated information directly onto paper for a permanent record, but for many applications they are too slow, they have no interactive capabilities, and they lack the resolution achievable in CRT display systems. Processed computer display information can be displayed directly on CRT tubes and light-pen devices or can be filmed in various size and color formats depending on the desired user output form. The actual display tube can be a storage tube, a television monitor screen, a plasma matrix panel, or a refresh-random tube, each with its own advantages and disadvantages. The refresh-random system is easily coupled for use with a light pen but is performance limited by the flicker speed of the display tube. Storage tube computer displays have large flick-free data capabilities but are difficult to partially erase or to use to show dynamically changing data. Television monitor screens are difficult to use with light-pen interactive devices but have the advantage of full color display and potentially high resolution, such as in screen converter systems. Figure 95 shows the basic block diagram of a computer television display system using either the digital or scan converter method.

BIBLIOGRAPHY

Backler, Jordon, Digital Display-CRT Terminals, September, 1975.

Craford, N. G., Recent developments in light-emitting diode technology, Proc. SPID, vol. 18, no. 2, 1977.

Dalisa, A. L., Electrophoretic display technology, Proc. Soc. Information Display, vol. 18, no. 1, 1977.

Fugate, K. O., High display viewability provided by thin-film EL, black layer and TFT drive, Proc. SPID, vol. 18, no. 2, 1977.

Little, Arthur D., World Electro-Optics Industry 1975-1980, E-O Components Update, Acorn Park, Mass., May, 1976.

Newman, W. M., and R. F. Sproull, Principles of Interactive Computer Graphics, McGraw-Hill, New York, 1973.

Optoelectronics Data Book for Design Engineers, 4th ed., Texas Instruments Incorporated, Semiconductor Group, Dallas, Tex., 1976.

Optoelectronics Designers Catalog 1977, Hewlett-Packard Components, Palo Alto, Calif., 1977.

Sobel, A., Gas-discharge displays: the state of the art, Proc. Soc. Information Display, vol. 18, no. 1, 1977.

Chapter 12

COMMUNICATIONS SYSTEMS

Atmospheric Systems

The most common atmospheric electro-optical communications system is the laser link, capable of video transmission with multiple audio channels or high data code digital transmission. The maximum range of the receiver-transmitter distance depends on both the desired signal quality (SNR) and the average atmospheric transmittance. Under average conditions a typical 5-mW He-Ne laser system has a range of 5 km to maintain the desired minimum audio-visual SNR of 41 dB. With clear atmosphere, the transmission range can increase to 12 km. The He-Ne laser communications system uses the 632.8-nm optical carrier wavelength which is amplitude modulated by a frequency-modulated RF subcarrier, typically with a bandwidth ranging from 15 to 30 MHz. The bit error rate (BER) is usually about 10^{-8} depending upon atmospheric conditions and transmitter-receiver design.

For daylight transmission atmospheric laser communications systems must use a narrow bandpass filter centered around the transmission wavelength to eliminate background noise (extraneous light). During nighttime transmission the filter can sometimes be removed to increase the received power to the photodetector-amplifier assembly.

This type of atmospheric communications link can be used for high bit rate computer data transmission over short building-to-building or intracity links, for PCM telephone communications links, and for mixed types of communications, particularly in areas that are rather congested in the RF region or have high levels of electromagnetic interference.

Optical Repeaters

The basic optical repeater consists of a photodetector, amplifier, equalizer, and signal regenerator coupled to an optical driver and light source.

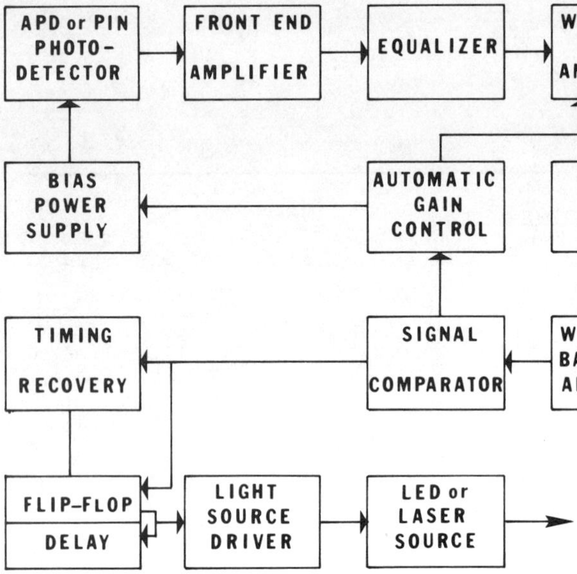

Fig. 96 Basic block diagram for an optical PCM receiver.

Figure 96 shows the basic block diagram for a typical PCM receiver. The equalizer in the circuit compensates for the effects of the photodetector and the frequency-time response of the optical fiber line. The objective of the equalization unit in the receiver is to eliminate or reduce the intersymbol interferences. Equalization imposes a power penalty on the total receiver unit, since the equalizing filter is not optimal in terms of noise performance. The equalizer response increases with frequency, enhancing the noise level. The types of equalizers that can be used are LC (passive or transversal), RC (active or transversal), or distributed transversal equalizers. If the communications system is attenuation limited, the equalizer is not needed. In systems limited by dispersion, however, the use of the equalizer can extend repeater spacings.

The main difference in the design and analysis of receivers for optical systems and conventional wire systems is the receiver noise characteristics which for optical repeaters can be rather complex since both Gaussian and Poisson noise components are present, some of which are dependent upon the transmitted signal. The front-end amplifiers in optical repeaters must be carefully designed to function with the detectors (capacitive devices), and the optical driver must be designed for the light sources (low-impedance, high-current devices). The light-source driver must be designed to deal with the transient response characteristics of the light source to reproduce the desired amplified and shaped pulses coming from the rest of the repeater. This is accomplished by internally correcting for the light-source response or by a feedback mechanism which responds to the output light signal. The amplifiers used in the repeater are either FET or bipolar discrete or some hybrid combination.

Repeaters have been built and tested for various types of transmission signals (analog and digital) at bit rates ranging up to 800 Mbps. The higher bit rate repeaters (>25 Mbps) use APD detectors with automatic gain control circuitry. The automatic gain control section of the receiver compensates for changes in the incoming signal level, amplifier gains, and the dramatic effects of temperature in the APD device. Analog repeaters are more complex in design compared to digital repeaters where the information signals can be completely and clearly regenerated in each successive repeater with little signal deterioration (very small BERs). In analog transmission each repeater-amplifier adds additional DP, DG, and noise to the signal which limits the number of repeaters that can be used in the transmission line as a function of the input SNR, the associated repeater noise level, and the specification limit on the final received SNR to the end-device. The minimum required received power (receiver sensitivity) is a function of bit rate in optical repeaters and varies greatly depending upon the optical component and amplifier designs and the use of detector gain. Figure 97 shows the minimum required receiver power versus bit rate for APD and PIN devices using integrating bipolar amplifiers and fibers with nominal pulse dispersion,

For an optical receiver using an APD followed by a baseband amplifier with a noise figure N_f, the SNR for a sinusoidal signal is given by

$$\mathrm{SNR} = \frac{(QeGmP_o)^2}{2(hvi_n)^2}$$

where

P_o = average received optical power
m = modulation index of the signal (-1 for 100% modulation)
i_n^2 = mean square noise current after avalanche gain

The mean square noise current is given by

$$i_n = i_q + i_T + i_d + i_l + i_g + i_b$$

the sum of the quantum, thermal, dark, leakage, background, and beat noise currents, respectively. These currents are defined as follows:

$$i_q^2 = \frac{2e^2}{hv} Qp_o G^2 N_d B$$

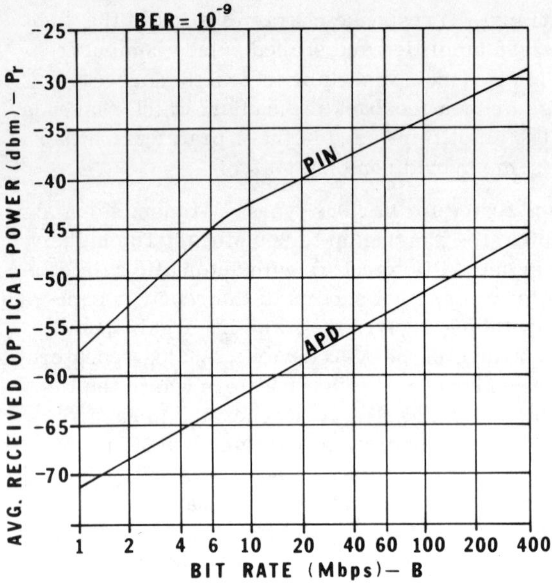

Fig. 97 Average received optical power vs. bit rate for PIN and APD detectors. PIN: $P_r = 16 \log B - 51$ for $B>7$; $P_r = 9 \log B - 58$ for $B<7$. APD: $P_r = 9 \log B - 71$.

$$i_T^2 = \frac{4KT}{R} N_t B$$

$$i_d^2 = 2eI_d G^2 N_d B$$

$$i_e^2 = 2eI_1 B$$

$$i_g^2 = \frac{2e^2}{hv} Qp_g G^2 N_d B$$

$$i_b = \frac{2B}{JW}\left(\frac{eGQP_o}{hv}\right)^2 \left(1 - \frac{B}{2w}\right)$$

where

N_d = noise figure for the random avalanche process
N_t = noise figure for the baseband amplifier
B = information bandwidth
I_d = detector dark current
W = spectral width of the source

J = number of spatial modes of the source
P_g = average background radiation power

The two dominant noise currents are i_q and i_T for systems with laser light sources. Since the avalanche gain process is seen to introduce gain-dependent noise, there is an optimal gain value for each given set of operating conditions.

The degree of pulse spreading in the transmission lines effects the repeater spacing. If the pulse spacing or RMS pulse width is less than 1/4, the effects of dispersion are negligible on repeater design. For widths greater than 1/2 the intersymbol interference (ability to distinguish successive pulses) becomes severe and the receiver sensitivity rapidly degrades, with an increase in the system noise level. When the fiber impulse response is such that the pulse width is less than 1/4, the allowable transmission cable loss is the difference between the transmitted power and the receiver sensitivity, and the system is considered loss limited. When the transmission cable pulse spreading is excessive, the cable length must be made shorter than the loss limited value to reduce intersymbol interference to an acceptable level, and the system is considered to be dispersion limited.

Optical repeaters must be designed to have low power consumption to minimize operational costs while maintaining sensitivity to completely regenerate the received signal. To reduce power consumption the amplifier bandwidths in the repeater are restricted so as to be slightly greater than the desired maximum bit rate. The power consumption of the transmitter section of the repeater depends upon the threshold current of the light source (usually a laser for high data rate systems), the drive current, and the source duty cycle. To minimize the power consumption of the overall fiber optic system beyond that of the repeater alone, an optimal light-source drive level must be used. This optical light-source drive level for long-distance runs represents the trade-off between the individual repeater power consumption and the total number of repeaters used in the transmission system.

Optical Repeater Spacing

The maximum spacing between repeaters is determined by many factors including the dispersion characteristics of the fiber cable and the electrical characteristics of the repeater unit. Given the relationships between required receiver power for various data rates, the receiver-dependent component of the repeater spacing is given by

$$S_{max} = \frac{P_c - P_r}{A}$$

where

S_{max} = maximum repeater spacing, km
P_c = power coupled from the source to the fiber, dBm W
P_r = required optical power (dBm W) for a given BER
A = attenuation loss of the fiber, dB/km

For low-loss cables presently available a reasonable attenuation figure is 3 dB/km. It should be noted from the equation to determine repeater spacing that the attenuation figure is in the denominator. Since the values for the attenuation losses are low, typically 2-5 dB/km, repeater spacing becomes a rather sensitive function of the attenuation A. For example, the difference of an attenuation loss of 3 dB/km versus 6 dB/km could translate to a repeater spacing difference of 2 km, which for long-distance cables can make a very significant impact on the total installation and operational costs.

The repeater spacing is also dependent upon the fiber cable dispersion characteristics. The greater the material and modal dispersion of the fibers, the shorter the repeater spacing will have to be to maintain signal integrity (BER). The material dispersion of optical fibers is given by

$$\sigma_{mat} = \lambda_o \frac{\delta\lambda}{c} \frac{d^2 n_1(\lambda)}{d\lambda^2}$$

where

λ_o = nominal source optical wavelength
$\delta\lambda$ = RMS spectral source width
$n_1(\lambda)$ = index of refraction of the fiber core material

For a laser source, λ_o is typically 800-1060 nm and $\delta\lambda$ is 1-20 Å. For a LED source, λ_o is typically 800-900 nm and $\delta\lambda$ is 370 Å. For silica fibers, n_1 is typically 1.5 and $d^2n_1(\lambda)/d\lambda^2$ is 1.52 X 10^{31} ns/km^2. Thus, σ_{mat} with a LED is typically 1.65 ns/km and 0.09 ns/km for laser sources. The modal dispersion for step index fibers can be approximated by $n_1\Delta/4c$, where $\Delta = 1 - [1 - (NA)^2]^{1/2}$. For graded index fibers, the modal dispersion is approximately given by $n_1\Delta/80c$. For a fiber NA of 0.16, the modal dispersions of silica fibers with $n_1 = 1.5$ would be σ_{mod}(SI) = 16 ns/km and σ_{mod}(GI) = 0.8 ns/km. The total RMS pulse width contribution from both modal and material dispersions is given by $\sigma = (\sigma^2_{mod} + \sigma^2_{mat})^{1/2}$. For a LED source, with SI fibers, $\sigma \simeq 16.1$ ns/km and 1.8 ns/km for GI fibers. For a laser source with SI fibers, $\sigma = 16.0$ ns/km and 0.8 ns/km for GI fibers. For step-index (SI) fibers the dispersion limitation on repeater spacing is essentially independent of the light source due to the dominance of the modal dispersion factor.

Using the typical fiber and light-source values given above Figs. 98 and 99 show the repeater spacing as a function of bit rate for dispersion limitations

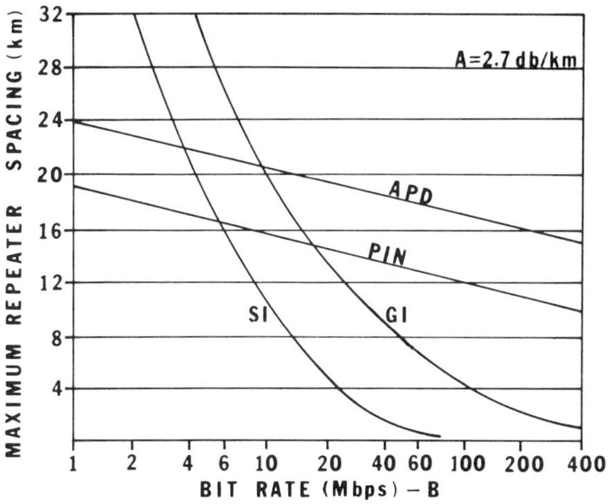

(SI)σ_{mod} – 10 ns/km λ_0 – 840 nm

(GI)σ_{mod} – 0.5 ns/km $\delta\lambda$ – 300 Å

σ_{mat} – 1.2 ns/km P_c – –10 dbm

Fig. 98 Maximum repeater spacing vs. bit rate for a LED light source.

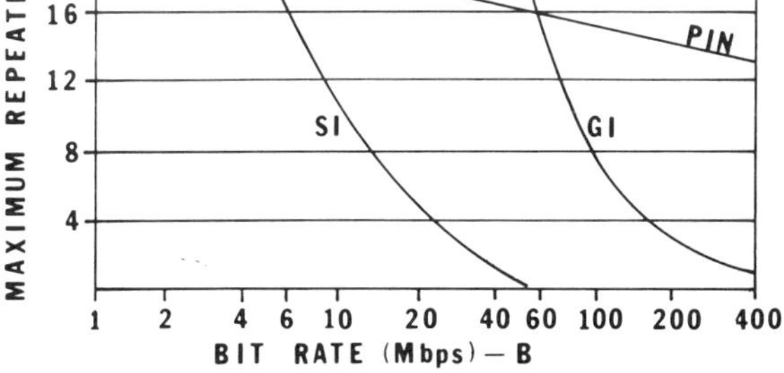

(SI)σ_{mod} – 10 ns/km λ_0 – 840 nm

(GI)σ_{mod} – 0.5 ns/km $\delta\lambda$ – 20Å

σ_{mat} –.06 ns/km P_c – 3 dBm

Fig. 99 Maximum repeater spacing vs. bit rate for a laser diode light source.

(SI and GI fibers) and for power loss limitations (APD and PIN detectors with integrating bipolar preamplifiers) for LED and laser light sources, respectively. The curves for the fiber dispersion limitations assume mode-mixing in both the step- and graded index fibers.

Fiber Optic System Design

The design of fiber optic communications systems is determined and based upon four basic user requirements: the desired data rate (bandwidth), the SNR (or BER), the distance between terminals (or end-devices), and the type of source information (digital or analog). Once these basic requirements are established, the design must consider numerous external variables, such as physical and chemical environmental conditions, cost, reliability, upgradability, size, weight, power requirements, and effects of electromagnetic interference (EMI) on transmitters, repeaters, substations, and end-devices. Many of the variables in optical system design are interrelated. In some designs, some parameters are predetermined by the nature of the specific application, such as environmental conditions.

The first step in the design process usually is to determine if the dispersion bandwidth of the fiber cable and light source combination are suitable to be used for the required data rate (information bandwidth) at the given terminal or end-device spacings. Since the dispersion bandwidth is a function of distance and cable and light-source parameters, it is possible that in some cases the best available cables and lasers cannot meet the required data rates without using an inordinately large number of repeaters, exceeding the alotted budget of the system. If this is the case, ways can sometimes be found to shorten the total link distance or to reduce the required data rates at minor sacrifice to end-device service quality, thus permitting the use of an optical system, within present budget limitations, which can be upgraded in the future. Upgradability is a general term applied to fiber optic communications systems meaning the improvement of the available bandwidth (data rate), extending the lifetime and reliability of components, particularly the light sources, improving the attenuation or dispersion properties of the cable, or reducing the losses at splices, connectors, and couplings with new equipment or designs.

The optical communications system designer must make decisions in terms of planning for upgradability of light sources, receivers, transmitters, modulators, and couplers, since the prices and quality of future components cannot be precisely determined. Given that the cost of the initial purchase and installation of fiber cable is rather significant in terms of total system cost, it is usually best to install the lowest attenuation cable possible for applications where upgradability is a factor, so that the original cable will not have to be replaced in the near future.

Once a fiber cable, light source, and modulation scheme have been found that meet the bandwidth requirements, giving consideration to upgradability, then the rest of the system parameters must be determined if the desired performance levels in terms of SNR and BER can be achieved. The major parameters to be considered here are the coupling light-source power levels, fiber attenuation and dispersion at the source wavelength, and the receiver sensitivity. When short-length analog links are used, an important parameter to consider is the level of sign distortion, which usually severely affects performance quality in analog transmission.

Once the rough system design has been completed with a set of specifications for all components, the next step in the overall design process might be aptly described by interactive perturbation and response analysis. Here, each separate component or system element is slightly altered (specifications changed) to see what effect the small change will have on the total system design and component choice. The procedure establishes a reasonable range of component specifications to meet the job requirements within the given budget. If certain system elements have definite or narrow ranges of choices available, it is usually easier to start with these elements in the design, since much unnecessary analysis can be eliminated.

It should be evident that no "optimal" design is really achievable in a practical fiber optic system. The reason for this is that practical designs are severely limited by the constraints of time and cost. Even if cost were not a constraint, an optimal system designed today using the best available components will be outdated (nonoptimal) in a rather short period of time.

Fiber Cable Choice. There are many types of fibers and cables available for fiber optic systems, making the options in design appear to be rather large. However, system limitations and requirements can rapidly narrow the choice of possible fibers and cables. The first choice to make is whether to use single fibers or fiber bundles. For single-strand fiber cables a large amount of power emitted by the source is lost in coupling the fiber. Fiber bundles have larger NA values and have fewer losses in coupling than single fibers. By using multifiber bundles and lenses, it is possible to reduce the coupling losses to be approximately equal to the packing fraction loss. Over short lengths where attenuation limitations are not critical, bundles are often used if the required data rate is relatively low. For long lengths, single fibers are usually used to make the cable bandwidth capabilities as large as possible.

Once the decision has been made to use bundles or single fibers the next step is to calculate the total system losses starting from coupling losses at the light source to coupling losses at the end-device. For these calculations the following cable parameters must be specified: effective fiber

NA, index profile, attenuation at the chosen light-source wavelength, and the length of cable between each repeater or terminal. When fiber bundles are used, the cross-sectional area and packing fraction must also be specified. The total losses that can occur over the transmission pathway include losses in the fiber, fiber splices, connectors, and detector-source coupling losses. To minimize input coupling losses the light source and detector packaging designs should match the fiber or fiber bundles to which they are attached. By the use of antireflection coatings or index-matching materials and properly matching the areas of the components to the cable, the coupling losses can be kept within an acceptable level for must communication system requirements. Whether or not the total loss is acceptable depends mostly upon the receiver sensitivity.

After an acceptable cable has been chosen, dispersion limitations must be considered to determine if the dispersion bandwidth is compatible with the desired data rate. Once a cable or set of cables has been found that meets the transmission requirements, the environmental and strength factors must then be considered. The major concerns here are the required cable strength for the specific application, whether or not internal conductors are needed for repeater power, temperature range of the application, the effect of water on the cable lifetime, and the total cable lifetime under the anticipated environmental stresses. From these final considerations a cable choice can usually be made. It will sometimes be the case that no existing cable satisfies the system requirements. If the system requirements cannot be made less stringent, an optical fiber cable transmission system is not feasible for the given application until improved cables are available. Using the lowest attenuation cables available today, most known communication systems can be reasonably converted to fiber optic transmission.

A flow chart of the fiber design choice process is given in Fig. 100. In the future, flow charts such as these for each system element can be translated into computer programs based on data banks of the existing commercial components. It may then be possible to interlink the individual element flow charts to obtain a totally computerized design based on existing components and equipment. The major advantage of computer designs is the ability to examine a large number of component variations and modulation schemes.

Receiver Design. There is an ever increasing number of prepacked receivers for fiber optic links, but most existing units have relatively low bandwidth capabilities (<30 MHz) and are designed for indoor applications. As a result, most receivers, particularly for high data rates, are designed to accommodate a specific application. Once the system requirements and constraints have been established the first step in the design process is the decision to use analog or digital modulation of the light

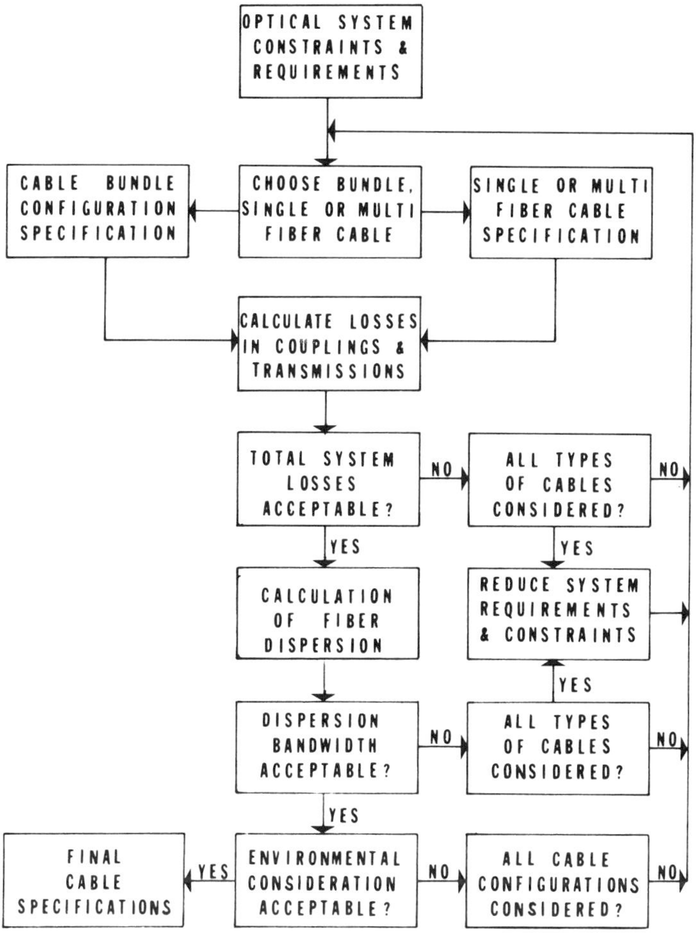

Fig. 100 Generalized fiber cable design flow chart.

source. Most of the first applications and installations of fiber optic units will use digital. If digital modulation is used the next step is to decide upon a particular modulation signal format such as PCM or PPM. The next step for either modulation scheme is to calculate the receiver noise equivalent bandwidth (NEB), which is determined by the desired information bandwidth, the noise introduced by various receiver components, and the possible use of filters to remove noise outside the transmissions signal bandwidth. Once this is established, the input SNR can be calculated from the BER and NEB.

The next step in the receiver design process is to select a particular transmission wavelength from the existing light sources available and a particular photodetector (PIN or APD). Needless to say the source wavelength must be compatible with the optical responsivity of the detector. Given that there are numerous light sources, detectors, and detector modules available, this step in the design process has a large range of possible choices. In high data rate systems APD detectors are usually chosen for laser light sources.

Once the light source and photodetector have been chosen, the type of preamplifier must then be determined, selected from one of four types: resistive, loading, integrating (FET or bipolar), or transimpedance feedback. Hybrid combinations can also be used. At low data rates the integrating FET preamplifier is often used. At high data rates integrating bipolar preamplifiers are most often used. After the choice of the preamplifier has been made, the next step in the receiver design is to calculate the minimal detectable signal level which is given by the minimal received optical power after coupling losses to generate a current equal to the NEB noise current. The required signal power to achieve the desired SNR is then calculated. If this power level cannot be achieved, the photodetector selection will have to be upgraded, or else the type of signal format (for digital transmission) will have to be changed to make the required signal bandwidth smaller.

Once the minimal received signal criterion has been satisfied the next step is to consider the dynamic range of the repeater. The term dynamic range means to encompass a variety of potential system variables including environmental changes (particularly temperature), different path lengths between repeaters or terminals, and changes in component characteristics with time. A properly designed receiver must anticipate the worst component responses as a function of potential component degradation, particularly the light source and detector and potential temperature excursions in the environment. If the anticipated temperature range has too large an effect on the receiver, temperature-compensation circuitry (either precision control or feedback types) must be used both on the receiver unit and on the light-source unit as well.

Finally, the packaging of the receiver must be considered and must account for environmental factors of temperature, humidity, and corrosion as well as protecting the detector from possible stray light sources. In addition, some applications will have to consider physical stresses applied directly to the receiver such a low-frequency large-amplitude vibrations encountered in aircraft systems.

The flow chart depicting the receiver design process is shown in Fig. 101. The data bases for component choices and the calculations involved in receiver design are much more complex than those for fibers and cables. The cost of a repeater unit at present is to a rather significant degree

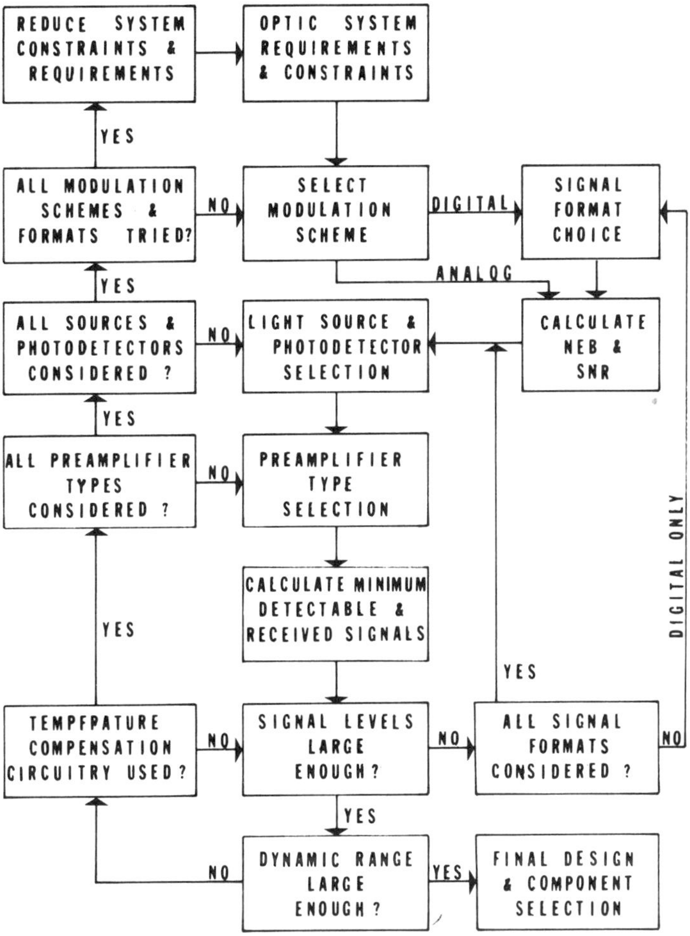

Fig. 101 Generalized receiver design flow chart.

determined by the choice of using a PIN or APD photodetector, thus upgradability considerations may weigh rather heavily for some borderline designs.

Transmitter Design. There are about as many transmitters on the market as receivers, designed for reasonably small bandwidth applications (<20 MHz). Transmitters come in a variety of prepackaged forms, some equipped with threaded socket connectors ready to attach to a fiber cable, often in bundle format, since most are made for short links. The transmitter design centers basically around the choice of an adequate light source and modulation scheme.

The choice of the light source must consider the total optical power required between terminals or repeaters, the output wavelength and spectral width, the speed and linearity of response, and the use of temperature-compensation circuitry (if needed). The choice of wavelength depends in part on the attenuation characteristics of the chosen fiber cable and in part on the spectral response of the photodetector. The spectral width depends mostly upon the restrictions on dispersion bandwidth within the transmission line. The required optical output power is determined by the required received signal at the receiver unit after considering and accounting for all of the transmission and coupling losses.

The choice of source wavelength for silica and glass fibers is usually between 800 and 900 nm but can be as high as 1060–1520 nm where fiber attenuations reach relative minima. For plastic fibers used with short links a wider variety of wavelengths (630 to 830 nm) can be used and the major consideration becomes required optical power, since attenuations in plastic fibers are so high.

The next step in the design process for optical transmitters is the choice of modulation schemes, i.e., digital or analog. If analog modulation is used, the information bandwidth and source power in terms of distortion levels must be considered. If digital modulation is used, the response speed of the light source and choice of signaling format must be considered. Once the modulation scheme has been determined, the next step is to calculate all the coupling losses and determine whether or not the optical coupled power is adequate. If it is not sufficient, a different signaling format could be tried for digital modulation or space division multiplexing could be considered.

Once the source and modulation have been selected, several parameters should be calculated and specified for design information on the receiver and fiber. These parameters should include the transmitted SNR, required electrical power consumption, and the effects of temperature on the transmitted signal. If temperature too greatly affects the output, then temperature-compensation circuitry should be used on the light-source drive current through suitable feedback circuitry or precise drive controllers.

A flow chart for the transmitter design process, similar to that depicted for receiver design, is shown in Fig. 102. The chart is again suitable for translation into computer format for use in single or multielement computer design.

Table 71 shows in a brief format the various component choices for different ranges of link length or data rate (bandwidth). The table is meant

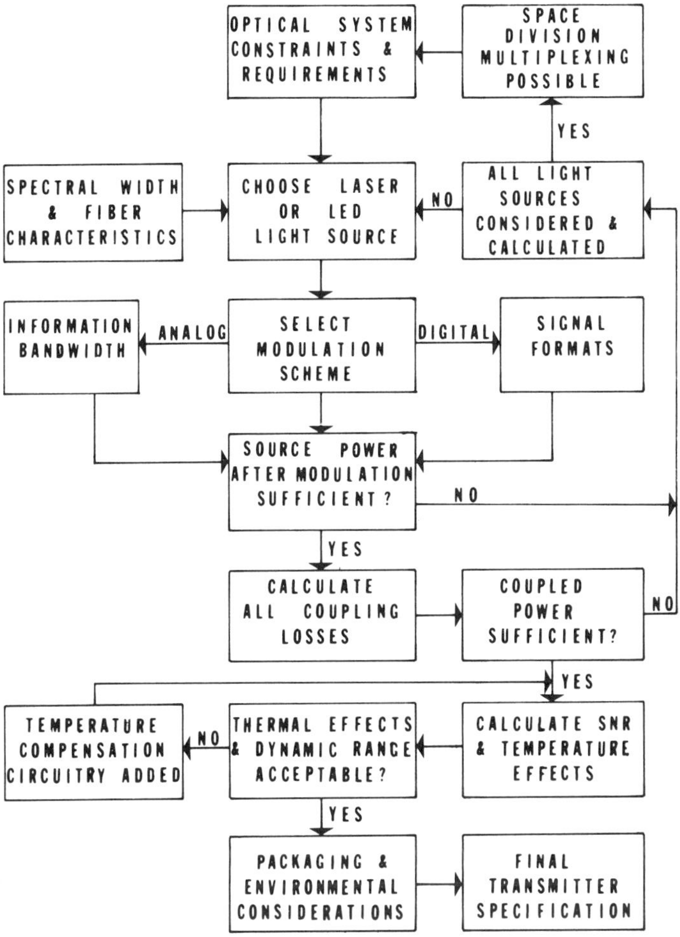

Fig. 102 Basic transmitter design flow chart.

to show sample choices of drivers, sources, fibers, detectors, and amplifiers for three basic groupings of fiber optical data rates. For intermediate values, careful analysis must be done as outlined in the design flow charts.

Multiterminal Networks. Fiber optic communication systems can be used to distribute various types of information signals to a number of remote terminals all interconnected in a multidirectional data distribution system. These networks can be used for information transfer in aircraft, ships,

Table 71 Component Choices for Different Data Rates

Component type	Short lengths <40 m Low data rates <2 Mbps	Medium lengths <400 m Medium data rates <25 Mbps	Long lengths >1 km High data rates >30 Mbps
Driver	Single IC or transistor	Controlled current source	Matched current source with pulse shaping
Source	LED (red, GaAs)	LED (red, GaAs) LED (GaAlAs, GaAsInP)	LED (GaAlAs, GaAsInP) DH laser
Fiber	Plastic or glass high loss bundle	Glass step index medium loss bundle or fiber	Graded index low loss fiber
Detector	Phototransistor, PIN	Silicon PIN	Si APD, GaAs APD
Amplifier[a]	Integrating FET	Transimpedance	Integrating bipolar

[a]Typical preamplifier types are resistive (voltage amplifier with low noise and resistive front end), integrating (integrating front end with bandwidth compensation circuitry), and transimpedance (current feedback amplifier with low bias current).

computers, interbuilding communications, manufacturing and power plants, and cable TV systems. These multiterminal networks use the data bus system where a single transmission line can carry numerous multiplexed signals spatially distributed to a number of terminals and end-devices. The design of multiterminal networks must consider the following:

1. The amount of optical power normally emitted from the original light source, the peak power available, the expected intensity degradation rate, and the expected source lifetime
2. The coupling bases from the light source to the transmission cable and from the cable to the various network components
3. The BER and SNR required to maintain signal integrity and end-device quality
4. The amount of optical power required to properly activate the photodetector (PIN or APD) and repeaters, if used, to maintain the desired BER and SNR

5. The maximum possible system losses from coupling, splicing, repeaters, fibers, signal splitters, and temperature and degradation effects
6. The most efficient and economical network distribution system for the particular application
7. The total system upgradability given future technological advances in light-source intensity, speed, and lifetime, increases in repeater-amplifier bit rate capacity, and decreases in system losses with new couplers, fibers, splices, and signal splitters

To estimate the amount of optical power available for distribution of the data in the multiterminal system, a cable and light source must be chosen. Given the reflection and insertion losses, packing fraction, numerical aperture and fiber diameters in the cable, and the radial directionality of the light source, the amount of light coupled into the fiber can be determined. For fiber bundles with large NA values this coupling of optical power can be as high as 10 dBm using laser sources. For single fibers with relatively low NA values this coupled optical power can be lower than -12 dBm for LED sources.

Data Bus Design. The two common types of optical distribution systems for communication systems are the series distribution system using access couplers and the parallel distribution system using star couplers. A system which combines both of these types of access coupling and signal divisions is called a hybrid configuration. Figure 103 shows a simple schematic diagram for an N-terminal system for series, parallel, and hybrid data bus design.

The alternatives in data bus design are quite numerous, increasing rapidly with the number of individual transmitters, repeaters, substations, terminals, and end-device units, and must consider which units are to have the ability to transmit or receive information from the other units in the system. For example, in a cable TV system the decision must be made as to whether or not the user terminal should be equipped to receive only now and later to expand to an interactive system. The design, construction, and performance of an optical data bus system depend on many factors including system transmission requirements, the number of present terminals to be installed, future expansion access, transmission line and coupling bases, environmental considerations, and the maximum dynamic range needed between any pair of terminals.

The dynamic range (DR) in the star data bus for an N-terminal system is given by

$$DR = A_{avg}^{(2-N)/2}$$

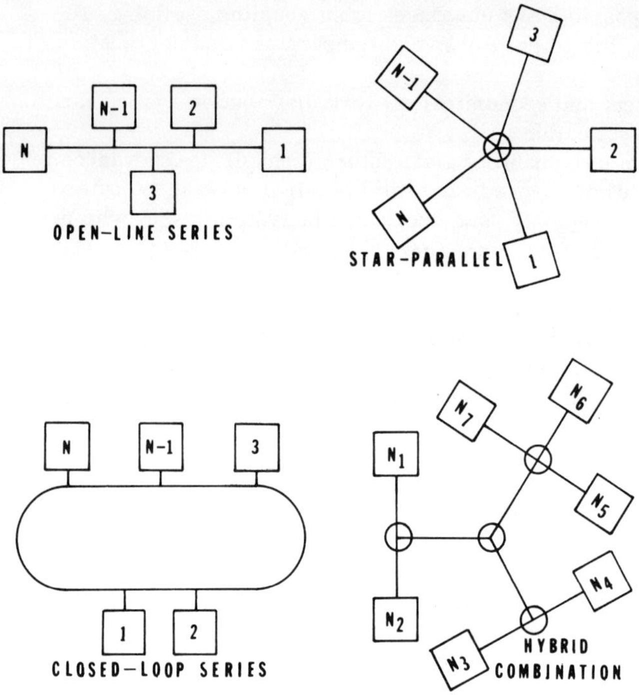

Fig. 103 Diagrams for various N-terminal data bus distribution systems.

where A_{avg} is the average attenuation between terminals or stations given by $A_{avg} = C^{-aL/(N-1)}$, where a is the attenuation coefficient of the fiber of length L. For the series bus system the dynamic range is given by

$$DR = \left(A_{avg} L_c^{\,2} L^t \frac{N - 2}{N - 1}\right)^{2-N}$$

where L_c is the connection insertion loss and L_t is the total transmission loss in the coupler given by $L_t = Tt/(1 - 2C_t)$, where T_t is the fraction of power at the input coupled to the output of the coupler and C_t is the fraction of the total power removed from the bus. The average value for $L_t L_c^{\,2}$ is about 1.5 dB.

The power ratio (PR) of the optical power in the transmission cable at the input to one terminal to the optical power at the output to another terminal is determined by the sum of the various system losses. For the radial system PR is given by

$$PR = \frac{A_{avg}^{N-1} L_c^{4} L_{is} L_s}{N}$$

where L_{is} is the star coupler insertion loss and L_S is the bidirectional coupler splitting factor, usually 1/2 (3 dB). For the series system the value of PR is given by

$$PR = 0.5\, A_{avg}^{N-1} L_c^{2N-2} L_t^{N-2} C_t (1 - 2C_t)^{N-2}$$

The total distribution system losses (TSL) in decibels excluding fiber losses can be expressed as follows for the series network:

$$TSL = L_{sf} + L_{tr} + L_{ct} + 2L_{cl} + (N - 3)(2L_{cl} + L_{ct} + L_{it})$$

where

L_{sf} = bidirectional splitting factor (3 dB)
L_{tr} = tap ratio of the T-coupler (usually about 10 dB)
L_{ct} = internal insertion loss of the T-coupler (usually about 1 to 2 dB)
L_{ci} = connector insertion loss (usually about 0.5 to 1.5 dB)
L_{it} = insertion loss from the T-coupler tapping given by 10 log $(1 - L_t)$

For the radial network TSL is given by

$$TSL = 4L_{cl} + L_{cs} + L_{is} + L_{sf}$$

where L_{is} is the insertion loss from the star coupler tapping given by 10 log N and L_{cs} is the internal loss of the star coupler (usually about 2-8 dB).

Using the typical values given above, the equation for the series TSL as a function of N is simply

$$TSL \simeq 4.5N + 3.5$$

For the radial network,

$$TSL \simeq 10 \log N + 13$$

These simplified equations are useful to demonstrate the difference between using star and series networks. For $N \leq 5$ the total distribution system

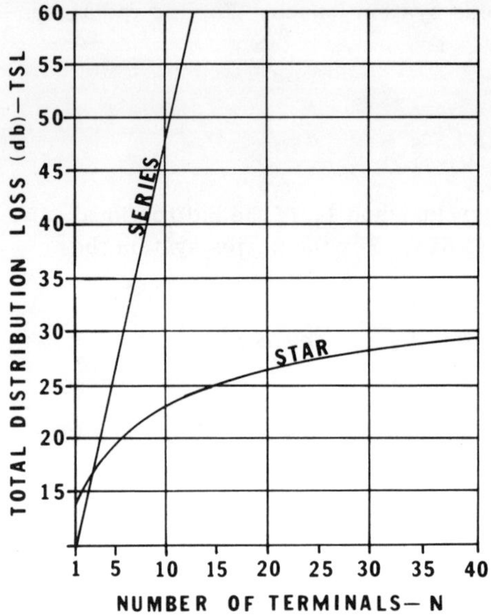

Fig. 104 Total optical distribution system losses vs. number of terminals. Series: TSL = 4.5N + 3.5. Star: TSL = 10 log N + 13.

losses are about the same for both networks, but for a larger number of terminals the difference rapidly increases. For N = 10, the TSL for the series network is about 48 dB and for the star network TSL is about 23, or approximately a twofold loss difference. Figure 104 shows the total system losses for star and series networks using the above parameters.

In some cases it may be required or desired to keep the received detector signals approximately the same for all transmission sources. This can be accomplished by using a different coupling factor at each terminal to provide a uniform response at each detector. Here, the same set of coupling factors must be used for one side of each coupler from terminal 1 to N as is used on the other side of each coupler from terminal N to 1. This scheme can get to be rather burdensome as reflected in the complexity of the power ratio given by

$$PR = \frac{A_{avg} L_c^{\,2} (A_{avg} L_t L_c^{\,2})^{N-2} (1 - A_{avg} L_t L_c)^2}{2(1 - A_{avg} L_t L_c^{\,2})^{N-1}}$$

This type of data bus system is sometimes called the <u>tapered duplex system</u>. The dynamic range for the tapered duplex data bus network is given by

$$DR = 1 + A_{avg} L_t L_c^{\,2}$$

In the radial system the power ratio varies as 1/N and as the exponent of N for the series network. The tapered duplex data bus power ratio is also not as favorable as the radial bus, as well as having the disadvantage of being rather inflexible in terms of future expansion, since the tapered duplex system must be designed exactly for N terminals.

The star (parallel) system has a signal level advantage over the series system especially for a large number of terminals (>10). Since the parallel system has only one mixer, it does not have the receiver problems associated with series systems which must use large dynamic range automatic gain control devices to process large signals from nearby terminals and small signals from distant terminals. The disadvantage of the parallel system is that the length of each terminal extension is rather large meaning that more cable must be used to accommodate an equal number of terminals than would be needed for a series system. It should be noted that many applications do indeed use only a small number of terminals in the region where the distribution system losses for series and parallel systems are about equal. In these cases the decision to choose either type of system must be based on other factors such as cabling costs and weight and whether future upgrading and terminal additions are planned.

The radial data bus configuration also has an advantage over the series (T-coupler) system in terms of dynamic range, since power can be distributed uniformly to all terminals, making dynamic range only a function of the distance to each individual terminal. If the distances are all equal, assuming equal coupling losses can be attained, the dynamic range in the star system can approach the theoretical value of 1 (0 dB) when cables of equal attenuation are used. The disadvantage of using this system is that the total length of cable is longer than used in the series configuration. This disadvantage can in part be compensated for by running a single multifiber cable part of the way between terminals and then splitting off the separate fibers to the individual terminals, and is best accomplished by choosing a centrally located transmitting substation to service a number of nearby end-terminals.

Hybrid data bus configurations which can combine some elements of all types of bussing configurations (radial, series, and tapered) are usually used for large networks. It is possible in some hybrid data bus designs to approach the dynamic ranges achievable in the radial systems with total optical power losses that are only slightly larger. Using hybrid designs it is possible to achieve greater flexibility for future expansion and alterations and use less total cabling than in the simple radial system.

Simplified Design Procedures. The complete design of a complex multiterminal fiber optic communications system can be very detailed and complicated, since the number of variables to consider is rather large. The decisions on cost limitations, pay-back periods (if any), and user requirements including bandwidth, SNR, BER, number of terminals, and terminal spacing are usually made on a managerial or corporate level.

The basic engineering decisions to make on a user specified system center about component choices: APD, PIN, LED, laser, fiber cable, repeaters, modulators, signal format encoders, multiplexers, and demultiplexers. The two major factors to consider are the optical power requirements and the response time requirements of the total system.

The optical power requirements and calculations are shown in a rather simplified format in Fig. 105. The first steps in the calculation procedure are to specify the required bandwidth or data rate (step 1A), the required SNR or BER (step 1B), the signal format and number of terminals (step 1C), and the data bus type and terminal spacing (step 1D). These are determined mainly by the application specifications of the user but can sometimes be altered by the systems designer to optimize power distributions or to minimize cabling between terminals. The next step is to choose an appropriate fiber cable (step 2), which can be selected from the fiber product literature. Actually, in the design process this step of the fiber choice can be saved until the end of the calculation by merely assuming some attenuation and rise time (dispersion) values and later determining if these assumed values were sufficient for the particular design.

The choice of the light source is next in the calculation procedure, which is usually specified by manufacture device number and the rated average output power in dBm (step 3A) which can be found in the light source product literature.

If the signal format is NRZ, 3 dBm is added to the output power value, and -6 dBm for RZ formats (step 3B). In many cases it is desirable to extend the lifetime of the light source to lower maintenance and replacement costs. To do this the peak light output power is limited. In the calculation this is accounted for by adding -3 dBm (step 3C) to the specified average output power. The total available light power P_S is then determined by the sum of steps 3A, 3B, and 3C.

The basic choices of components of course depends on many variables. To prevent excessive reiterations on the design procedure, Table 68 outlines some basic component choices depending upon the bandwidth or transmission length of the given fiber optic communications system.

The detector choice follows after the light source calculations. The type of detector used (APD or PIN) must be specified by the manufacturer and device number, which can be found in the detector product literature (step 4A). The required received optical power P_r can be determined from the

		Information source
1A.	Required bandwidth = ___ MHz or bit rate = ___ Mbps	Ap. Sp.
1B.	Required SNR = ___ dB or BER = ___	Ap. Sp.
1C.	Signal format: ___, terminal spacing L = ___ km	Ap. Sp.
1D.	Data bus type: ___, number of terminals N = ___	Ap. Sp.
2.	Fiber type: ___, attenuation A = ___ dB/km	P.L.
3A.	Light source type: ___, average output power ___ dBm	P.L.
3B.	If digital: NRZ add -3 dB, RZ add -6 dBm	
3C.	If source output halfed to extend lifetime, add 3 dBm	
3D.	Total available light power: Ps = 3A + 3B + 3C = ___ dBm	
4A.	Detector type and manufacturer specifications: ___	P.L.
4B.	Required received optical power Pr = ___ dBm	Figs. 106 and 107
4C.	Total power margin = Ps - Pr = ___ dB	
5.	Total fiber loss = $A\Sigma L_N$ = ___	
6.	Source coupling loss = ___ dB	P.L., E.D.
7.	Detector module coupling loss X N = ___ dB	P.L., E.D.
8.	Number of splices ___ X loss/splice ___ = ___ dB	Table 71
9.	Allowance for temperature variations = ___ dB	Table 72
10.	Allowance for component time degradation = ___ dB	Table 73
11.	Number of extra connectors ___ X loss/connector ___ = ___ dB	Table 71
12.	Data bus attenuation allowance = ___ dB	Fig. 104
13.	Total attenuation = 5 + 6 + 7 + 8 + 9 + 10 + (11 or 12) + ___ dB	
14.	Excess power = EP = EP = 4C - 13 = ___ dB If EP ___ 0 upgrade components or use repeaters	

Fig. 105. Basic optical power calculations. Ap. Sp. = application specifications; P.L. = product literature; E.D. = experimental data.

required data rate or bandwidth and SNR or BER from Figs. 106, 107, 108, or 97, depending upon the signal format used (step 4B). When computer designs are used the questions to use in place of the figures are given where applicable on the individual figures and curves. Once the required received optical power has been determined, the total available power margin can be determined (step 4C).

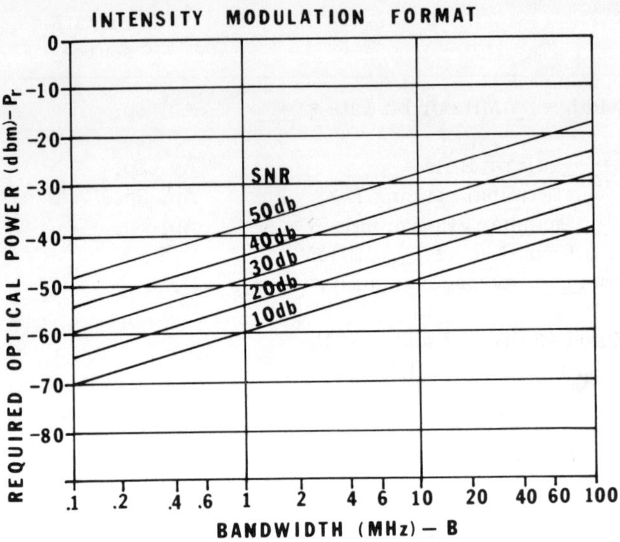

Fig. 106. Required optical power vs. bandwidth for analog receiver with PIN detector. P_r = 10 log B - 61 + (SNR - 10) (0.55).

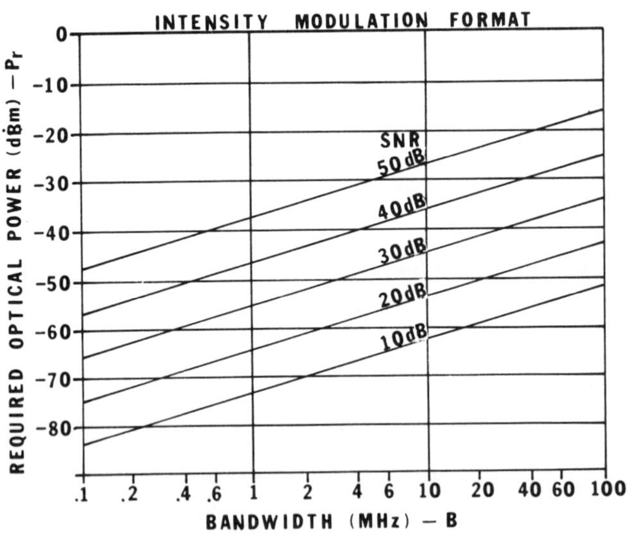

Fig. 107 Required optical power vs. bandwidth for analog receiver with APD detector. P_r = 10 log B - 73 + (SNR - 10) (0.88).

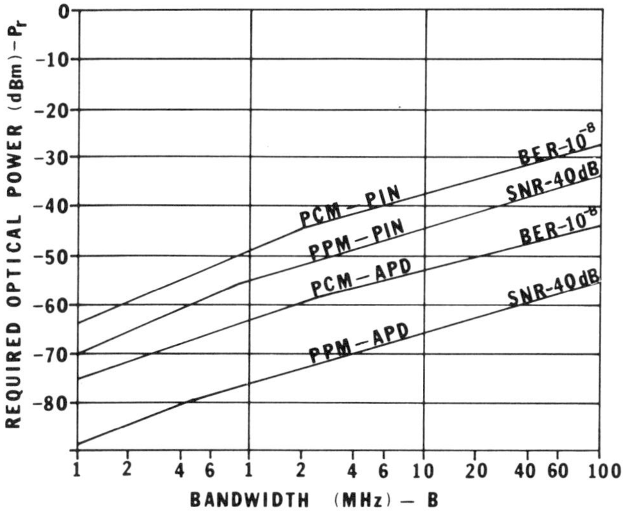

Fig. 108 Required optical power vs. bandwidth for analog receiver using PPM and PCM.

The next steps in the optical power calculations deal with various fiber, coupling, and splicing losses. First (step 5) the total fiber loss is determined by $A \sum_{N-1}^{N} L_N$. If the system has terminals which are all equally spaced, the fiber loss is given by A(L)(N). The next loss to consider is from light source to fiber coupling (step 6) which can be determined from the product literature, from the experimental data, or by calculation. This loss is often an important consideration in the choice of fiber NA and whether or not to use bundle or single fiber cables. The next loss to consider is the detector coupling loss which can be determined from the product literature, from the experimental data, or by calculation (step 7). When larger area devices are used, this coupling loss is usually rather small (<2 dB).

The next loss concerns cable splicing losses (step 8), which may or may not be used in a given system. The use of splices is often determined by the terminal spacing requirements and the maximum available lengths of the chosen fiber cable. Table 72 briefly summarizes the expected splicing losses depending upon the fiber type.

In any practical fiber optic link, allowances must be made for the degradation of components with time, particularly the detectors and light sources.

Table 72 Expected Connector and Splice Losses

Fiber type	Demountable connector loss (dB)	Fixed splice loss (dB)
Step	1.0	0.3
Graded	1.4	0.5
Bundle	2.0	1.0

Table 73 Temperature-Variation Loss Allowance

Compensation circuitry	Temperature range 10-30° C	Allowance loss (dB)
No	No	4
No	Yes	2
Yes	No	1
Yes	Yes	0

Table 74 Component-Degradation Loss Allowances

Component combination	Allowance loss (dB)
LED + PIN	2
LED + APD	3
Laser + PIN	4
Laser + APD	5

Allowances must also be made for variations in the environmental temperature. In some indoor applications the variation of temperature is nominal and there is little concern for allowances to compensate for detector sensitivity shifting and light-source drive current changes. However, for most outdoor applications temperature range is an important consideration especially when no compensation circuitry is used for the light source and detector modules. Table 73 lists the allowances that should be made for temperature variations (step 9) depending upon the temperature range and the use of compensation circuitry. Table 74 shows the loss allowances for component time degradation (step 10) depending upon the light source-detector combination.

If there are extra connectors in the fiber optic link, the losses from these various connectors are the next losses to be accounted for in the optical power calculations (step 11). Table 72 summarizes the expected connector losses depending upon the type of fiber cable used. For multiterminal networks the total data bus attenuation must be considered (step 12). This can be evaluated by independent calculation or by using Fig. 75. The equations for the series and star TSL can be reevaluated using the actual chosen component losses for signal splitters, connectors, etc., to give a more accurate estimate of expected network losses. Losses using the series network can become very high above seven terminals, necessitating the use of repeaters between terminals.

The total system attenuation can then be determined (step 13) by adding all of the above individual losses. The excess power EP is then calculated (step 14) by subtracting the total attenuation from the power margin. EP will often be a negative value. If EP is between 0 and -20 dB, the use of an APD instead of a PIN detector, the use of a laser instead of an LED light source, the use of lower loss fiber cable, or the use of a star versus series networks can sometimes shift EP to a positive value, meaning the power requirements have been satisfied. If not, repeaters will have to be used. For digital transmission the use of repeaters is essentially limited only by total cost and BER considerations. For analog transmission, however, the use of repeaters becomes rather limited due to signal distortion. Often the designer will come to the conclusion that for certain user applications there will be no alternatives (in analog transmission) except to convert the analog signal to some digital format and use repeaters to meet the terminal spacing and SNR user requirements. Figures 98 and 99 show the expected repeater spacings using LED and laser light sources with APD and PIN detectors for various data rates with low attenuation cables. These figures apply for single links and do not account for the losses encountered in multiterminal networks.

Once the optical power requirements have been satisfied, the system rise time must be calculated to determine whether the chosen components and fibers can respond fast enough to handle the given data rate or bandwidth. The various steps in the calculation of the rise time are shown in Fig. 109. The first step is to determine the total allowable system rise signaling formats (step 1). Table 75 shows the total rise time in units of nanoseconds (ns) for various modulation and signal formats as a function of the bit rate or bandwidth. For the PPM format the total rise time is a function of the pulse separation to pulse width ratio as shown in Fig. 110. At higher SNR values commonly encountered in optical systems the PIN and APD curves in Fig. 81 can be put into a simple equation form as shown on the curves for computer design calculations.

Once the allowable system rise time has been determined the response time of the light source unit must be specified (step 2) from the light source

	Information source
1. Total allowable system rise time = ____	Table 74
2. Light source (module) rise time = ____	P.L.
3. Photodetector (digital) rise time = ____	P.L.
4. Receiver module (analog) rise time = ____	P.L.
5. Fiber: NA = ___, n = ___	P.L.
6. Source: λ_0 = ___ nm, $\delta\lambda$ = ___ nm	P.L.
7. Modal dispersion rise time = TMOD = MD X L = ____ SI: $MD = 600n[1 - (1 - NA^2)^{1/2}]$; GI: $MD_{SI}/20$ Series: $TMOD = MD \Sigma L_N$, Star: $TMOD = MD \times L_{max}$	P.L. or Calcul. or E.D.
8. Material dispersion rise time = TMAT = MT X L = ____ $MT = \lambda_0 \times \delta\lambda \times 10^{-4}$ (SI or GI) Series: $TMAT = MT \Sigma L_N$, Star: $TMAT = MT \times L_{max}$	P.L. or Calcul. or E.D.
9. Rise time squares sum = S = $\underline{2}^2 + (\underline{3}^2 \text{ or } \underline{4}^2) + \underline{7}^2 \times \underline{8}^2$ = ____	
10. System rise time = SRT = $1.11 \times S^{1/2}$ = ____ If SRT > AST, choose faster detector and/or light source and/or lower dispersion cable	

Fig. 109 Basic rise time calculations. P.L. = product literature; E.D. = experimental data; Calcul. = calculation.

Table 75 Allowable System Rise Time Calculations

Modulation and signal formats	Total rise time (ns)[a]
Digital NRZ	0.7/bit rate
Digital RZ	0.35/bit rate
Intensity Modulation	0.35/bandwidth
Pulse Position Modulation	$(SR \times PS/PW \times \text{bandwidth})^{-1}$
Pulse Code Modulation	$(SR \times \text{bits/sample} \times \text{bandwidth})^{-1}$

[a]PS/PW = Pulse Separation-to-Pulse Width ratio (see Fig. 81). SR = sampling rate = 2.5 unless otherwise specified bandwidth units in MHz; bit rate units in Mbps. Bit/sample = (SNR/6) X 1.2.

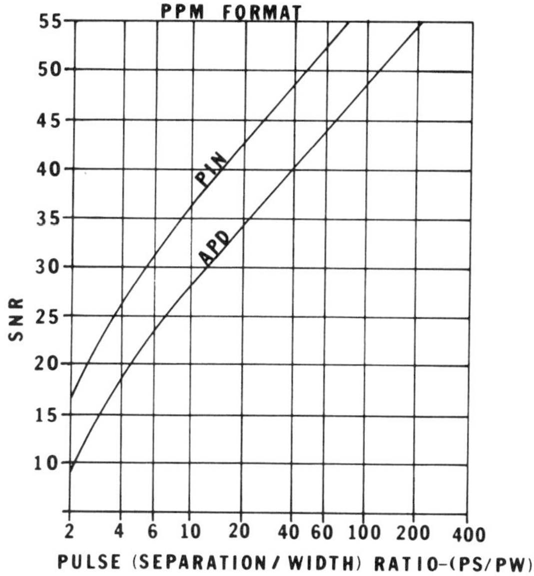

Fig. 110 SNR vs. pulse separation-width ratio. PIN; SNR > 31: SNR = 20 log (PS/PW) + 16. APD; SNR > 25: SNR = 20 log (PS/PW) + 7.

product literature. Similarly the response time of the photodetector or detector module is specified (step 3). For analog transmission the response time of the complete receiver module must be considered, usually determined from the product literature (step 4). The next step is to calculate the modal and material dispersion rise times from the fiber NA and refractive index values (n_{clad}) and the light source emission wavelength (λ_0) and spectral width ($\delta\lambda$) as shown in steps 5 to 8. The calculated fiber rise times (ns/km) are multiplied by the appropriate fiber length between terminals depending upon the type of network used to link all of the user terminals. For star networks, L_{max} is defined as the maximum terminal-to-terminal distance in the system.

Once all of the individual rise times have been determined, the sum of the squares of the rise times S is then calculated (step 9). The actual expected system rise time (SRT) is then calculated (step 10). If the value of SRT is greater than AST, which is often the case, faster detectors or light sources, or cable with smaller dispersion values, will have to be used. It should be noted that steps 5 to 8 are only used as a means of approximating the dispersion rise times when they have not been determined experimentally. To be safe, especially with large networks, it is best to determine total cable rise time from experimental data or from the manufacturer's specifications to evaluate the accuracy and applicability of the given dispersion value.

BIBLIOGRAPHY

Akiyama, S., et al., 3-Channel TV transmission on optical fiber, International Conference on Integrated Optics and Optical Fiber Communications, July, 1977, Tokyo.

Aoki, F., et al., Practical use of optical fiber communications for electric power companies, ibid.

Banks, F.M., et al., An experimental 45Mb/s digital transmission system using optical fibers, Proc. ICC, 1974, Minneapolis, Minn.

Boerner, M., et al., Single-made transmission systems for civil telecommunications, Proc. IEEE, vol. 123, 1976.

Bouillie, R., et al., Influence of baseband frequency responses of optical components on the transmission system design, Topical Meeting on Optical Fiber Transmission, February, 1977, Williamsburg, Va.

Brackett, C.A., et al., Optical data links for short-haul high level performance at 16 and 32 Mbps, Ibid.

Callan, R. R., et al., A duct installation of two fiber optical cables, IEE Conf. Publ., vol. 132, 1975.

Daikoku, K., et al., A proposal on optical fiber transmission systems in a low-loss 1.0-1.4 μm wavelength region, Optical and Quantum Electronics, vol. 9, no. 1, 1977.

Electro-Optic Telecommunications, Arthur D. Little, August, 1975.

Elion, G. R., and H. A. Elion, Fiber Optics in Communications Systems, Marcel Dekker, New York, 1978.

Eppes, T. A., et al., Long distance optical communications, Proceedings of the International Conference on Communcations, 1975.

Eppes, T. A., et al., Low-power fiber optic repeater design, Electro-Optical Systems Design, May, 1977.

Goell, J. E., Input amplifiers for optical PCM receivers, Bell System Tech. J., vol. 53, no. 9, 1974.

Gravel, R. L , et al., Star repeaters for fiber optic links, Appl. Opt., vol. 16, no. 2, 1977.

Hiramatsu, T., et al., An experimental 400 Mbps digital transmission system using selfoc fiber, Topic Meeting on Optical Fiber Transmission, February, 1977, Williamsburg, Va.

Kimura, T., et al., Considerations on Optical fiber transmission systems and problems associated with their system design, Rev. Electrical Commun. Lab., vol. 24, 1976.

Markstein, H. W., Fiber optics for electronic interconnection, Electronic Packaging and Production, April, 1977.

Miki, T., et al., Hybrid digital transmission system over optical fiber cable, International Conference on Integrated Optics and Optical Fiber Communications, July, 1977, Tokyo.

Mochida, Y., et al., Low noise 100 Mb/s optical receiver, Second European Conference on Optical Fiber Communication, September, 1976, Paris, France.

Nawata, K., et al., Studies on PCM-IM 100 Mbps optical transmission experiments, Electronics Communications Engineers of Japan, vol. 75, 1975.

Personick, S. D., Comparison of equalizing and non-equalizing repeaters for optical fiber systems, Bell System Tech. J., vol. 55, no. 7, 1976.

Uchida, T., et al., An experimental 123 Mb/s fiber optic communication system, Topical Meeting on Optical Fiber Transmission, January, 1975, Williamsburg, Va.

Chapter 13

SPECIALIZED ELECTRO-OPTIC COMPONENTS

Optoelectronic Devices

The basic types of optoelectronic devices are the interrupter-reflector modules (which detect objects modifying the given radiation pathway), isolator-coupler devices (which transmit low-noise electrical signals without using a common ground), and discrete detectors and emitters. The standardized symbols used for various electro-optical components are shown in Fig. 111. Interrupter-reflector modules are specified as a coupled pair, defined by the input current, the type of object that will pass through the interrupted gap (visible or infrared, opaque or reflective), and the desired monitor and electrical outputs.

Some typical applications of opto-couplers include a transmission line isolator to isolate chopper circuits, solid-state relays and pulse amplifiers, and effective replacement pulse transformers and mechanical relays. Figure 112 shows the accepted schematic symbols for various types of optically coupled isolators. Note that either the rectangular or semicircular end forms can be used in a given circuit diagram as long as one or the other is used consistently throughout.

Opto-Couplers

The main purpose of opto-couplers is to couple two or more electrical systems together without the use of a common ground, to effectively isolate the individual switching circuits. The basic opto-coupler often consists of a GaAs infrared-emitting diode (IRED) as the input stage and a silicon npn phototransistor as the output stage coupled together with either a gas-filled gap or some infrared-transmitting optical material (glass, etc.). The spectral emission of the IRED should be carefully matched to the spectral response of the phototransistor to achieve optional performance. In many opto-couplers the active area of the sensor is rather large. To

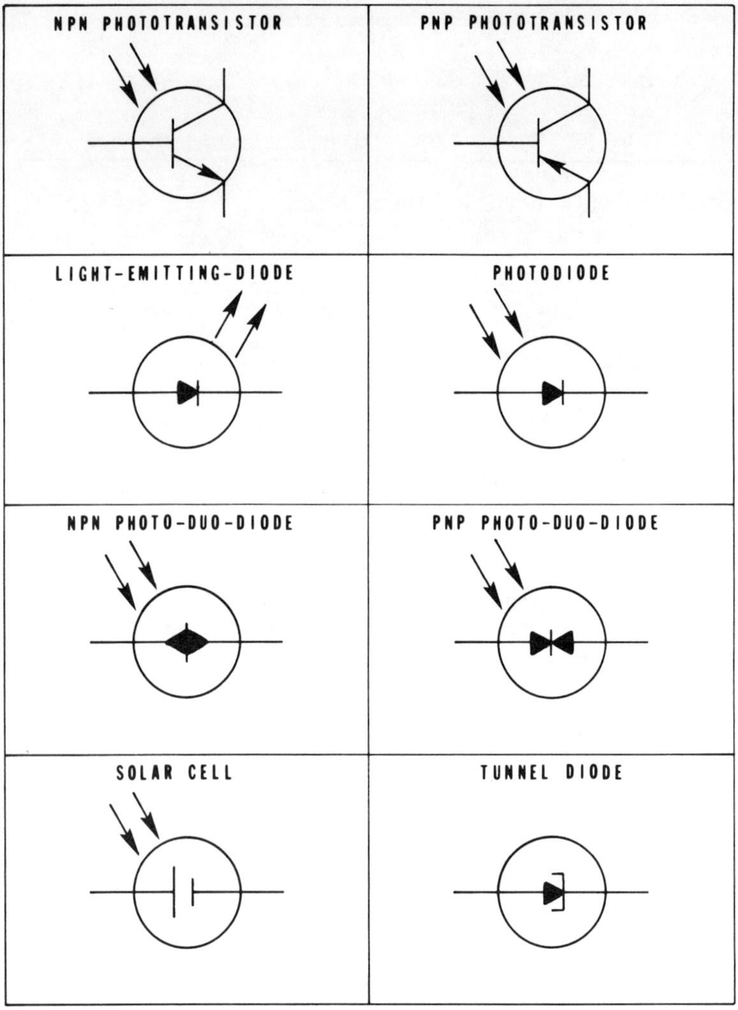

Fig. 111 Electro-optic component schematic symbols.

increase the response efficiency further causes a correspondingly large collector-base junction capacitance. The isolation capacitance through the dielectric from the input to output typically ranges from 1 to 3 pF. The frequency response is in the range of dc to 150 kHz.

The overall size of most opto-couplers is equivalent to a power transistor, enabling them to be used on standard printed circuit (pc) boards.

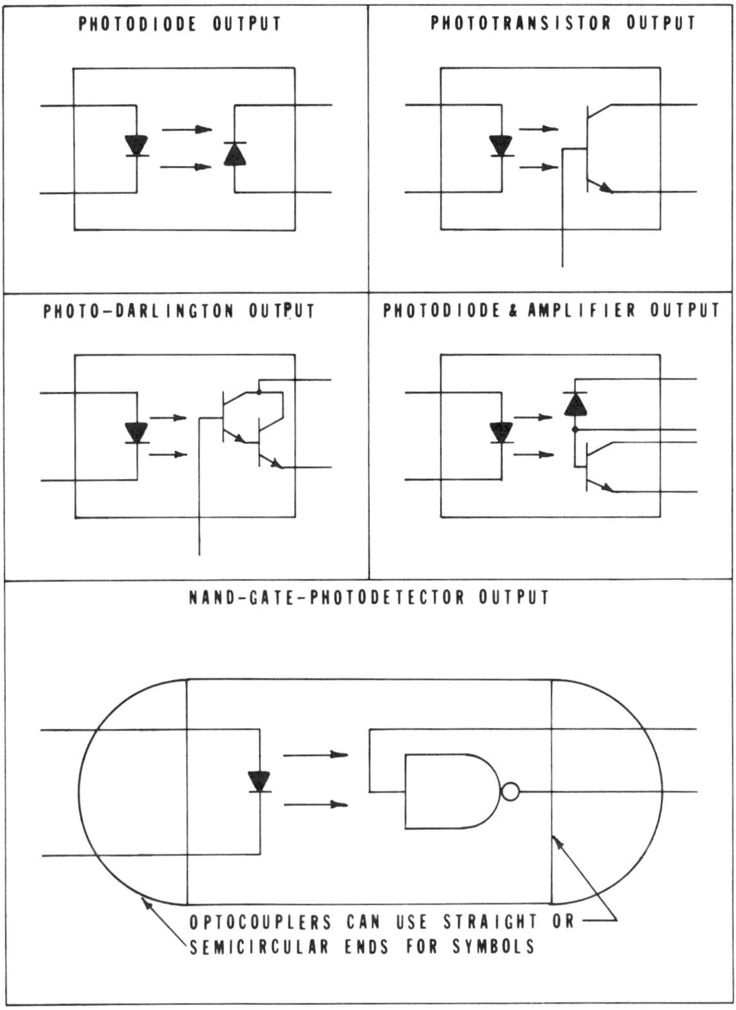

Fig. 112 Optically coupled isolators schematic symbols.

Opto-couplers are often preferred over isolation transformers, line drivers, and relays due to their fast switching times and high-voltage isolation properties. Noise received at the input of an opto-coupler is essentially isolated from the output signals by the coupling medium, and since the input circuit is basically a photodiode, common-mode noise is rejected. The current gain in opto-couplers varies widely depending on the phototransistor efficiency and the coupling medium transmittance at the IRED emission wavelength. The current gain is usually stable over a wide temperature range due to the offsetting temperature coefficients of the IRED and the silicon phototransistor. Figure 84 shows the collector current as a function of the collector-emitter voltage and the forward current for a typical opto-coupler. The ratio of the output to input current usually lies within the range of 0.1 to 1.0.

The leakage resistance is usually in the range of 10^{11} to 10^{13} Ω. The high-voltage isolation between the input and output is one of the main advantages of opto-couplers in circuit system coupling. The degree of isolation depends on the electrical properties of the coupling material and the unit construction. The isolation voltage rating depends on the time derivative factor dV/dT and the signal waveform. Typical breakdown voltages for nitrogen-filled (or inert gas-filled) or glass separator opto-couplers is in the range of 600 to 1200 V.

Specialized LEDs and Fiber Coupling Devices

The packaging design of LEDs and laser diodes involves the consideration of electrical and optical requirements as well as various environmental considerations including temperature, chemical environment, and mechanical constraints. In most designs reflections within the emitter can cause power losses. Total internal reflection can be eliminated in LEDs when the ratio of the junction radius to the dome radius is equal to the sine of the critical angle:

$$\sin\theta = \frac{n_a}{n_e}$$

where n_a is the index of refraction of air (or the surrounding gas) and n_e is the index of refraction of the top LED material (epoxy $\simeq 1.5$ or semiconductor $\simeq 3.6$). The angle θ_c is usually in the range of 14 to 24°, respectively, when semiconductor material and epoxy are used as the top material.

Doping the basic $GaAs_{1-x}P_x$ LED with nitrogen improves the brightness qualities for the visible spectrum devices as well as the quantum efficiency, especially in the region for $x \geq 0.44$. Doping with nitrogen, usually in the form of Si_3O_4, also increases the peak emission wavelength as a function of the alloy composition x.

High-radiance light-emitting diodes are commonly used for fiber optic communications despite the intrinsic disadvantage of having a relatively large angular radiation profile as compared to semiconductor lasers. Since most individual optical fibers of low attenuation have a small acceptance angle the coupling losses can be high. To decrease the coupling

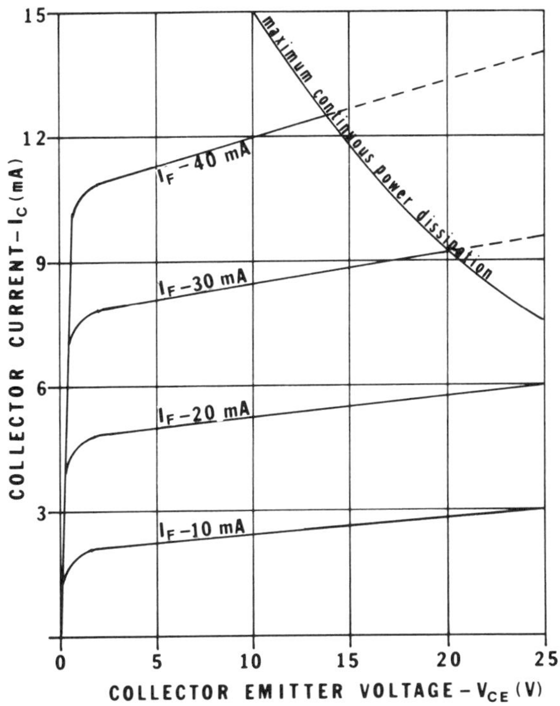

Fig. 113 Typical opto-coupler collector current vs. collector-emitter voltage.

losses two types of specialized electro-optical components can be used, both of which are based on the same principle of focusing light into the optical fiber. One method is to use an optical fiber pigtail (short fiber segment) which has a relatively spherical end. The pigtail is permanently attached to the LED, and installation simply requires the coupling to the flat end of the long communications fiber. The spherical end of the pigtail is attached using epoxy and is very near the LED surface to focus emitted radiation into the optical fiber segment. This method can achieve coupling efficiencies in the range of 7 to 9% depending on the LED surface area. Another method is to use a spherical lens directly on the LED active surface which is held in place by epoxy. The LED surface can be pre-etched to fit the surface curvature of the lens, thus providing excellent contact and radiation concentration abilities. This focusing method can achieve 9 to 11% coupling efficiencies depending on the active-surface diameter, assumed to be about 25 to 35 μm. The total optical power transmitted in the beam of the collection optics (P_t) is given

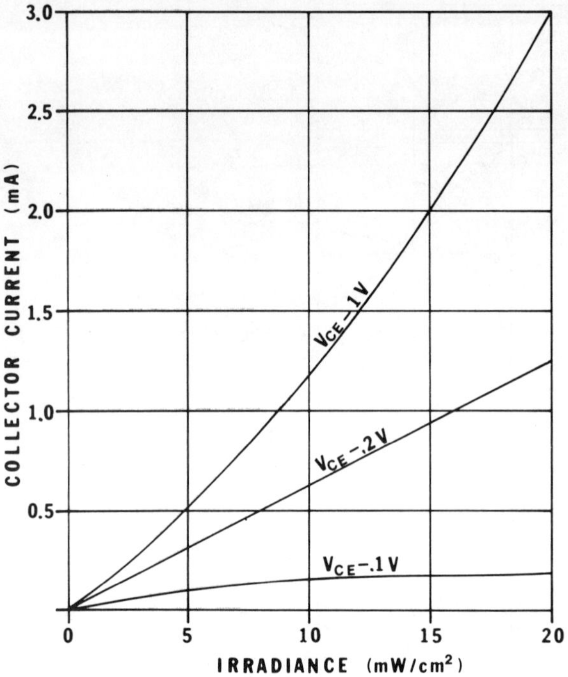

Fig. 114 Typical silicon phototransistor collector current vs. irradiance.

by $P_t = P_0(1 - \cos\theta)T$, where P_0 is the total radiated optical power of the semiconductors, θ is the half-angle of the collection core, and T is the transmittance of the lens including material losses and reflectance losses by index of refraction mismatching.

Card and Tape Readers

Solid-state read heads for punched cards and tapes usually used with computer terminals basically consist of infrared-emitting diodes (IREDs) as the light sources and silicon phototransistor sensors. The electrical and especially the optical properties of the IREDs and phototransistors are carefully matched for optimal performance. Figure 114 shows the typical silicon phototransistor collector current as a function of irradiance and the collector-emitter voltage. Figure 115 shows the effect of temperature on the relative output and dark current of a typical silicon phototransistors.

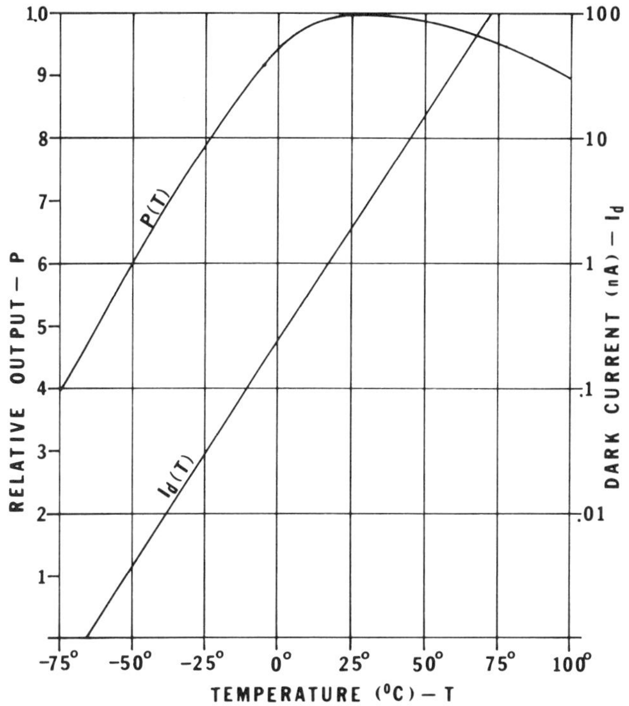

Fig. 115 Typical silicon phototransistor temperature vs. relative output and dark current.

The solid-state readers effectively replace mechanical readers and filament lamp types. The IREDs used in the reader are compatible with DTL and TTL circuits commonly encountered in computer systems and permit the use of opto-couplers to eliminate system noise and grounding problems. In the design of optical readers one of the most important parameters is the system contrast ratio defined as the ratio of the hole to no-hole output current of the phototransistors, which should be made as high as possible (50:1 or greater) to minimize potential reading errors. Contrast ratios for optical solid-state readers range from about 40 to 40 X 10^4. High-contrast ratios for low-opacity tapes is achieved using an opaque aperture plate between the source and detector units which usually has holes with a diameter in the range of 1 mm. The actual contrast (current) ratio is affected by numerous factors including the tape opacity, the degree of optical cross talk between channels, the use of an aperture plate and its hole and reflectance parameters, and the sensor-detector spacing (usually 6 to 10 mm) and the electro-optical parameters.

Acousto-Optic Devices

In a typical acousto-optic device a piezoelectric transducer of dimensions H X L is attached to a crystal of specified orientation. The transducer can be controlled electrically by various modulation schemes to produce acoustic waves that have wavelengths much smaller than H and L. When a light beam is transmitted through the material the ultrasonic waves produced at the transducer cause the beam to be deflected by an angle ϕ. The acoustic wave power decreases exponentially as a function of distance from the H X L element, thus limiting the effective beam diameter that can be adequately modulated. The fraction of incident light intensity that is actually deflected by the device is given by

$$\sin^2 \frac{\phi}{2}$$

where

$$\phi = \pi \frac{2LM_2P}{\lambda^2 H^{1/2}}$$

where

λ = incident beam optical wavelength
P = acoustic power
M_2 = acousto-optic figure of merit

The three figures of merit for these materials are given by

$$M_1 = \frac{n^7 p^2}{\rho V} \qquad M_2 = \frac{n^6 p^2}{\rho V^3} \qquad M_3 = \frac{n^7 p^2}{\rho V^2}$$

where

n = index of refraction at the incident wavelength
ρ = strain-optic component
p = material density
V = acoustic velocity

Since n, p, and V vary with crystal orientation, the figure of merit must be specified along a given crystal direction.

Acousto-optic materials are preferred to have a high figure of merit, low optical and acoustic losses, and high mechanical integrity to withstand the ultrasonic waves for extended periods of time. The figures of merit of these materials are compared to SiO_2 as a reference material with $M_1 = M_2 = M_3 = 1.0$. Table 76 lists some typical acousto-optic material properties.

Table 76 Acousto-Optic Material Properties

Material	Range (μm)[a]	n(0.633 nm)[b]	A.W.V.[c]	M_1	M_2	M_3
GaP	0.6–10.0	3.31	6.32	75	30	69
$LiNbO_3$	0.5–4.5	2.20	6.57	8.3	4.6	7.5
$PbMoO_4$	0.4–5.5	2.39	3.66	15	24	25
SiO_2	0.2–4.5	1.46	5.96	1.0	1.0	1.0
TeO_2	0.4–5.0	2.27	4.26	18	23	26
TiO_2	0.5–5.5	2.58	7.86	7.9	2.6	6.2

[a]Approximate effective optical transmission range.
[b]Index of refraction measured at 0.633 μm.
[c]Acoustic wave velocity X 10^5 cm/s in the longitudinal polarization direction.
Source: R. J. Pressley (Ed.), CRC Handbook of Lasers with Selected Data on Optical Technology, Cleveland, Ohio, 1971.

Highly efficient acousto-optic modulators can be obtained using doubly confined optical and acoustic waves where the guided optical waves and the surface acoustical waves are simultaneously confined in the transverse direction. The three types of structures used to attain highly efficient mode conversions are the strip loaded, metal clad, and rectangular wave-guide structures. Typical doubly confined structures require less than 2 mW of acoustic power to achieve 100% mode conversion. Such devices can also be used as tunable optical filters, since the acoustic wavelength determines the wavelength of light for which the mode conversion can occur. The bandwidth of these filters is given by $0.5/(n_i - n_d)L$, where n_i is the refractive index of the incident light, n_d is the index of the diffracted light, and L is the interaction length. Filter bandwidths of less than 5 Å are thus achievable.

Optical Taps

Often in optical transmission links it is necessary to tap (partially or selectively remove) some of the light in the wave-guide pathway. For short links the insertion loss and cross talk requirements for these devices are sometimes not very critical, and various types of beam-splitter configurations can be used. However, for some applications the required insertion loss and cross talk must be minimized to increase the transmission pathway without the use of repeaters or to increase the number of terminals that can be connected to one original light source. To minimize insertion

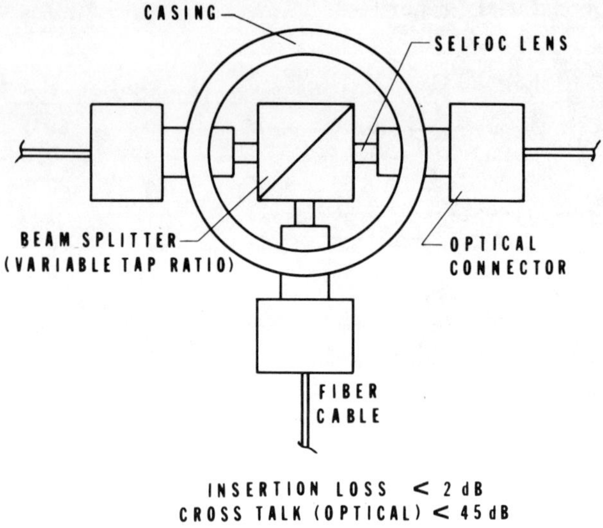

Fig. 116 Low-loss optical beam splitter with a variable tap ratio.

losses, the coupling of the optical fibers must be designed very carefully, since the core diameters are often less than 60 μm, making any physical misalignments costly in terms of optical power loss.

One practical design is shown in Fig. 116 which uses optical fiber connectors to couple the input fiber to the tapped fiber (to the device or terminal) and to the output (downstream) fiber. Using this method the terminal device may be disconnected without interruption of the downstream operation. To minimize cross talk, the number of reflecting surfaces must be minimized. Thus, the rod lenses or slab lenses are directly attached to the beam-splitter surface using index-matching epoxy. The opposite end of the lens is coated with an antireflection coating to reduce losses at the fiber-air-lens interface. For some applications where cross talk must be extremely small this interface can be eliminated by direct attachment of the fiber to the lens by epoxy or by using index-matching oil (silicone) in the optical connector. Another method is to use slab lenses directly attached to the optical wave guide. Here one end of two of the slab lenses is cut at a 45° angle and coated with a partially reflective coating, and the two ends are attached together via index-matching epoxy. By controlling the coating reflectivity at the central wavelength, the beam-splitter tap ratio can be controlled to any desired value. Tap ratios of 1:1 to 1:10 are most common in optical fiber systems, largely determined by the number of terminals needed, the maximum light source output, and the minimum

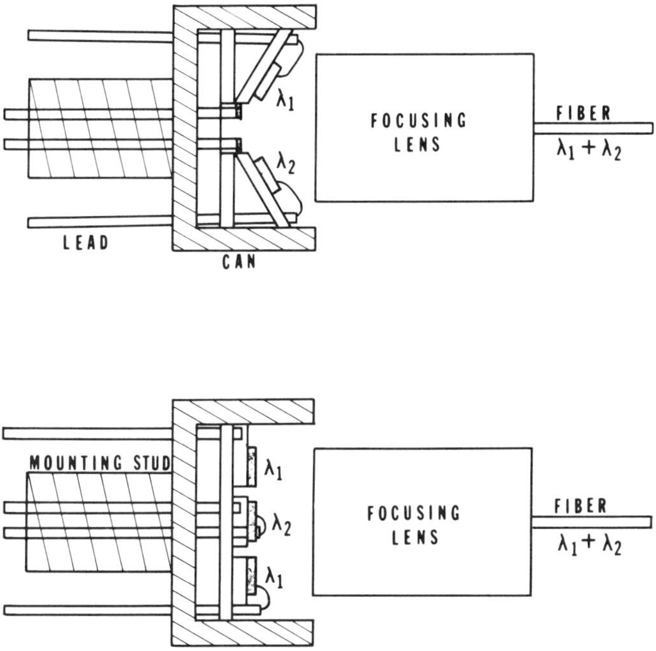

Fig. 117 LED or LD configurations for direct light source wavelength division multiplexing.

required received signal power at each terminal device. One method of reducing cross talk in most devices coupled to optical transmission wave guides is to use fiber terminations that have off-angle (nonperpendicular to the fiber axis) end surfaces, to reduce reflection cross-talk components.

Wavelength Division Multiplexing Devices

The general term wavelength division multiplexing devices applies to optical devices capable of coupling into or splitting out of an optical wave guide two or more wavelengths of transmission information. Devices which can both couple and split two or more wavelengths into or from an optical wave guide are called wavelength division multiplexers and splitters, or WDM mixer-splitters.

Light sources can be designed to transmit two or more wavelengths by using chips mounted in the source "can" which emit at different wavelengths. Figure 117 shows two configurations for such WDM devices utilizing two wavelengths. The top of the figure shows the off-angle type where chips of differing wavelengths are tilted at off-normal angles to more efficiently couple into the focusing lens connected to the optical wave guide (fiber). Rod lenses, spherical lenses, and elliptical lenses are commonly used to couple the light source to the fiber. Using this type of spatial configuration for the semiconductor chips, 2, 3, 4, 5, or more wavelengths may be coupled with equal power into the optical fiber. The bottom of Fig. 117 shows the concentric circle-type WDM light source. Here, the semiconductor materials are made in concentric ring formats, each ring with a different wavelength, each usually of a differing width and optical power. The problem with these devices is that they are expensive to fabricate and if one chip fails, the entire device must be replaced, which is relatively expensive and inefficient. For LEDs the practical number of wavelengths that can be added (multiplexed) into one fiber is usually considered to be three to minimize cross talk by spectral overlap. For digital transmission four wavelengths can be used, where electrical subtraction techniques are used in the detection circuits to eliminate cross talk. Given the fabrication complexity of such light sources and a finite mean time before failure (MTBF) for each semiconductor chip, the practical limit and use of such devices is two or three wavelengths.

The detectors used for WDM optical systems can take several forms. One is the common PIN or APD type with a relatively broadband responsivity. Another type is the narrow-band responsivity detector where the peak responsivity is shifted to be in coincidence with the incident wavelength peak. Another type is the coated detector, where various types of optical filters or coatings are added to the detector window to limit the incident wavelength regions. Another type is the two-layer detector where two separate semiconductor layers are used, each sensitive (optimized) for only one wavelength region. These two-layer detectors usually have a low quantum efficiency for the top layer which must be kept relatively transparent (physically thin) to the second (bottom layer) wavelength.

One method of using one device as both a multiplexer and a splitter is shown in Fig. 118. This particular configuration is applicable to using two-way transmission on a single optical fiber. The central component for these devices is a dichroic filter whose spectral properties (wavelength vs. transmission %) are shown in Fig. 119. The cutoff and cut-on wavelengths of each dichroic filter can be shifted to accommodate the peak wavelength of the light sources. The same device can be used to couple two LEDs or LDs onto a single fiber or to split the incoming optical signal into two wavelength regions for the PIN or APD detectors. Multiple (>2) wavelengths can be multiplexed using this method by using a series of dichroic filters. As the number of wavelengths increases, the insertion loss, especially for the last wavelength in the filtered sequence, also increases.

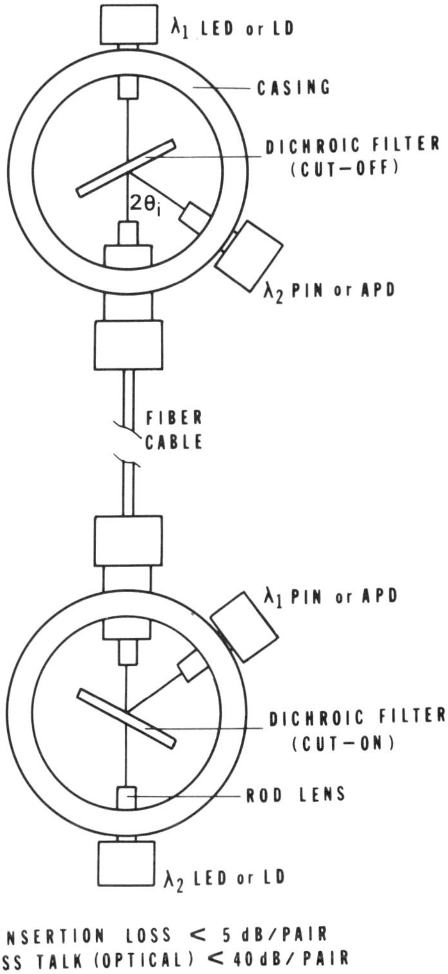

Fig. 118 Wavelength division multiplexing and splitting component schematic for two-way on one fiber.

The cross talk in such devices is caused by many sources. The fiber-to-lens interface at the optical connectors is one source of cross talk caused by reflection. These reflection sources occur at both ends of the fiber. The longer the fiber, the higher its attenuation or the higher the modulation frequency of the signal, the lower the cross talk by fiber length reflections.

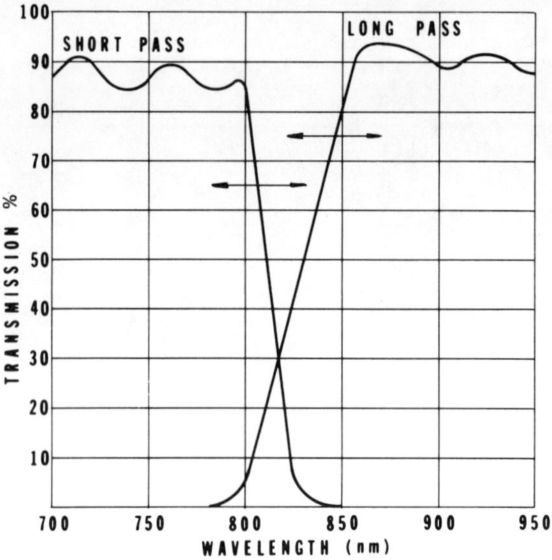

Fig. 119 Dichroic filter combinations used for wavelength division multiplexing.

These reflections can be reduced by direct attachment of the fiber to the lens or by using index-matching oil in the connector which is subsequently sealed with heat-shrink tubing or wax. Each surface in the optical pathway sequence is a source of reflection and thus a possible source of cross talk. One method of reducing cross talk is to cut the lens to the desired incidence angle and use the polished face as the dichroic filter substrate. Another method of reducing cross talk is to add other dichroic filters in the pathway between the central dichroic filter and the detector or to coat the detector window with a multilayer dichroic filter.

When LD light sources are used, cross talk can easily be made to be very small. Here cross talk by spectral overlap usually does not occur. The source of cross talk occurring mainly by reflections can usually be limited to less than 60 dB. When LED sources are used, a significant portion of the cross talk is caused by the spectral overlap of the individual sources. Cross talk can be reduced in these cases by using sources with peak wavelengths that are far apart or by using sources with narrow spectral half-widths. There is an approximately twofold difference between optical and electrical cross talk, given in decibels. For WDM devices used for analog video systems the cross talk must usually be kept below 45 dB, which can easily be achieved even using LEDs with spectral half-widths of 40 nm.

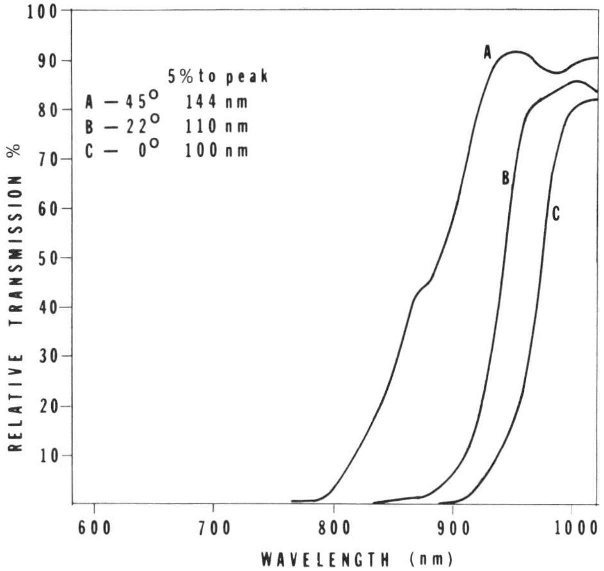

Fig. 120 Dichroic filter relative transmission vs. wavelength and incidence angle.

Another source of cross talk in such devices is caused by the divergence of the light source beam beyond the end of the lens. The greater the divergence, the greater the chance of light entering or reflecting in or on an undesired location. To restrict this type of optical interference, opaque coated metal or plastic tubes are sometimes attached to the end of the lens and extend to the filter interface, effectively blocking any divergent light or entering stray light. This permits the incidence angle of the beam onto the filter to be relatively small (20°). To minimize the insertion loss caused by losses in the lens collimating system, the optical pathway from the input-light lens to the other lenses must be kept small (usually less than 20 mm), which is accomplished by using a large incidence angle to the dichroic filter or by using small lenses and connectors. However, as the incidence angle increases the desired dichroic filter properties are diminished. The best dichroic filter properties for such applications have steep slopes for the wavelength-vs.-transmission plots. Figure 120 shows the transmission properties of a typical dichroic filter at various incidence angles showing the sequential degradation of the slope with increasing angle. A compromise of incident angle versus insertion loss and cross talk must be considered for each transmission case using different information bandwidths. The compromise angle is usually in the range of 18 to 22°. By combining FDM and WDM a device configuration utilizing two wavelengths and three frequency bands can transmit six channels on one fiber.

BIBLIOGRAPHY

General Electric Optoelectronics Manual, Semiconductor Products, Syracuse, N. Y., 1976.

Horiuchi, S., et al., A new LED structure with a self-aligned spherical lens for efficient coupling to optical fibers, Proc. SPID, vol. 18, no. 2, 1977.

Optoelectronics Designers Catalog 1977, Hewlett-Packard Components, Palo Alto, Calif., 1977.

The Optoelectronics Data Book For Design Engineers, 4th ed., Texas Instruments Incorporated, Semiconductor Group, Dallas, Tex., 1976.

Chapter 14

COMPUTER DESIGN OF ELECTRO-OPTIC COMPONENTS AND SYSTEMS

There are numerous applications of computers in electro-optics including the design of individual components or elements, the design and performance optimization of multielement systems, as well as for real-time on-line types of computer functions and calculations for measurement and control systems. In addition, interactive computer graphics terminals (using light pens and two-way keyboards) are often used to design, optimize, or test various electrical circuits used to control, drive, or interact with optical components such as LEDs, LDs (light sources), PINs, APDs (photodetectors), KDP, ADP (E-O crystals) or with complete networks such as data and video transmission links, optical switching telecommunications networks, and other applications.

Probably one of the most commonly known uses of computers in the design of optical systems is for lenses and multielement lens systems. The purpose of such computer programs and designs is to minimize or maximize or compromise various optical parameters such as effective focal length, resolution, chromatic and other types of aberrations, effective usable wavelength region, reflection, transmission, as well as physical parameters such as size, weight, temperature variations of optical parameters, element thickness, refractive index, and other properties. The major application area for this type of computer design is with camera and telescope lenses and mirrors.

For multiple-element optical systems, the design of the individual elements is often complicated by the number of system variables. In addition to those previously stated, complete designs must consider the selection of materials used for the optical coatings on each element and the optimal design of each of the lens coatings, which are often multilayer (2 to 30 are common) in configuration. It is evident from the nature and complexity of such designs that the term <u>optimal design</u> has little meaning, since not all or even two or three parameters can be "optimized" for any given lens system function. Thus, the selection of multielement lenses for technical application is often specified by optimization of the most desirable optical parameters such as resolution or minimal chromatic aberration (such as for laser lenses).

Integrated optical circuits often use computer calculations to design components which have a broad spectrum of possible input parameters, electrical, optical, or magnetic. Integrated optical circuits (IOCs) like integrated electrical circuits (ICs) have stringent physical (spatial) requirements and demand precise designs to achieve accurate chemical etchings or electron-beam etchings. The design of these devices requires accurate measurement of the material properties including the electro-optical parameters as well as such physical parameters as the index of refraction over the usable wavelength region.

An example of a simple multielement design for an electro-optic component is shown in Fig. 121. This figure shows one section of a multiple wavelength mixer and splitter used in optical fiber transmission systems. The component base material could be metal (aluminum or stainless) for direct machining applications or a silicon wafer for photoetching techniques or a coated metal base for electron-beam etching methods. The problem with such multielement optical devices is the precise placement of each optical element on the component base. The absolute precision of such placements is determined basically by the geometric variations of the optical elements (filters and lenses in this case), which are placed on the supporting base, sometimes called a micro-optical bench. For example, if the rod lens axial eccentricity is ± 0.1 mm, the placement precision required for the V-groove is about the same magnitude as the eccentricity or the overall variation for all elements used on the bench. In this case several (two or more) wavelengths from a single beam enter a rod lens and strike the surface of a dichroic filter. One or more wavelengths may be reflected from the first or second filter surface, which is tilted at an angle θ from the incident beam. The optical insertion loss of such devices is highly dependent upon the alignment of the lenses to the incident beam and filter surfaces. For example, an offset of 0.2 mm might be equivalent to a 1-dB increase in the insertion loss. Thus, the placement of the lens or lenses used after the beam has transmitted through the filter must take into account the small off-axis translation of the central beam caused by the refraction in the filter. This off-axis distance is dependent upon the filter substrate refractive index at the wavelengths of the transmitted beam, the effective index of refraction of the coated filter, the angle (θ) of the filter to the incident beam, and the filter thickness (T). A computer program to calculate the relative lens and filter positions needs the physical variables defined and usually internally varies the incidence angle and filter thickness. For such calculations the refractive index profile should be defined to ± 0.0001. If a medium other than air is used between the optical elements, it is sometimes necessary to define the interelement material (gas, liquid, or solid) refractive index spectral profile to ± 0.00001.

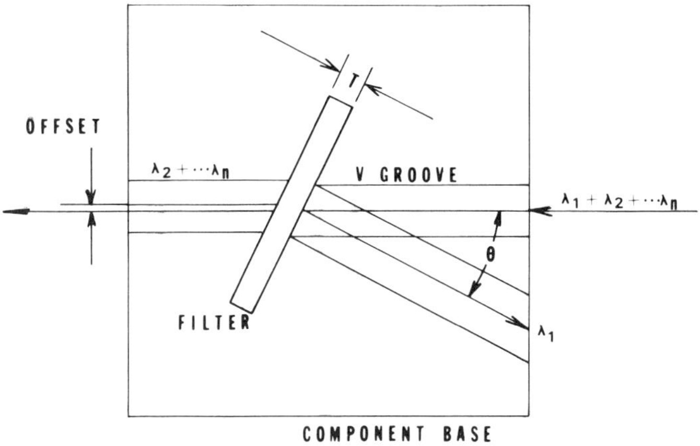

Fig. 121 Multiple wavelength mixer and splitter example of computer-designed E-O components.

In optical or electro-optical components such as these (especially IOC types), the distances between all elements are made as small as possible to minimize coupling and insertion losses.

An example of computer designs in electro-optical systems is shown in Fig. 122. This figure shows two LEDs and two dichroic filters which are used for the two-way video or data transmission link on a single optical fiber such as is shown in Fig. 118. Such a system must consider the following parameters:

1. The LED relative power and coupled optical power as a function of temperature and the circuit-driving current and voltage
2. The peak LED wavelength and the spectral power distribution as a function of temperature and driving current
3. The dichroic filter spectral transmission properties
4. The required coupled optical power to the detectors
5. The total insertion loss of the complete transmission link (used to determine the maximum repeaterless distance)
6. The optical cross talk caused by reflections and the spectral overlap of the LEDs as a function of temperature, LED peak wavelengths, and filter characteristics
7. The spectral attenuation properties of the transmission line (fiber) and the transmission bandwidth (or data rate)

A sample computer program used for such calculations is shown in Fig. 123 along with the typical calculation results. The calculations require

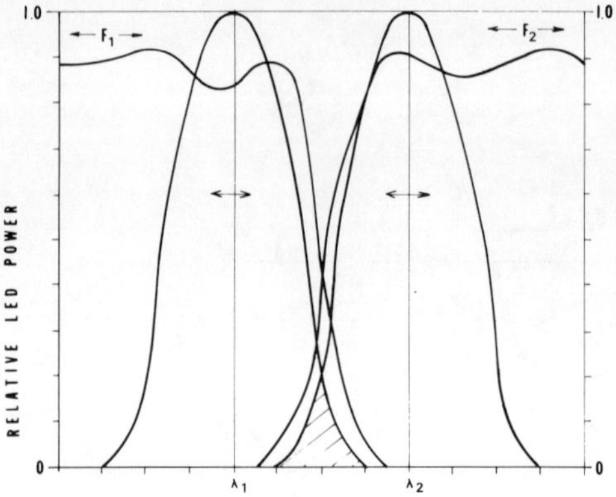

Fig. 122 Two-way LED transmission on one fiber example of computer-designed E-O systems. F_1 short-pass dichroic filter; F_2 long-pass dichroic filter.

accurate numerical data on the filter transmission properties and LED spectral power distribution shape. The filter characteristics must be determined experimentally as a function of the incidence angle. The LED data must also be determined experimentally, usually in the format of optical power versus wavelength as a function of temperature and circuit-driving current and voltage. For both the filter and LED data the digitized format for accurate computer calculations requires the wavelength increment (DLAMDA) to be about 2 nm. In actual systems the optical cross talk is found to be a function of the modulation frequency and fiber length due to the Rayleigh scattering properties of the fiber. In addition, minor changes in the calculation may be required to account for the detector responsivity changes as a function of the LED peak wavelength. For broadband response detectors these effects are usually small and negligible on the overall system design. However, for some specialized detectors which have narrow peak responsivities the detector response effect must be considered in the calculations. Cross talk in such devices can be reduced by using LEDs that have peak wavelengths that are further apart, by using index-matching fluid in the optical connectors to reduce the number of reflection surfaces, by shifting the cutoff and cut-on wavelengths of the filters, by using dichroic filter coatings on the detectors, by using an additional filter in the optical pathway before the detector, and by using other methods which may be accounted for in the overall computer design calculations. The degree of reflected power at the fiber ends is a function of the fiber attenuation, which is a function of the LED spectral power distribution and peak wavelength. For medium loss fibers ($A > 5$ dB/km) at low modulation frequencies this effect can become significant and must also be considered in the calculation.

```
      IMPLICIT DOUBLE PRECISION(A-H,O-Z)
      DIMENSION IGRAPH(71,32),YLED(2,200),YFILTR(2,200)
      DIMENSION SCALE(2),TPLED(2),TFPLED(2),YFLED(2,200)
      DIMENSION CROS(200),ITEMP(8),LSHIF(2,8)
      CHARACTER BLANK,EXCLA,PERIOD,STER,UNDER,EFF,IGRAPH
      DATA BLANK/' '/,EXCLA/'+'/,PERIOD/'.'/,STER/'*'/
      DATA UNDER/'-'/,EFF/'F'/
C*********************************************************************
C     INITIALIZE SCALE FACTORS FOR FILTER AND LED'S ; SCALE
C     FACTORS FOR THE LED'S CORRESPOND TO CHANGING THE DRIVING
C     CURRENT ; SCALE FACTORS FOR THE FILTERS CORRESPOND TO
C     CHANGING THE RELATIVE ABSOLUTE VALUES OF TRANSMISSION
C     FOR CHANGES IN FILTER QUALITY OR INCIDENCE ANGLE
C*********************************************************************
      DLAMDA=2.0
      DO 50 M=1,8
      DO 2 N=1,2
      DO 1 LAMDA=1,200
      YFILTR(N,LANDA)=0.0
      YFLED(N,LANDA)=0.0
      YLED(N,LANDA)=0.0
      CROS(LAMDA)=0.0
    1 CONTINUE
    2 CONTINUE
C*********************************************************************
C     DEFINE FILTER 1 TRANSMISSION PROPERTIES VERSUS WAVELENGTH
C     BY SHIFTING CURVE ; EACH INCREMENT IS DLAMDA NANOMETERS
C*********************************************************************
      ISHIF1=16
      YFILTR(1,22+ISHIF1)=0.001
      YFILTR(1,23+ISHIF1)=0.0015
      YFILTR(1,24+ISHIF1)=0.002
      YFILTR(1,25+ISHIF1)=0.0025
      YFILTR(1,26+ISHIF1)=0.003
      YFILTR(1,27+ISHIF1)=0.0035
      YFILTR(1,28+ISHIF1)=0.004
      YFILTR(1,29+ISHIF1)=0.006
      YFILTR(1,30+ISHIF1)=0.010
      YFILTR(1,31+ISHIF1)=0.015
      YFILTR(1,32+ISHIF1)=0.020
      YFILTR(1,33+ISHIF1)=0.025
      YFILTR(1,34+ISHIF1)=0.030
      YFILTR(1,35+ISHIF1)=0.040
      YFILTR(1,36+ISHIF1)=0.050
      YFILTR(1,37+ISHIF1)=0.060
      YFILTR(1,38+ISHIF1)=0.080
      YFILTR(1,39+ISHIF1)=0.10
      YFILTR(1,40+ISHIF1)=0.13
      YFILTR(1,41+ISHIF1)=0.16
      YFILTR(1,42+ISHIF1)=0.20
      YFILTR(1,43+ISHIF1)=0.26
      YFILTR(1,44+ISHIF1)=0.32
      YFILTR(1,45+ISHIF1)=0.38
      YFILTR(1,46+ISHIF1)=0.44
      YFILTR(1,47+ISHIF1)=0.50
      YFILTR(1,48+ISHIF1)=0.56
      YFILTR(1,49+ISHIF1)=0.62
      YFILTR(1,50+ISHIF1)=0.68
```

Fig. 123 Typical computer program for WDM-LED device calculations.

```
      VFILTR(1,51+ISHIF1)=0.74
      VFILTR(1,52+ISHIF1)=0.80
      VFILTR(1,53+ISHIF1)=0.815
      VFILTR(1,54+ISHIF1)=0.83
      VFILTR(1,55+ISHIF1)=0.84
      VFILTR(1,56+ISHIF1)=0.85
      VFILTR(1,57+ISHIF1)=0.84
      VFILTR(1,58+ISHIF1)=0.83
      VFILTR(1,59+ISHIF1)=0.815
      VFILTR(1,60+ISHIF1)=0.80
      VFILTR(1,61+ISHIF1)=0.79
      VFILTR(1,62+ISHIF1)=0.78
      VFILTR(1,63+ISHIF1)=0.77
      VFILTR(1,64+ISHIF1)=0.77
      VFILTR(1,65+ISHIF1)=0.78
      VFILTR(1,66+ISHIF1)=0.79
      VFILTR(1,67+ISHIF1)=0.795
      VFILTR(1,68+ISHIF1)=0.80
      VFILTR(1,69+ISHIF1)=0.805
      VFILTR(1,70+ISHIF1)=0.81
      VFILTR(1,71+ISHIF1)=0.815
      VFILTR(1,72+ISHIF1)=0.82
      VFILTR(1,73+ISHIF1)=0.825
      VFILTR(1,74+ISHIF1)=0.83
      VFILTR(1,75+ISHIF1)=0.84
      VFILTR(1,76+ISHIF1)=0.85
      VFILTR(1,77+ISHIF1)=0.86
      VFILTR(1,78+ISHIF1)=0.87
      VFILTR(1,79+ISHIF1)=0.88
      VFILTR(1,80+ISHIF1)=0.89
      VFILTR(1,81+ISHIF1)=0.90
      VFILTR(1,82+ISHIF1)=0.91
      VFILTR(1,83+ISHIF1)=0.92
      DO 3 LAMDA=84,180
      VFILTR(1,LAMDA+ISHIF1)=0.92
    3 CONTINUE
C**********************************************************************
C     DEFINE FILTER 2 TRANSMISSION PROPERTIES VERSUS WAVELENGTH
C     BY SHIFTING CURVE ; EACH INCREMENT IS DLAMDA NANOMETERS
C**********************************************************************
      ISHIF2=-44
      DO 4 LAMDA=1,58
      IF(LAMDA+ISHIF2.LT.1) GO TO 4
      VFILTR(2,LAMDA+ISHIF2)=0.92
    4 CONTINUE
      VFILTR(2,59+ISHIF2)=0.92
      VFILTR(2,60+ISHIF2)=0.91
      VFILTR(2,61+ISHIF2)=0.90
      VFILTR(2,62+ISHIF2)=0.89
      VFILTR(2,63+ISHIF2)=0.88
      VFILTR(2,64+ISHIF2)=0.87
      VFILTR(2,65+ISHIF2)=0.86
      VFILTR(2,66+ISHIF2)=0.85
      VFILTR(2,67+ISHIF2)=0.84
      VFILTR(2,68+ISHIF2)=0.83
      VFILTR(2,69+ISHIF2)=0.825
      VFILTR(2,70+ISHIF2)=0.82
      VFILTR(2,71+ISHIF2)=0.815
      VFILTR(2,72+ISHIF2)=0.81
      VFILTR(2,73+ISHIF2)=0.805
```

Fig. 123-Continued

```
      YFILTR(2,74+ISHIF2)=0.80
      YFILTR(2,75+ISHIF2)=0.795
      YFILTR(2,76+ISHIF2)=0.79
      YFILTR(2,77+ISHIF2)=0.78
      YFILTR(2,78+ISHIF2)=0.77
      YFILTR(2,79+ISHIF2)=0.77
      YFILTR(2,80+ISHIF2)=0.78
      YFILTR(2,81+ISHIF2)=0.79
      YFILTR(2,82+ISHIF2)=0.80
      YFILTR(2,83+ISHIF2)=0.815
      YFILTR(2,84+ISHIF2)=0.83
      YFILTR(2,85+ISHIF2)=0.84
      YFILTR(2,86+ISHIF2)=0.85
      YFILTR(2,87+ISHIF2)=0.84
      YFILTR(2,88+ISHIF2)=0.83
      YFILTR(2,89+ISHIF2)=0.815
      YFILTR(2,90+ISHIF2)=0.80
      YFILTR(2,91+ISHIF2)=0.74
      YFILTR(2,92+ISHIF2)=0.68
      YFILTR(2,93+ISHIF2)=0.62
      YFILTR(2,94+ISHIF2)=0.56
      YFILTR(2,95+ISHIF2)=0.50
      YFILTR(2,96+ISHIF2)=0.44
      YFILTR(2,97+ISHIF2)=0.38
      YFILTR(2,98+ISHIF2)=0.32
      YFILTR(2,99+ISHIF2)=0.26
      YFILTR(2,100+ISHIF2)=0.20
      YFILTR(2,101+ISHIF2)=0.16
      YFILTR(2,102+ISHIF2)=0.13
      YFILTR(2,103+ISHIF2)=0.10
      YFILTR(2,104+ISHIF2)=0.08
      YFILTR(2,105+ISHIF2)=0.06
      YFILTR(2,106+ISHIF2)=0.05
      YFILTR(2,107+ISHIF2)=0.04
      YFILTR(2,108+ISHIF2)=0.03
      YFILTR(2,109+ISHIF2)=0.025
      YFILTR(2,110+ISHIF2)=0.020
      YFILTR(2,111+ISHIF2)=0.015
      YFILTR(2,112+ISHIF2)=0.010
      YFILTR(2,113+ISHIF2)=0.006
      YFILTR(2,114+ISHIF2)=0.004
      YFILTR(2,115+ISHIF2)=0.0035
      YFILTR(2,116+ISHIF2)=0.003
      YFILTR(2,117+ISHIF2)=0.0025
      YFILTR(2,118+ISHIF2)=0.002
      YFILTR(2,119+ISHIF2)=0.0015
      YFILTR(2,120+ISHIF2)=0.001
C*********************************************************************
C     SHIFT LED PEAK WAVELENGTHS FOR  DIFFERENT TEMPERATURES,
C     GIVEN IN UNITS OF DEGREES CENTIGRADE ; TOTAL SHIFT IS
C     EQUIVALENT TO 0.229 NM/DEGREE CENTIGRADE FOR BOTH LED'S
C*********************************************************************
      ITEMP(M)=-20+((M-1)*10)
      DO 7 N=1,2
      LSHIF(1,M)=M-22
      LED1=909+(DLAMDA*LSHIF(1,M))
      LSHIF(2,M)=M-62
      LED2=909+(DLAMDA*LSHIF(2,M))
      IF(N.EQ.2) GO TO 5
C
```

```
      YLED(N,40+LSHIF(N,M))=0.0001
      YLED(N,41+LSHIF(N,M))=0.0006
      YLED(N,42+LSHIF(N,M))=0.0012
      YLED(N,43+LSHIF(N,M))=0.0019
      YLED(N,44+LSHIF(N,M))=0.0027
      YLED(N,45+LSHIF(N,M))=0.0036
      YLED(N,46+LSHIF(N,M))=0.0046
      YLED(N,47+LSHIF(N,M))=0.0057
      YLED(N,48+LSHIF(N,M))=0.0069
      YLED(N,49+LSHIF(N,M))=0.0082
      YLED(N,50+LSHIF(N,M))=0.0096
      YLED(N,51+LSHIF(N,M))=0.0111
      YLED(N,52+LSHIF(N,M))=0.0127
      YLED(N,53+LSHIF(N,M))=0.0144
      YLED(N,54+LSHIF(N,M))=0.0162
      YLED(N,55+LSHIF(N,M))=0.0181
      YLED(N,56+LSHIF(N,M))=0.0210
      YLED(N,57+LSHIF(N,M))=0.0222
      YLED(N,58+LSHIF(N,M))=0.0244
      YLED(N,59+LSHIF(N,M))=0.0267
      YLED(N,60+LSHIF(N,M))=0.0291
      YLED(N,61+LSHIF(N,M))=0.0331
    5 CONTINUE
      YLED(N,62+LSHIF(N,M))=0.0376
      YLED(N,63+LSHIF(N,M))=0.0426
      YLED(N,64+LSHIF(N,M))=0.0481
      YLED(N,65+LSHIF(N,M))=0.0541
      YLED(N,66+LSHIF(N,M))=0.0606
      YLED(N,67+LSHIF(N,M))=0.0676
      YLED(N,68+LSHIF(N,M))=0.0751
      YLED(N,69+LSHIF(N,M))=0.0836
      YLED(N,70+LSHIF(N,M))=0.0936
      YLED(N,71+LSHIF(N,M))=0.1056
      YLED(N,72+LSHIF(N,M))=0.1196
      YLED(N,73+LSHIF(N,M))=0.1356
      YLED(N,74+LSHIF(N,M))=0.1536
      YLED(N,75+LSHIF(N,M))=0.1726
      YLED(N,76+LSHIF(N,M))=0.1926
      YLED(N,77+LSHIF(N,M))=0.2126
      YLED(N,78+LSHIF(N,M))=0.2326
      YLED(N,79+LSHIF(N,M))=0.2528
      YLED(N,80+LSHIF(N,M))=0.2735
      YLED(N,81+LSHIF(N,M))=0.2983
      YLED(N,82+LSHIF(N,M))=0.3333
      YLED(N,83+LSHIF(N,M))=0.3750
      YLED(N,84+LSHIF(N,M))=0.4188
      YLED(N,85+LSHIF(N,M))=0.4565
      YLED(N,86+LSHIF(N,M))=0.4957
      YLED(N,87+LSHIF(N,M))=0.5420
      YLED(N,88+LSHIF(N,M))=0.5897
      YLED(N,89+LSHIF(N,M))=0.6314
      YLED(N,90+LSHIF(N,M))=0.6752
      YLED(N,91+LSHIF(N,M))=0.7338
      YLED(N,92+LSHIF(N,M))=0.7949
      YLED(N,93+LSHIF(N,M))=0.8451
      YLED(N,94+LSHIF(N,M))=0.8974
      YLED(N,95+LSHIF(N,M))=0.9263
      YLED(N,96+LSHIF(N,M))=0.9573
      YLED(N,97+LSHIF(N,M))=0.9788
      YLED(N,98+LSHIF(N,M))=1.0000
```

Fig. 123–Continued

```
      YLED(N,99+LSHIF(N,M))=0.9830
      YLED(N,100+LSHIF(N,M))=0.9658
      YLED(N,101+LSHIF(N,M))=0.8881
      YLED(N,102+LSHIF(N,M))=0.8120
      YLED(N,103+LSHIF(N,M))=0.7387
      YLED(N,104+LSHIF(N,M))=0.6667
      YLED(N,105+LSHIF(N,M))=0.5638
      YLED(N,106+LSHIF(N,M))=0.4615
      YLED(N,107+LSHIF(N,M))=0.4010
      YLED(N,108+LSHIF(N,M))=0.3419
      YLED(N,109+LSHIF(N,M))=0.2900
      YLED(N,110+LSHIF(N,M))=0.2393
      YLED(N,111+LSHIF(N,M))=0.1875
      YLED(N,112+LSHIF(N,M))=0.1367
      YLED(N,113+LSHIF(N,M))=0.1107
      YLED(N,114+LSHIF(N,M))=0.0855
      YLED(N,115+LSHIF(N,M))=0.0726
      YLED(N,116+LSHIF(N,M))=0.0598
      YLED(N,117+LSHIF(N,M))=0.0465
      YLED(N,118+LSHIF(N,M))=0.0342
      YLED(N,119+LSHIF(N,M))=0.0272
      YLED(N,120+LSHIF(N,M))=0.0216
      YLED(N,121+LSHIF(N,M))=0.0179
      YLED(N,122+LSHIF(N,M))=0.0154
      YLED(N,123+LSHIF(N,M))=0.0112
      YLED(N,124+LSHIF(N,M))=0.0077
      YLED(N,125+LSHIF(N,M))=0.0058
      YLED(N,126+LSHIF(N,M))=0.0044
      YLED(N,127+LSHIF(N,M))=0.0036
      YLED(N,128+LSHIF(N,M))=0.0028
      YLED(N,129+LSHIF(N,M))=0.0022
      YLED(N,130+LSHIF(N,M))=0.0016
      YLED(N,131+LSHIF(N,M))=0.0012
      YLED(N,132+LSHIF(N,M))=0.0007
      YLED(N,133+LSHIF(N,M))=0.0004
      YLED(N,134+LSHIF(N,M))=0.0001
    7 CONTINUE
C*********************************************************************
C     CALCULATE TRANSMISSION CURVE AFTER FILTERING EQUIVALENT
C     TO MULTIPLYING THE RELATIVE LED POWER TIMES THE FILTER
C     TRANSMISSION AT EACH CORRESPONDING WAVELENGTH
C*********************************************************************
      DO 9 N=1,2
      DO 8 LAMDA=1,200
      YFLED(N,LAMDA)=YLED(N,LAMDA)*YFILTR(N,LAMDA)
    8 CONTINUE
    9 CONTINUE
C*********************************************************************
C     CALCULATE THE TOTAL RELATIVE POWER OF THE LED'S BEFOR &
C     AFTER FILTERING I USE ACALE FACTORS TO ADJUST ANY DIFFERENCES
C     IN THE RELATIVE DRIVE CURRENTS OF THE LED'S OF FILTER CHANGES
C*********************************************************************
      TPLED(1)=0.0
      TPLED(2)=0.0
      SCALE(1)=1.0
      SCALE(2)=1.0
      DO 11 N=1,2
      DO 10 LAMDA=2,200
      XPLED=SCALE(N)*DLAMDA*(YLED(N,LAMDA)+
     1YLED(N,LAMDA-1))/(2.0*45.69)
```

```
      TPLED(N)=XPLED+TPLED(N)
   10 CONTINUE
   11 CONTINUE
      TFPLED(1)=0.0
      TFPLED(2)=0.0
      DO 13 N=1,2
      DO 12 LAMDA=2,200
      XFPLED=SCALE(N)*DLAMDA*(YFLED(N,LAMDA)+
     1YFLED(N,LAMDA-1))/(2.0*45.69)
      TFPLED(N)=XFPLED+TFPLED(N)
   12 CONTINUE
   13 CONTINUE
      WRITE(6,14) ITEMP(M),LED1,TPLED(1),LED2,TPLED(2),LED1,
     1TFPLED(1),LED2,TFPLED(2)
   14 FORMAT(1H1,' LED OPERATIONAL TEMPERATURE = ',I4,' DEGREES',
     1' CENTIGRADE'/1H0,' TOTAL RELATIVE POWER OF LED AT',I4,'NM =',
     2F6.3/1H ,' TOTAL RELATIVE POWER OF LED AT',I4,'NM ='F6.3/
     31H0,' TOTAL RELATIVE FILTERED POWER OF LED AT',I4,'NM =',F6.3/
     41H ,' TOTAL RELATIVE FILTERED POWER OF LED AT',I4,'NM =',F6.3/)
C
      RFP1=(TPLED(1)-TFPLED(1))*100.0/TPLED(1)
      RFP2=(TPLED(2)-TFPLED(2))*100.0/TPLED(2)
      WRITE(6,15) LED1,RFP1,LED2,RFP2
   15 FORMAT(1H ,' PERCENT LOSS OF LED AT',I4,'NM BY FILTER =',F6.2,'%'/
     11H ,' PERCENT LOSS OF LED AT',I4,'NM BY FILTER =',F6.2,'%'/)
      IF(RFP1.LE.0.00001) GO TO 17
      IF(RFP2.LE.0.00001) GO TO 17
C
      FILOS1=-10.0*DLOG10((100.0-RFP1)/100.0)
      FILOS2=-10.0*DLOG10((100.0-RFP2)/100.0)
      WRITE(6,16) LED1,FILOS1,LED2,FILOS2
   16 FORMAT(1H ,' INSERTION LOSS FROM FILTER FOR LED AT',I4,'NM =',F5.2
     1,'DB'/1H ,' INSERTION LOSS FROM FILTER FOR LED AT',I4,'NM =',F5.2,
     2'DB'/)
   17 CONTINUE
C***********************************************************************
C     CALCULATE THE THEORETICAL CROSSTALK CAUSED BY SPECTRAL
C     OVERLAP OF THE LED'S AT 0 MODULATION - DC WORST CASE
C***********************************************************************
      DO 19 LAMDA=1,200
      IF(YFLED(1,LAMDA).LT.0.00001) GO TO 19
      IF(YFLED(2,LAMDA).LT.0.00001) GO TO 19
      IF(YFLED(1,LAMDA).LT.YFLED(2,LAMDA)) GO TO 18
      CROS(LAMDA)=YFLED(2,LAMDA)
      GO TO 19
   18 CROS(LAMDA)=YFLED(1,LAMDA)
   19 CONTINUE
      CROST=0.0
      DO 20 LAMDA=2,200
      CROSA=DLAMDA*(CROS(LAMDA)+CROS(LAMDA-1))/(2.0*45.69)
      CROST=CROSA+CROST
   20 CONTINUE
      IF(TFPLED(1).LE.TFPLED(2)) GO TO 21
      CROSTL=-2.0*10.0*DLOG10(CROST/TFPLED(2))
      GO TO 22
   21 CROSTL=-2.0*10.0*DLOG10(CROST/TFPLED(1))
   22 CONTINUE
      REFL=-12.0*10.0*DLOG10((100.0-4.0)/100.0)
      CROSTC=CROSTL-REFL
      TINLOS=REFL+FILOS1+FILOS2
```

Fig. 123-Continued

```
C***********************************************************************
C       CROSSTALK ISOLATION INCREASE FROM MODULATION IS GIVEN
C       BY : 10*LOG(81HZ/F) FOR 5MHZ>F>51KHZ ; FOR FIBER VIDEO
C       TRANSMISSION THE WEIGHTED AVERAGE FREQUENCY USING THE
C       PRE-EMPHASIS CURVE AND HUMAN RESOLUTION RESPONSE IS 1 MHZ
C***********************************************************************
        CROST1=CROSTC-(10.0*DLOG10(0.81/10000.0))-28.0
        WRITE(6,23) REFL,TINLOS,CROSTL,CROSTC,CROST1
     23 FORMAT(1H ,' LOSSES BY FIBER + LENS + FILTER INTERFACES =',
       1F5.2,'DB'/
       21H ,' TOTAL EXPECTED MINIMUM INSERTION LOSS/PAIR =',F5.2,'DB'/
       31H0,' CROSS-TALK CAUSED BY LED SPECTRAL OVERLAP =',F6.2,'DB'/
       41H ,' TOTAL EXPECTED MAXIMUM CROSSTALK AT 0 MOD.=',F6.2,'DB'/
       51H ,' TOTAL EXPECTED MAXIMUM CROSSTALK AT 1 MHZ =',F6.2,'DB'//
       61H ,9X,' RELATIVE LED + FILTER SPECTRAL CURVES'/)
        DO 25 IX=1,70
        DO 24 IY=1,31
        IGRAPH(IX,IY)=BLANK
     24 CONTINUE
     25 CONTINUE
C***********************************************************************
C       CALCULATE THE GRAPH POSITIONS OF THE FILTERS AND THE LED'S
C***********************************************************************
        DO 31 N=1,2
        DO 30 LAMDA=2,120,2
        DO 27 IY=1,29
        LX=(LAMDA+12)/2
        DIF=VLED(N,LAMDA)-(IY*0.033333)
        IF(VLED(N,LAMDA).LE.0.0166667) GO TO 26
        IF(DIF.LE.0.0167) GO TO 28
     26 IGRAPH(LX,30)=STER
     27 CONTINUE
        GO TO 29
     28 IGRAPH(LX,31-IY)=STER
     29 CONTINUE
     30 CONTINUE
     31 CONTINUE
        DO 37 N=1,2
        DO 36 LAMDA=2,120,2
        DO 33 IY=1,29
        LX=(LAMDA+12)/2
        DIF=VFILTR(N,LAMDA)-(IY*0.0333333)
        IF(VFILTR(N,LAMDA).LE.0.0166667) GO TO 32
        IF(DIF.LE.0.0167) GO TO 34
     32 CONTINUE
     33 CONTINUE
        GO TO 35
     34 IGRAPH(LX,31-IY)=EFF
     35 CONTINUE
     36 CONTINUE
     37 CONTINUE
C***********************************************************************
C       INITIALIZE GRAPH Y CORDINATES (0 TO 1 IN 1/3 INCREMENTS)
C       AND X COORDINATES (WAVELENGTH UNITS OF DLAMDA INCREMENTS)
C***********************************************************************
        DO 38 IA=1,10
        IAA=1+(3*(IA-1))
        IGRAPH(5,IAA)=UNDER
     38 CONTINUE
        DO 39 IB=1,30
```

```
      IGRAPH(6,IB)=EXCLA
   39 CONTINUE
      A=1.1
      DO 60 L=1,30,3
      A=A-0.1
      WRITE(6,600) A,(IGRAPH(IX,L),IX=5,71)
  600 FORMAT(1H ,1X,F3.1,67A1)
      L1=L+1
      L2=L+2
      DO 61 J=L1,L2
      WRITE(6,610) (IGRAPH(IX,J),IX=5,71)
  610 FORMAT(1H ,' * *',67A1)
   61 CONTINUE
   60 CONTINUE
      WRITE(6,44)
   44 FORMAT(1H ,' 0.0-+-------------------------------------------------------
     1-----------------')
      WRITE(6,45)
   45 FORMAT(1H ,5X,'+',12(6X,'+'))
      WRITE(6,46)
   46 FORMAT(1H ,6X,'710',2X,'730',2X,'750',2X,'770',2X,'790',2X,
     1'810',2X,'830',2X,'850',2X,'870',2X,'890',2X,'910',2X,
     2'930',2X,'950'/1H ,26X,'WAVELENGTH(NM)')
   50 CONTINUE
      STOP
      END
```

Fig. 123-Continued

LED OPERATIONAL TEMPERATURE = 20 DEGREES CENTIGRADE

TOTAL RELATIVE POWER OF LED AT 875NM = 1.000
TOTAL RELATIVE POWER OF LED AT 795NM = 0.996

TOTAL RELATIVE FILTERED POWER OF LED AT 875NM = 0.731
TOTAL RELATIVE FILTERED POWER OF LED AT 795NM = 0.762

PERCENT LOSS OF LED AT 875NM BY FILTER = 26.92%
PERCENT LOSS OF LED AT 795NM BY FILTER = 23.29%

INSERTION LOSS FROM FILTER FOR LED AT 875NM = 1.36DB
INSERTION LOSS FROM FILTER FOR LED AT 795NM = 1.15DB

LOSSES BY FIBER + LENS + FILTER INTERFACES = 2.13DB
TOTAL EXPECTED MINIMUM INSERTION LOSS/PAIR = 4.66DB

CROSS-TALK CAUSED BY LED SPECTRAL OVERLAP = 44.83DB
TOTAL EXPECTED MAXIMUM CROSSTALK AT 0 MOD.= 42.70DB
TOTAL EXPECTED MAXIMUM CROSSTALK AT 1 MHZ = 55.62DB

RELATIVE LED + FILTER SPECTRAL CURVES

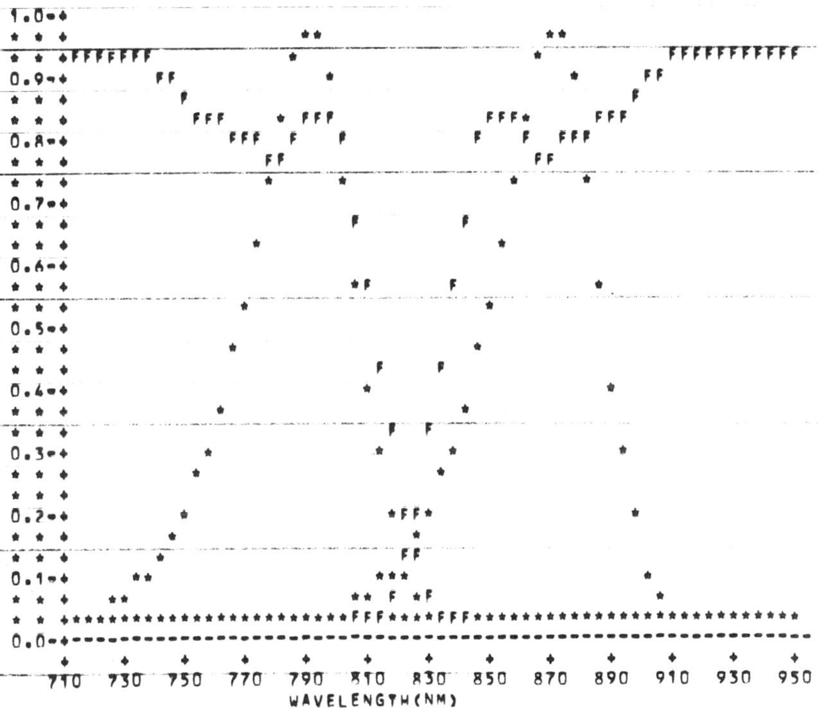

PROPERTIES OF THE ELEMENTS

Element symbol	Ionization potentials (V) I	II	III	Specific heat (cal/g K)	Melting point (°C)	Boiling point (° C)
Ar	15.75	27.62	40.90	0.124	-189	-186
Ag	7.57	21.48	34.82	0.056	962	2212
Al	5.98	18.82	28.44	0.215	660	2467
As	9.81	18.63	28.34	0.079	817	613^{sub}
Au	9.22	20.51		0.031	1064	2807
B	8.30	25.15	37.92	0.245	2300	2550^{sub}
Ba	5.21	10.00	35.50	0.046	725	1640
Be	9.32	18.21	153.85	0.436	1278	2970
Bi	7.29	16.68	25.56	0.029	271	1560
Br	11.84	21.60	35.91	0.110	-7	59
C	11.26	24.38	47.87	0.170	3550	4827
Ca	6.11	11.87	51.21	0.156	839	1484
Cd	8.99	16.90	37.47	0.055	321	765
Cl	13.01	23.80	39.91	0.114	-101	-35
Co	7.86	17.05	33.49	0.108	1495	2870
Cr	6.76	16.49	30.96	0.107	1857	2672
Cs	3.89	25.10	35.02	0.057	28	678
Cu	7.72	20.29	36.83	0.092	1083	2567
F	17.42	34.98	62.65	0.197	-219	-188
Fe	7.87	16.18	30.64	0.107	1535	2750
Ga	6.02	20.57	30.76	0.089	30	2403
Ge	7.88	15.93	34.21	0.077	937	2830
H	13.59			3.415	-259	-253
He	24.48	54.40		1.241	-272	-269
Hf	6.99	14.91	23.20	0.033	2227	4602
I	10.45	19.13		0.009	113	184
In	5.78	18.86	28.03	0.056	157	2080

Properties of the Elements–Continued

K	4.34	31.81	45.98	0.180	64	774
Kr	14.00	24.56	36.91	0.059	-157	-152
La	5.61	11.43	19.17	0.048	920	3454
Li	5.39	75.62	127.42	0.835	181	1347
Mg	7.64	15.03	80.14	0.242	649	1090
Mn	7.43	15.64	33.69	0.115	1244	1962
Mo	7.10	16.15	27.13	0.059	2617	4612
N	14.53	29.59	47.43	0.249	-210	-196
Na	5.14	47.29	71.72	0.293	98	883
Nb	6.88	14.32	25.04	0.064	2468	4742
Ne	21.56	41.07	63.50	0.246	-249	-246
Ni	7.63	18.15	35.16	0.106	1453	2732
O	13.61	35.11	54.89	0.219	-218	-183
Os	8.49	16.98		0.031	3045	5027
P	10.48	19.72	30.16	0.171	44	280
Pb	7.42	15.03	31.93	0.032	328	1740
Pd	8.33	19.42	32.92	0.058	1552	3140
Pt	9.00	18.56		0.032	1772	3827
Rb	4.18	27.49	40.03	0.086	39	688
Re	7.87	16.59		0.033	3180	5627
Rh	7.46	18.07	31.05	0.058	1966	3727
Ru	7.36	16.76	28.46	0.057	2310	3900
S	10.36	23.41	35.03	0.176	112	445
Sb	8.64	16.49	25.30	0.050	631	1750
Sc	6.54	12.81	24.75	0.120	1539	2832
Se	9.75	21.50	32.01	0.076	217	685
Si	8.15	16.34	33.49	0.168	1410	2355
Sn	7.34	14.63	30.49	0.052	232	2270
Sr	5.69	11.03		0.072	769	1384
Ta	7.88	16.20		0.033	2996	5425
Tc	7.28	15.26	29.54	0.058	2172	4877
Te	9.01	18.60	31.02	0.048	450	990
Ti	6.82	13.57	27.47	0.124	1660	3287
Tl	6.11	20.42	29.81	0.031	304	1457
V	6.74	14.65	29.31	0.115	1890	3380
W	7.98	17.71		0.032	3410	5660
Xe	12.13	21.20	31.30	0.038	-112	-107
Y	6.38	12.23	20.52	0.069	1523	3337
Zn	9.39	17.96	39.70	0.092	420	907
Zr	6.84	13.13	22.98	0.067	1852	4377

INDEX